Analytical Inventory Management and Optimization

In this vital resource with discussion of and applicability to various industries, Ali, Rabiu, and Ismail guide readers through the challenging environment of inventory management and enable them to successfully balance the demand and supply of items in stock, a critical task in any field.

The book covers a wide range of topics related to inventory management and optimization. After introducing the key concepts and principles of inventory management, such as inventory analytics, optimization, and models, the authors provide a comprehensive and in-depth understanding of various inventory control techniques that are essential for effective inventory management, such as ABC analysis, EOQ model, safety stock, and reorder point. They then introduce various mathematical models and optimization techniques such as system-level and item-level inventory analysis and show how to perform sensitivity analyses to test the robustness of these models. They then look at the role of inventory management in various industries such as supply chain management and logistics, manufacturing, and more; addressing the effects and integration of cutting-edge technologies like artificial intelligence, machine learning, and robotics to conventional inventory management practices. Additional topics include inventory forecasting, inventory management systems, inventory auditing and control, and risk management. In combining mathematical underpinnings in the area with practical case studies throughout, readers will gain a solid understanding of the real-world applications of these different techniques so that they can apply inventory management and optimization best practices in their workplaces.

The comprehensive coverage makes the book a valuable reference for practitioners and students, particularly postgraduate and MBA students who require such insights to improve business functions and make informed decisions. Because it provides the foundational mathematical knowledge required to comprehend any chapter, it is also accessible for readers without a strong background in mathematics.

Analytical Inventory Management and Optimization

Theories, Methods and Applications

Majid Khan Majahar Ali
Sani Rabiu
Mohd Tahir Ismail

CRC Press
Taylor & Francis Group
Boca Raton London New York

CRC Press is an imprint of the
Taylor & Francis Group, an **informa** business

First edition published 2025
by CRC Press
2385 NW Executive Center Drive, Suite 320, Boca Raton FL 33431

and by CRC Press
4 Park Square, Milton Park, Abingdon, Oxon, OX14 4RN

CRC Press is an imprint of Taylor & Francis Group, LLC

ISBN: 978-1-032-86036-7 (hbk)
ISBN: 978-1-032-87866-9 (pbk)
ISBN: 978-1-003-53486-0 (ebk)

DOI: 10.1201/9781003534860

Typeset in CMR10 font
by KnowledgeWorks Global Ltd.

Dedication

To our families, for their unwavering support and encouragement throughout this journey. To our mentors and colleagues, whose guidance and insights have been invaluable. To all the students, researchers, and practitioners who strive to advance the field of inventory management. This book is dedicated to you.

Contents

Foreword xxi

Preface xxiii

Acknowledgements xxv

About the Authors xxvii

Contributors xxix

1 Inventory Management 1
1.1 Introduction to Inventory Management 1
1.2 Key Concepts and Principles 2
1.3 Principles of Inventory Management 3
1.3.1 Inventory Control Systems 3
1.3.2 Demand Forecasting 4
1.3.3 Inventory Planning 4
1.3.4 Supplier Management 4
1.3.5 Inventory Optimization 4
1.3.6 Inventory Auditing 5
1.3.7 Safety Stock Management 5
1.3.8 Inventory Reporting and Analytics 5
1.3.9 Order Management 5
1.3.10 Return Management 5
1.3.11 Cycle Counting 6
1.3.12 Inventory Valuation Methods 6
1.3.13 Warehouse Layout and Design 6
1.3.14 Inventory Segmentation 6
1.3.15 Sustainability and Environmental Considerations . . . 6
1.3.16 Collaboration and Integration With Other Business Functions 7
1.3.17 Technology and Automation 7
1.3.18 Lead Time Management 7
1.3.19 Product Life Cycle Management 7
1.3.20 Service Level Agreements and Performance Metrics . . 7
1.3.21 Risk Management and Contingency Planning 8
1.3.22 Inventory Accounting and Financial Integration 8

1.3.23 Regulatory Compliance 8
1.3.24 Customer Relationship Management Integration 8
1.3.25 Distribution and Transportation Management 8
1.3.26 Vendor Managed Inventory and Consignment Stock . . 8
1.3.27 Inventory Replenishment Strategies 9
1.4 Objectives of Inventory Management 9
1.4.1 Ensuring Sufficient Supply to Meet Demand 9
1.4.2 Minimizing Costs Associated With Holding and Managing Inventory 10
1.4.3 Balancing Between Overstocking and Stockouts 10
1.5 Costs Associated with Inventory Management 11
1.5.1 Holding Costs 11
1.5.2 Ordering Costs 12
1.5.3 Stockout Costs 13
1.6 Problems and Challenges in Inventory Management 13
1.6.1 Demand Variability 14
1.6.2 Strategies to Overcome Demand Variability 14
1.6.3 Lead Time Variability 15
1.6.4 Strategies to Overcome Lead Time Variability 15
1.6.5 Other Common Challenges 16
1.6.6 Strategies to Overcome Other Common Challenges . . 16
1.7 Types of Inventories 17
1.7.1 Raw Materials 17
1.7.2 Work-in-Progress 17
1.7.3 Finished Goods 18
1.7.4 Maintenance, Repair, and Operations Supplies 18
1.8 Inventory Control and Optimization 19
1.8.1 Definition of Inventory Optimization 19
1.8.2 What Is Inventory Optimization 19
1.8.3 Importance of Inventory Optimization 20
1.8.4 Goals of Inventory Optimization 20
1.9 Relevance to Organizations 21
1.9.1 Impact on Customer Satisfaction and Profitability . . . 21
1.9.2 Importance of Maintaining Optimal Inventory Levels . 21
1.9.3 Inventory Control Systems 22
1.9.4 Inventory Optimization Methods 23
1.10 Fundamental Concepts and Practices 25
1.10.1 Inventory Analytics 25
1.10.2 Inventory Models 26
1.11 Impact of Inventory Management on Organizational Operations 34
1.11.1 Optimized Inventory Levels 34
1.11.2 Reduced Costs 34
1.11.3 Improved Supply Chain Collaboration 34
1.11.4 Enhanced Customer Service 35

1.11.5 Toyota Production System 35
1.11.6 Amazon's Fulfillment Centers 35
1.12 Practical Applications 35
1.12.1 Practical Tips for Practitioners 36
1.12.2 Tools and Technologies 36
1.13 Conclusion 37

Exercises **38**

2 Inventory Control Techniques **40**
2.1 Introduction to Inventory Control Techniques 40
2.1.1 Definition of Inventory Control Techniques 41
2.1.2 Importance of Inventory Control Techniques 41
2.1.3 Contribution to Effective Inventory Management . . . 41
2.2 Just-In-Time 42
2.2.1 Key Principles of JIT 43
2.2.2 Benefits of JIT 43
2.2.3 Challenges of JIT 44
2.2.4 Applications of JIT 45
2.3 ABC Analysis 46
2.3.1 Categories in ABC Analysis 46
2.3.2 Methodology for Categorizing Inventory Items 47
2.3.3 Optimization Strategies 48
2.3.4 Apply Control Policies 49
2.3.5 Benefits 50
2.3.6 Challenges of ABC Analysis 50
2.3.7 Limitations 51
2.4 VED Analysis 56
2.4.1 Vital (V) 57
2.4.2 Essential (E) 57
2.4.3 Desirable (D) 57
2.4.4 Integration with ABC Analysis 58
2.4.5 Advantages of VED Analysis 58
2.4.6 Disadvantages of VED Analysis 59
2.5 Safety Stock Management 60
2.5.1 Key Points About Safety Stock Management 60
2.5.2 Methods for Calculating Safety Stock Levels 61
2.5.3 Advantages of Safety Stock 62
2.5.4 Disadvantages of Safety Stock 63
2.5.5 Balancing Safety Stock 63
2.6 Inventory Turnover 64
2.6.1 Benefits of Calculating Inventory Turnover 64
2.7 Reorder Point (ROP) 65
2.7.1 Definition of Reorder Point and Its Role in Inventory Management 66

2.7.2 Calculation of Reorder Point Based on Lead Time and Demand . . . 66
2.7.3 Impact of Reorder Points on Inventory Control and Cost Management . . . 68
2.8 Economic Order Quantity (EOQ) Model . . . 69
2.8.1 Mathematical Formulation and Assumptions of the EOQ Model . . . 69
2.8.2 Economic Production Quantity (EPQ) . . . 70
2.8.3 Advantages . . . 81
2.8.4 Disadvantages . . . 81
2.9 Comparison of Inventory Control Techniques . . . 82
2.9.1 ABC Analysis . . . 82
2.9.2 Economic Order/Production Quantity (EOQ/EPQ) . . 83
2.9.3 Safety Stock . . . 83
2.9.4 Reorder Point (ROP) . . . 83
2.10 Criteria for Selecting Appropriate Techniques Based on Organizational Needs . . . 84
2.10.1 Integrated Inventory Management in a Retail Chain . . 84
2.10.2 Manufacturing Company Using Multiple Techniques . . 85
2.11 Importance of Inventory Control from an Organizational Perspective . . . 86
2.12 Functioning of Inventory Management Systems Within Organizations . . . 88
2.12.1 Role of Coordination Among Different Departments in Inventory Management . . . 88
2.12.2 Strategies for Optimizing Expenditure on Materials and Inventory . . . 89
2.13 Coordination and Optimal Expenditure on Materials . . . 90
2.13.1 Techniques for Optimizing Material Expenditure . . . 91
2.14 Multi-Echelon Inventory Optimization (MEIO) . . . 92
2.14.1 Overview of MEIO . . . 92
2.14.2 Key Benefits of MEIO . . . 92
2.14.3 Components of MEIO Models . . . 93
2.14.4 MEIO Techniques and Approaches . . . 93
2.14.5 Challenges in MEIO Implementation . . . 94
2.15 Conclusion . . . 94

Exercises **96**

3 Advanced Inventory Optimization Models **103**
3.1 Introduction . . . 103
3.1.1 Multi-Echelon Inventory Model . . . 103
3.2 Sensitivity Analysis . . . 106
3.2.1 Conducting Sensitivity Analysis . . . 106
3.3 Inventory Optimization for Seasonal Demand . . . 107

3.3.1 Seasonal Forecasting Models . . . 108
3.3.2 Dynamic Safety Stock Strategies . . . 109
3.3.3 Flexible Capacity Planning . . . 110
3.4 Applications of Advanced Inventory Optimization Techniques . 113
3.4.1 Retail Industry . . . 114
3.4.2 Manufacturing Industry . . . 115
3.4.3 Healthcare Industry . . . 116
3.4.4 E-commerce Industry . . . 117
3.5 Conclusion . . . 118

Exercises **119**

4 Inventory Management in the Manufacturing Sector **122**
4.1 Introduction to Inventory Management in Manufacturing . . . 122
4.1.1 Definition and Scope of Inventory Management Specific to the Manufacturing Sector . . . 123
4.1.2 Importance of Inventory Management in Manufacturing Operations . . . 123
4.1.3 Impact on Production . . . 126
4.2 Significance of Inventory Management in Manufacturing . . . 126
4.3 Role of Inventory Management . . . 127
4.3.1 Strategizing Inventory Policies . . . 127
4.3.2 Allocating Resources Efficiently . . . 128
4.3.3 Ensuring the Availability of Materials for Continuous Production . . . 130
4.3.4 Practical Applications . . . 132
4.4 Inventory Analytics . . . 135
4.4.1 Descriptive Analytics . . . 135
4.4.2 Predictive Analytics . . . 136
4.4.3 Key Performance Indicators (KPIs) and Metrics for Inventory Performance in Manufacturing . . . 137
4.4.4 Techniques for Forecasting Demand and Lead Times Specific to Manufacturing . . . 140
4.5 Conclusion . . . 144

Exercises **145**

5 Emerging Trends in Inventory Management **146**
5.1 Introduction . . . 146
5.2 Integration of Cutting-Edge Technologies . . . 146
5.2.1 Artificial Intelligence and Machine Learning in Inventory Management . . . 147
5.2.2 Enhanced Data Analysis Capabilities . . . 147
5.2.3 Inventory Optimization Using Machine Learning Algorithms . . . 148
5.2.4 Linear Regression . . . 148

5.2.5 Decision Trees 149
5.2.6 Classification and Regression 150
5.2.7 Visualization 150
5.2.8 Neural Networks 154
5.2.9 Robotics in Warehouse Operations 160
5.2.10 Automated Picking and Packing 160
5.2.11 Autonomous Mobile Robots 161
5.2.12 Collaborative Robots (Cobots) 163
5.3 Cloud-Based Inventory Systems 164
5.3.1 Real-Time Monitoring and Decision-Making 164
5.3.2 Enhanced Collaboration and Flexibility 166
5.3.3 Flexibility and Collaboration Across Multiple Locations 168
5.4 Conclusion 170

Exercises **172**

6 Inventory Forecasting **174**
6.1 Introduction to Inventory Forecasting 174
6.1.1 Historical Background 174
6.2 Expectation 175
6.3 Variance 176
6.4 Covariance 176
6.5 Exponential Smoothing 177
6.6 MAPE (Mean Absolute Percentage Error) 178
6.7 Linear Regression 178
6.7.1 SSE (Sum of Squared Errors) 179
6.7.2 Ridge Regression 179
6.8 LASSO (Least Absolute Shrinkage and Selection Operator) . . 180
6.9 Elastic Net 182
6.10 Random Forest 183
6.11 Support Vector Machine (SVM) 186
6.12 Role of Inventory Forecasting in Inventory Management 189
6.12.1 Predicting Future Inventory Requirements 190
6.12.2 Time Series Analysis 190
6.12.3 Causal Models 192
6.12.4 Implications for Supply Chain Management 196
6.12.5 Bullwhip Effect Reduction 196
6.12.6 Supplier Relationship Management 197
6.12.7 Inventory Optimization 198
6.12.8 Logistics and Operational Efficiency 199
6.12.9 Transportation Planning 199
6.12.10 Warehouse Space Utilization 201
6.12.11 Labor Planning 204
6.13 Inventory Forecasting Techniques 206

6.13.1 Demand Planning 206
6.13.2 Demand Forecasting 207
6.13.3 Comparative Analysis of Techniques 209
6.14 Practical Applications of Inventory Forecasting 210
6.14.1 Impact on Organizational Operations 210
6.14.2 Best Practices and Implementation Strategies 211
6.15 Practical Applications of Inventory Forecasting 212
6.15.1 Impact on Organizational Operations 213
6.15.2 Best Practices and Implementation Strategies 213
6.16 Conclusion 214

Exercises 216

7 Inventory Auditing and Control 217
7.1 Introduction 217
7.1.1 Overview of Inventory Auditing and Control 217
7.1.2 Importance of Inventory Auditing in Inventory Management 218
7.2 Techniques of Inventory Auditing 219
7.3 Inventory Reconciliation 219
7.3.1 Steps Involved in Inventory Reconciliation 220
7.3.2 Importance of Inventory Reconciliation 220
7.3.3 Challenges in Inventory Reconciliation 220
7.3.4 Best Practices 221
7.3.5 Mathematical Formulation 221
7.4 Practical Applications of Inventory Auditing 222
7.4.1 Best Practices for Implementing Inventory Auditing . . 224
7.5 Impact on Supply Chain Management 224
7.5.1 Role of Inventory Auditing in Supply Chain Coordination 224
7.5.2 Integration With Modern Supply Chain Technologies . 225
7.6 Current Trends and Future Directions 226
7.6.1 Emerging Techniques in Inventory Auditing 226
7.6.2 Technological Advancements and Their Implications . . 227
7.6.3 Future Challenges and Opportunities 228
7.7 Conclusion 229

Exercises 231

8 Inventory Metrics and KPIs 233
8.1 Introduction 233
8.1.1 Importance of Inventory Management 233
8.1.2 Overview of Metrics and KPIs 234
8.2 Choosing Reliable and Efficient Inventory Metrics 235
8.2.1 Criteria for Selecting Metrics 235
8.2.2 Aligning Metrics With Business Goals 236

8.3 Key Inventory Metrics and KPIs 237
8.3.1 Inventory to Sales Ratio 237
8.3.2 Demand Forecast Accuracy 238
8.3.3 Inventory Turnover 240
8.3.4 Gross Margin Return on Investment (GMROI) 242
8.3.5 Days Sales of Inventory (DSI) 243
8.3.6 Sell-Through Rate 245
8.4 Using Metrics to Monitor Performance 246
8.4.1 Analyzing Inventory Data 247
8.4.2 Improving Sales and Merchandising Processes 250
8.5 Benefits of Inventory Metrics and KPIs 251
8.5.1 Measuring and Achieving Progress 251
8.5.2 Guiding Behaviour and Decision-Making 252
8.5.3 Enhancing Productivity 253
8.5.4 Industry Applicability 253
8.5.5 Practical Applications 254
8.5.6 Lessons Learned 254
8.5.7 Future Directions in Inventory Management 255
8.6 Conclusion 256

Exercises **257**

9 Inventory Risk Management **259**
9.1 Introduction 259
9.1.1 Overview of Inventory Risk Management 259
9.1.2 Importance of Managing Inventory Risks 260
9.2 Identifying Inventory Risks 261
9.2.1 Demand Risks 261
9.2.2 Supply Risks 262
9.2.3 Operational Risks 262
9.2.4 Financial Risks 263
9.3 Risk Assessment Models 263
9.3.1 Analytic Hierarchy Process (AHP) Method 264
9.3.2 Quantitative and Qualitative Assessment Techniques . 277
9.4 Mitigating Inventory Risks 288
9.4.1 Demand Risk Mitigation Strategies 288
9.4.2 Supply Risk Mitigation Strategies 289
9.4.3 Operational Risk Mitigation Strategies 289
9.4.4 Financial Risk Mitigation Strategies 290
9.5 Classifying Risks in Inventory Management 291
9.5.1 Categorization of Risks 291
9.5.2 Prioritizing Risks 292
9.5.3 Risk Matrix 294
9.5.4 Criticality Analysis 295
9.6 Implementing Risk Management Strategies 297

9.6.1 Developing a Risk Management Plan 298
9.6.2 Integrating Risk Management into Inventory Processes 298
9.6.3 Monitoring and Reviewing Risk Management Activities 299
9.7 Implications for Companies 300
9.7.1 Operational Efficiency 300
9.7.2 Financial Stability 300
9.7.3 Competitive Advantage 301
9.7.4 Future Directions in Inventory Risk Management . . . 302
9.8 Conclusion 303

Exercises **305**

10 Inventory Sustainability **308**
10.1 Introduction 308
10.1.1 Importance of Sustainable Inventory Management . . . 309
10.1.2 Waste Reduction Strategies 310
10.1.3 Minimizing Carbon Footprint 310
10.1.4 Promoting Ethical Sourcing and Production 311
10.1.5 Models of Sustainable Inventory Management 312
10.1.6 Current Trends in Sustainable Supply Chains 317
10.1.7 Emerging Topics in Sustainable Inventory Management 318
10.1.8 Technological Innovations and Sustainability 319
10.1.9 Theoretical Framework for Sustainable Inventory Control 320
10.1.10 Case Studies of Sustainable Inventory Practices 320
10.1.11 Implications for Practice and Research 322
10.2 Conclusion 323

Exercises **324**

11 Inventory Ethics and Social Responsibility **325**
11.1 Introduction 325
11.1.1 Importance of Social Responsibility in Inventory Management 326
11.1.2 Ethical Sourcing Practices 327
11.1.3 Fair Treatment in Production 327
11.1.4 Transparency in Distribution 328
11.1.5 Resource Allocation Ethics 329
11.1.6 Waste Reduction and Environmental Impact 329
11.1.7 Social Equity and Fairness 330
11.2 Fostering Ethical Decision-Making 330
11.2.1 Developing Ethical Frameworks 331
11.2.2 Training and Education for Ethical Practices 331
11.2.3 Creating a Culture of Responsibility 332

11.3 Socially Responsible Inventory Management Strategies 332
11.3.1 Sustainable Sourcing and Procurement 333
11.3.2 Ethical Inventory Policies 333
11.3.3 Community Engagement and Impact 334
11.4 Role of Technology in Enhancing Ethical Practices 335
11.4.1 Technological Innovations for Transparency 335
11.4.2 Digital Tools for Ethical Supply Chain Management . . 336
11.4.3 Monitoring and Reporting Ethical Compliance 336
11.4.4 Theoretical Approaches to Inventory Ethics 337
11.4.5 Future Directions in Inventory Ethics and Social Responsibility 338
11.5 Conclusion 339

Exercises **340**

12 Inventory Simulation and Gaming **341**
12.1 Introduction 341
12.1.1 Benefits of Simulation and Gaming Techniques 342
12.1.2 Objectives of Using Simulation and Gaming 342
12.1.3 Types of Simulation Models 343
12.1.4 Developing and Implementing Simulation Models . . . 343
12.1.5 Case Studies of Simulation in Practice 344
12.1.6 Types of Inventory Management Games 345
12.1.7 Designing Effective Inventory Management Games . . . 345
12.2 Educational and Strategic Dimensions 346
12.2.1 Strategic Applications of Simulation and Gaming . . . 347
12.2.2 Enhancing Decision-Making Skills Through Simulation and Gaming 347
12.3 Challenges and Limitations 348
12.3.1 Limitations and Considerations 349
12.3.2 Overcoming Challenges in Practical Applications . . . 349
12.4 Future Directions in Inventory Simulation and Gaming 350
12.4.1 Potential Areas for Research and Development 351
12.4.2 Integrating Simulation and Gaming With Other Tools 351
12.4.3 Implications for Practitioners and Researchers 352
12.4.4 Final Thoughts and Recommendations 353
12.5 Conclusion 353

Exercises **355**

13 Inventory Education and Training **357**
13.1 Introduction 357
13.1.1 Importance of Continuous Learning in Inventory Management 358
13.1.2 Objectives of Inventory Education and Training 358
13.2 Strategies for Effective Inventory Education 358

13.2.1 Traditional Training Methods 359
13.2.2 Modern Educational Approaches 359
13.2.3 Blended Learning Techniques 360
13.3 Methodologies for Teaching Inventory Management 360
13.3.1 Case-Based Learning 360
13.3.2 Simulation and Gaming Techniques 361
13.3.3 Hands-On Workshops and Practicals 361
13.3.4 Measuring Training Effectiveness 361
13.3.5 Feedback and Continuous Improvement 362
13.3.6 Best Practices in Training Evaluation 362
13.3.7 Technological Advancements in Training Tools 363
13.3.8 Online Learning Platforms and Resources 363
13.3.9 The Role of Artificial Intelligence and Machine Learning 363
13.4 Practical Implications of Inventory Education 364
13.4.1 Impact on Organizational Efficiency 364
13.4.2 Enhancing Employee Competence and Skills 364
13.4.3 Aligning Training With Organizational Goals 365
13.5 Challenges and Opportunities in Inventory Training 365
13.5.1 Common Challenges in Training Implementation 365
13.5.2 Opportunities for Improvement and Innovation 365
13.5.3 Addressing Diverse Learning Needs 366
13.6 Future Directions in Inventory Education and Training 366
13.6.1 Emerging Areas of Focus 366
13.6.2 Integration of New Technologies 367
13.6.3 Developing Comprehensive Training Programs 367
13.6.4 Recommendations for Practitioners and Educators . . . 368
13.7 Conclusion 369

Exercises **370**

14 Inventory Research and Development **371**
14.1 Introduction 371
14.1.1 Importance of Continuous Research in Inventory Management 372
14.1.2 Objectives of Research and Development in Inventory Management 372
14.1.3 Advancements in Inventory Optimization Techniques . 373
14.1.4 Innovations in Inventory Forecasting and Planning . . 374
14.1.5 Integration of Big Data and Analytics 374
14.2 Research Methodologies in Inventory Management 375
14.2.1 Quantitative Research Methods 375
14.2.2 Qualitative Research Approaches 376
14.2.3 Case Studies and Empirical Research 377
14.2.4 Practical Implications for Inventory Control 377

14.2.5 Impact on Supply Chain Efficiency 378
14.2.6 Enhancements in Inventory Decision-Making 379
14.2.7 Common Challenges in Inventory Research 379
14.2.8 Opportunities for Future Research 380
14.2.9 Addressing Gaps in Current Research 381
14.2.10 Technological Innovations in Inventory Management . . 381
14.2.11 The Impact of Artificial Intelligence and Machine Learning 382
14.2.12 Utilization of Simulation and Modeling Tools 383
14.2.13 Best Practices for Conducting Inventory Research . . . 383
14.2.14 Emerging Areas of Focus 384
14.2.15 Potential for Interdisciplinary Research 385
14.2.16 Developing Collaborative Research Opportunities . . . 385
14.2.17 Implications for Practitioners and Researchers 386
14.3 Conclusion 387

Exercises **388**

15 Inventory Career Opportunities **389**
15.1 Introduction 389
15.1.1 Importance of Inventory Professionals in Supply Chain Operations 390
15.1.2 Objectives of Exploring Inventory Careers 390
15.1.3 Entry-Level Positions 390
15.1.4 Mid-Level Roles 391
15.1.5 Senior and Specialized Positions 391
15.1.6 Career Pathways and Advancement Opportunities . . . 392
15.1.7 Skills and Qualifications for Inventory Careers 392
15.1.8 Educational and Professional Qualifications 393
15.1.9 Certifications and Training Programs 393
15.1.10 Soft Skills and Competencies 394
15.1.11 Impact of Technological Advancements on Inventory Careers 394
15.1.12 Skills Required for Technology-Driven Roles 395
15.1.13 Adapting to Technological Changes in Inventory Management 396
15.1.14 Global Market Trends and Their Effect on Inventory Careers 396
15.1.15 Regional Variations in Career Opportunities 397
15.1.16 Cultural and Market Trends Shaping Inventory Careers 398
15.2 Career Development and Professional Growth 398
15.2.1 Networking and Professional Associations 399
15.2.2 Continuing Education and Professional Development . 400
15.2.3 Navigating the Job Market 400
15.2.4 Preparing for Interviews and Assessments 401

15.2.5 Building a Strong Professional Resume and Portfolio . 402
15.2.6 Emerging Trends and New Opportunities 403
15.2.7 Potential Areas for Growth and Specialization 403
15.2.8 Preparing for Future Career Challenges 404
15.2.9 Recommendations for Career Seekers and Advisors . . 404
15.3 Conclusion 405

Exercises **406**

Solutions **407**

Glossary **467**

References **475**

Index **479**

Foreword

The field of inventory management has undergone significant transformations over the past few decades. From traditional methods to the integration of advanced technologies, the evolution of inventory management practices has been remarkable. As someone who has closely observed these changes, I am delighted to introduce this comprehensive resource that captures both foundational principles and emerging trends in inventory management. This book is an invaluable asset for anyone involved in supply chain management, logistics, and operations. It provides a thorough understanding of inventory management techniques, optimization models, and the latest advancements that are shaping the future of the field. The authors have done an excellent job of combining theoretical insights with practical applications, making this book a must-read for practitioners and scholars alike. I am confident that this book will serve as a vital resource for enhancing knowledge and improving practices in inventory management. It is a testament to the authors' dedication and expertise, and I am honored to present it to the readers.

Faculty of Sciences and Natural Resources, Universiti Malaysia Sabah
Prof. Jumat Sulaiman
August 2024

Preface

Inventory management is a crucial aspect of supply chain management and logistics. Effective inventory management enables organizations to meet demand, minimize costs, and optimize operations. This book aims to provide a comprehensive guide to inventory management, encompassing fundamental concepts, control techniques, optimization models, and emerging trends. The journey of writing this book has been both challenging and rewarding. It involved extensive research, collaboration with experts, and the application of various inventory management practices. Our goal was to create a resource that is both informative and practical, offering insights that can be applied in real-world scenarios. Each chapter of this book has been meticulously crafted to provide a deep understanding of the different facets of inventory management. From basic principles to advanced optimization techniques, the content is designed to cater to a diverse audience, including students, researchers, and practitioners.

I hope this book serves as a valuable resource for those looking to enhance their knowledge and expertise in inventory management. I extend our gratitude to everyone who supported us in this endeavor and look forward to the positive impact this book will have on its readers.

Sincerely,

School of Mathematical Sciences
Universiti Sains Malaysia (USM)
Majid Khan Majahar Ali, Sani Rabiu and Mohd Tahir Ismail
August 2024

Acknowledgements

On the momentous occasion of the 50th anniversary of the School of Mathematical Sciences, Universiti Sains Malaysia, we are deeply honored and profoundly grateful to acknowledge the institution that has been a pillar of academic excellence, a beacon of intellectual innovation, and a thriving hub for scholarly inquiry over the past five decades. The School's unwavering commitment to fostering a vibrant research culture and its dedication to nurturing the next generation of scholars have established it as a leading center of mathematical research and education.

This book, *Analytical Inventory Management and Optimization: Theories, Methods, and Applications*, is a direct reflection of the academic rigor and research tradition that the School has upheld. It is the product of a rich intellectual heritage that has been cultivated over the years, driven by the School's vision to advance mathematical sciences, not only in Malaysia but on a global scale.

We would like to extend our deepest appreciation to the leadership of the School of Mathematical Sciences, whose far-reaching vision and unwavering support have been integral to the success of this endeavor. The strategic foresight and academic stewardship provided by the administration have created an environment where research flourishes and new ideas are nurtured. Their support has allowed us to push the boundaries of knowledge and delve deeply into the complex subject matter that defines this book.

Special recognition is due to the faculty members, whose mentorship and intellectual contributions have been indispensable in shaping this work. Their tireless efforts, both in the classroom and in research collaborations, have significantly influenced the development of the concepts and methods presented in this book. Moreover, the collaborative spirit among colleagues and peers has enriched our research journey, fostering an environment where ideas are freely exchanged and refined, leading to the comprehensive treatment of analytical inventory management found in this volume.

We extend our gratitude to Universiti Sains Malaysia as a whole, whose continuous commitment to academic excellence and research advancement has been the foundation upon which generations of scholars have built their careers. The University's dedication to creating an intellectually stimulating and resource-rich environment has allowed countless researchers to thrive and make significant contributions to the global scientific community. That this book is being published during the School's golden jubilee celebration makes it all the more significant, as it stands as a proud testament to the achievements and scholarly pursuits that have been nurtured here.

We also wish to acknowledge and thank Routledge and CRC Press for their belief in the importance of this work. Their professionalism, dedication, and global reach have ensured that the ideas and insights contained in this book will have a far-reaching impact, extending to scholars, practitioners, and institutions worldwide. Your partnership has been invaluable in making this work accessible to a broader audience.

Lastly, our heartfelt appreciation goes to the students, researchers, and academic staff who have been a constant source of inspiration. Their dedication to learning, insatiable curiosity, and passion for knowledge have driven us to explore new frontiers in inventory management and optimization. This book is, in many ways, a tribute to their pursuit of excellence. It is through their enthusiasm and commitment that we have been motivated to continue contributing to the ever-evolving field of mathematical sciences.

In sum, this book is a reflection of the collective efforts, vision, and dedication of all those who have been part of this academic journey. It is our hope that the work presented here will contribute to the ongoing dialogue and advancement in inventory management and optimization, and serve as a lasting legacy in the field of mathematical sciences.

School of Mathematical Sciences
Universiti Sains Malaysia (USM)
Professor Dr. Hailiza Kamarulhaili
August 2024

About the Authors

Majid Khan Majahar Ali is a senior lecturer in the School of Mathematical Sciences at Universiti Sains Malaysia, where he joined in 2017 specializing in Operational Research. He earned his Ph.D. in Mathematics from Universiti Malaysia Sabah. His research spans seaweed cultivation, solar drying systems, and the use of big data for models and simulations to predict moisture loss under various environmental conditions.

Sani Rabiu is a lecturer at the Nigerian Defence Academy in Kaduna. He holds a B.Sc. and a Postgraduate Diploma in Education from Bayero University Kano, a Postgraduate Diploma from the Nigerian Defence Academy, an M.Ed. from the National Open University of Nigeria, and an M.Sc. from Aliko Dangote University of Science and Technology. Currently completing his Ph.D. in Operational Research at Universiti Sains Malaysia, his research focuses on mathematical modeling and optimization techniques to address real-world problems related to inventory management and optimization.

Mohd Tahir Ismail is an associate professor and researcher at Universiti Sains Malaysia. His research interests include financial time series, econometrics, categorical data analysis, and applied statistics. He is currently the Vice President of the Malaysian Mathematical Sciences Society and an active member of other scientific professional bodies.

Contributors

Mohd Tahir Ismail
School of Mathematical Sciences
Universiti Sains Malaysia (USM)
Gelugor, Penang, Malaysia

Majid Khan Majahar Ali
School of Mathematical Sciences
Universiti Sains Malaysia (USM)
Gelugor, Penang, Malaysia

Sani Rabiu
Nigerian Defence Academy Kaduna, Nigeria
School of Mathematical Sciences
Universiti Sains Malaysia (USM)
Gelugor, Penang, Malaysia

1

Inventory Management

This chapter presents the fundamental concepts and principles of inventory management, a crucial component of supply chain management and logistics. It covers the key objectives, costs, and challenges associated with managing inventory, along with the different types of inventories and strategies for controlling and optimizing them. The chapter delves into essential practices such as inventory analytics and optimization models, offering a comprehensive overview for understanding the role of inventory management in various industries. It serves as a foundational guide for practitioners, researchers, and students, preparing them for more advanced study in inventory optimization techniques and their applications in organizational operations.

1.1 Introduction to Inventory Management

Inventory management is a critical component of supply chain and logistics management. It involves overseeing the flow of goods from manufacturers to warehouses and from these facilities to points of sale. Inventory management aims to ensure the right quantity of goods is available at the right time to meet customer demand, while also minimizing the costs associated with holding inventory. At its core, inventory management involves the tracking and controlling of stock items, which can include raw materials, work-in-progress (WIP), and finished goods. Effective inventory management helps businesses maintain optimal inventory levels, which is essential for preventing overstocking and stockouts. Overstocking can lead to high holding costs and potential obsolescence, while stockouts can result in lost sales, customer dissatisfaction, and a damaged reputation. The importance of inventory management in ensuring the smooth operation of supply chains cannot be overstated. Here are some key reasons why inventory management is vital:

- **Demand Fulfillment:** Efficient inventory management ensures that products are available to meet customer demand without delay. This is crucial for maintaining customer satisfaction and loyalty.
- **Cost Control:** Through optimizing inventory levels, companies can reduce the costs associated with storing excess inventory, such as

DOI: 10.1201/9781003534860-1

warehousing costs, insurance, and spoilage. This contributes to overall cost efficiency.

- **Cash Flow Management:** Inventory is a significant component of a company's working capital. Effective inventory management frees up cash that would otherwise be tied up in unsold stock, allowing for better cash flow management and investment in other areas of the business.

- **Operational Efficiency:** Proper inventory management streamlines operations by ensuring that materials and products are available when needed, which minimizes production delays and disruptions in the supply chain.

- **Forecasting and Planning:** Inventory management provides valuable data that can be used for forecasting future demand and planning production schedules. This data-driven approach helps in making informed decisions and aligning inventory levels with market trends.

- **Risk Management:** Through maintaining appropriate inventory levels, companies can mitigate risks associated with supply chain disruptions, such as supplier delays, natural disasters, and geopolitical events. Having buffer stock can ensure business continuity in such situations.

Effective inventory management relies on a combination of strategies, techniques, and technologies. Companies often use inventory management systems (IMS) to automate and streamline inventory tracking, ordering, and replenishment processes. These systems can provide real-time visibility into inventory levels, helping businesses make data-driven decisions. Inventory management is an indispensable part of supply chain and logistics management. It plays a crucial role in ensuring that the supply chain operates smoothly, cost-effectively, and responsively to market demands. As businesses strive to meet customer expectations and remain competitive, the importance of effective inventory management will continue to grow.

1.2 Key Concepts and Principles

In this section, we discuss the basic terminology and principles guiding effective inventory management practices. Understanding these concepts is fundamental to mastering inventory management and ensuring the smooth operation of supply chains.

- **Inventory:** Inventory refers to the goods and materials that a business holds for the purpose of resale, production, or utilization in services. It includes raw materials, WIP and finished goods.

- **Stock:** Stock is often used interchangeably with inventory, though it specifically refers to the items that are available for sale or use.
- **Lead Time:** Lead time is the time interval between placing an order and receiving the goods. It includes the time taken for order processing, manufacturing, and shipping. Understanding lead time is crucial for planning and maintaining adequate inventory levels.
- **Reorder Point:** The reorder point is the inventory level at which a new order should be placed to replenish stock before it runs out. It is calculated based on lead time and average demand during that lead time.
- **Safety Stock:** Safety stock is the additional inventory held to mitigate the risk of stockouts caused by uncertainties in demand or supply. It acts as a buffer against variability in the supply chain.
- **Economic Order Quantity (EOQ):** EOQ is a model used to determine the optimal order quantity that minimizes the total cost of inventory, including ordering and holding costs.
- **Just-In-Time (JIT):** JIT is an inventory strategy where materials and products are produced or acquired only as needed for immediate use, reducing inventory holding costs and increasing efficiency.
- **Stock Keeping Unit (SKU):** SKU is a unique identifier for each distinct product and service that can be purchased. SKUs are used to track inventory and manage stock efficiently.
- **Inventory Turnover:** Inventory turnover is a ratio that measures how many times inventory is sold and replaced over a specific period. It is an indicator of inventory efficiency and product demand.

1.3 Principles of Inventory Management

Effective inventory management is guided by several key principles that ensure optimal inventory levels and smooth supply chain operations:

1.3.1 Inventory Control Systems

Inventory control systems are the backbone of inventory management, providing the tools and technologies necessary to track and manage inventory levels, orders, and deliveries. These systems can be manual or automated, with modern enterprises increasingly relying on sophisticated software solutions such as enterprise resource planning (ERP) systems and warehouse management systems (WMS). These technologies facilitate real-time inventory tracking, demand forecasting, and efficient order fulfillment.

1.3.2 Demand Forecasting

Accurate demand forecasting is essential for maintaining optimal inventory levels. It involves predicting future customer demand based on historical data, market trends, and other influencing factors. Effective demand forecasting helps prevent overstocking and stockouts, enabling organizations to adjust their inventory strategies proactively. Advanced analytical techniques, including statistical models and machine learning algorithms, are often employed to enhance the accuracy of demand forecasts.

1.3.3 Inventory Planning

Inventory planning involves determining the appropriate quantity and timing of inventory to be stocked. This component focuses on setting reorder points, calculating safety stock levels, and determining order quantities. Inventory planning aims to strike a balance between having sufficient inventory to meet demand and minimizing the costs associated with holding excess inventory. Techniques such as JIT inventory and EOQ are commonly used in this process.

1.3.4 Supplier Management

Effective supplier management is critical for ensuring a reliable supply of inventory. It encompasses selecting and evaluating suppliers, negotiating contracts, and establishing strong relationships with key suppliers. Supplier management aims to ensure timely delivery of high-quality materials and products, thereby reducing the risk of production delays and stockouts. Collaboration and communication with suppliers are enhanced through the use of supplier relationship management (SRM) systems.

1.3.5 Inventory Optimization

Inventory optimization focuses on finding the most efficient balance between inventory investment and service levels. It involves analyzing inventory data to identify opportunities for improvement and implementing strategies to optimize stock levels. Techniques such as ABC analysis, which categorizes inventory based on its importance, and multi-echelon inventory optimization (MEIO), which considers the entire supply chain network, are employed to enhance inventory efficiency.

1.3.6 Inventory Auditing

Regular inventory auditing is crucial for maintaining accurate inventory records. Auditing involves physically counting inventory and reconciling it with recorded data to identify discrepancies. This process helps detect issues such as shrinkage, theft, or administrative errors. Regular audits ensure data accuracy, which is vital for effective inventory management and decision-making.

1.3.7 Safety Stock Management

Safety stock acts as a buffer to protect against uncertainties in demand and supply. Determining the appropriate level of safety stock involves analyzing variability in demand and lead times. Proper safety stock management helps prevent stockouts and ensures that customer demand can be met even during unexpected disruptions. Balancing the cost of holding safety stock with the risk of stockouts is a key consideration in this component.

1.3.8 Inventory Reporting and Analytics

Comprehensive reporting and analytics are essential for monitoring and managing inventory performance. Inventory reports provide insights into key metrics such as inventory turnover, fill rate, and carrying costs. Advanced analytics tools enable organizations to identify trends, patterns, and areas for improvement. Data-driven insights support strategic decision-making and continuous improvement in inventory management practices.

1.3.9 Order Management

Order management involves the processes related to receiving, tracking, and fulfilling customer orders. Effective order management ensures that customer orders are processed efficiently and accurately, from order entry through to delivery. This component is crucial for maintaining customer satisfaction and operational efficiency. Integrated order management systems can automate many of these processes, reducing errors and improving order fulfillment times.

1.3.10 Return Management

Return management, or reverse logistics, deals with the handling of returned goods. This includes the processes for receiving, inspecting, restocking, or disposing of returned items. Effective return management helps minimize the costs associated with returns and maximizes the recovery of value from returned goods. It also plays a significant role in customer satisfaction and retention.

1.3.11 Cycle Counting

Cycle counting is a periodic auditing process where a subset of inventory is counted on a regular basis, rather than conducting a full physical inventory count at once. This method helps maintain accurate inventory records throughout the year and can identify and rectify discrepancies more quickly than annual audits.

1.3.12 Inventory Valuation Methods

Inventory valuation methods determine the cost assigned to inventory and how it impacts financial statements. Common methods include first-in-first-out (FIFO), last-in-first-out (LIFO), and weighted average cost. Choosing the right inventory valuation method is crucial for accurate financial reporting and tax compliance.

1.3.13 Warehouse Layout and Design

Efficient warehouse layout and design significantly impact inventory management. Proper design maximizes space utilization, facilitates easy movement of goods, and reduces handling times. Techniques such as slotting, where items are placed in the most optimal locations within the warehouse, and utilizing automated storage and retrieval systems (AS/RS), can greatly enhance warehouse efficiency.

1.3.14 Inventory Segmentation

Inventory segmentation involves classifying inventory into different categories based on criteria such as demand variability, value, and supply chain characteristics. Common segmentation strategies include ABC analysis, where items are classified into three categories (A, B, and C) based on their importance and value. This helps prioritize management efforts and resources.

1.3.15 Sustainability and Environmental Considerations

Increasingly, inventory management must incorporate sustainability and environmental considerations. This includes managing inventory in a way that minimizes waste, reduces carbon footprint, and adheres to regulatory and corporate sustainability goals. Practices such as recycling, reducing excess inventory, and choosing eco-friendly suppliers contribute to more sustainable inventory management.

1.3.16 Collaboration and Integration With Other Business Functions

Inventory management does not operate in isolation; it must be closely integrated with other business functions such as marketing, finance, and production. Collaboration across these functions ensures alignment of inventory strategies with overall business goals, enhances communication, and improves overall efficiency.

1.3.17 Technology and Automation

The role of technology and automation in inventory management cannot be overstated. Technologies such as RFID (radio frequency identification), IoT (Internet of Things) sensors, and blockchain can enhance visibility, accuracy, and traceability of inventory. Automation tools, including robotic process automation (RPA) and artificial intelligence (AI), can streamline and optimize various inventory management processes.

1.3.18 Lead Time Management

Lead time management involves monitoring and reducing the time taken from ordering inventory to its receipt. Effective lead time management helps in maintaining optimal inventory levels and reduces the risk of stockouts. This includes coordinating with suppliers to ensure timely deliveries and using techniques like lead time buffering.

1.3.19 Product Life Cycle Management

Inventory management strategies need to account for the various stages of a product's life cycle, from introduction and growth to maturity and decline. Different stages require different inventory approaches, such as ramping up production during growth phases and managing obsolescence during decline phases.

1.3.20 Service Level Agreements and Performance Metrics

Service level agreements (SLAs) with suppliers and performance metrics for inventory management processes help ensure that all parties meet agreed-upon standards. Key performance indicators (KPIs) such as order accuracy, fulfillment time, and inventory turnover rates are critical for measuring and improving performance.

1.3.21 Risk Management and Contingency Planning

Risk management involves identifying potential risks that could disrupt inventory flow, such as supplier failures, natural disasters, or market fluctuations, and developing contingency plans to mitigate these risks. This includes maintaining safety stock, diversifying suppliers, and having alternative logistics arrangements.

1.3.22 Inventory Accounting and Financial Integration

Inventory management has significant financial implications. Integrating inventory management with financial accounting ensures accurate reporting of inventory costs, valuation, and impacts on profitability. This includes adherence to accounting standards and regular financial audits.

1.3.23 Regulatory Compliance

Compliance with industry regulations and standards is essential for inventory management. This includes adherence to safety standards, environmental regulations, and trade compliance. Ensuring regulatory compliance helps avoid legal issues and potential fines.

1.3.24 Customer Relationship Management Integration

Integrating inventory management with customer relationship management (CRM) systems allows for better alignment with customer needs and preferences. This integration helps in demand planning and ensures that inventory strategies are aligned with customer service goals.

1.3.25 Distribution and Transportation Management

Effective inventory management also involves the efficient distribution and transportation of goods. This includes optimizing transportation routes, selecting appropriate transportation modes, and managing logistics partners to ensure timely and cost-effective delivery of inventory.

1.3.26 Vendor Managed Inventory and Consignment Stock

Vendor managed inventory (VMI) is a collaborative strategy where the supplier manages the inventory levels based on the agreed-upon metrics. Consignment stock involves suppliers placing inventory at the buyer's location but retaining ownership until the inventory is used. These strategies can improve inventory efficiency and reduce capital investment.

1.3.27 Inventory Replenishment Strategies

Replenishment strategies determine how and when inventory is restocked. Common strategies include continuous replenishment, periodic review, and fixed order quantity systems. Choosing the right replenishment strategy helps maintain optimal inventory levels and improve service levels.

Understanding the key concepts and principles of inventory management is fundamental to managing stock effectively. Mastering these concepts, businesses can ensure optimal inventory levels, reduce costs, and maintain a competitive edge in the market. Effective inventory management not only supports operational efficiency but also contributes to overall customer satisfaction and business success.

1.4 Objectives of Inventory Management

The primary objectives of inventory management are to ensure a sufficient supply to meet demand, minimize costs associated with holding and managing inventory, and maintain balance between overstocking and stockouts. Achieving these objectives is crucial for maintaining operational efficiency, customer satisfaction, and overall business profitability. As depicted in Figure 1.1: Inventory Management Stages, the dynamic relationship between inventory levels and time highlights critical phases such as production, demand fluctuations, lead time, replenishment, and backorder management. The figure also identifies key points like the reorder point and stockout point, which are essential for maintaining optimal inventory levels and avoiding disruptions. This section discuss each of these objectives in detail.

1.4.1 Ensuring Sufficient Supply to Meet Demand

One of the foremost objectives of inventory management is to ensure that there is enough inventory to meet customer demand at all times. This involves:

- **Demand Forecasting:** Accurate demand forecasting is essential for anticipating customer needs and planning inventory levels accordingly. Analyzing historical sales data, market trends, and seasonality, businesses can predict future demand more accurately and adjust their inventory levels to match.

- **Safety Stock:** Maintaining safety stock is a key strategy to protect against demand variability and supply chain disruptions. Safety stock acts as a buffer, ensuring that there is sufficient inventory to cover unexpected spikes in demand or delays in replenishment.

- **Reorder Point:** Setting appropriate reorder points ensures that new orders are placed in a timely manner before stock levels fall below the critical threshold. The reorder point is typically calculated based on lead time and average demand during that lead time.

1.4.2 Minimizing Costs Associated With Holding and Managing Inventory

Effective inventory management seeks to minimize the various costs associated with holding and managing inventory. These costs can be categorized into several types:

- **Holding Costs:** Holding costs, also known as carrying costs, include expenses related to storing and maintaining inventory. These costs encompass warehousing, insurance, depreciation, and opportunity costs of capital tied up in inventory. Reducing holding costs involves optimizing inventory levels and improving storage efficiency.

- **Ordering Costs:** Ordering costs are incurred each time an order is placed for replenishing inventory. These include administrative expenses, transportation costs, and handling fees. Using techniques such as EOQ can help determine the optimal order size that minimizes total ordering and holding costs.

- **Stockout Costs:** Stockout costs arise when inventory levels are insufficient to meet demand, leading to lost sales, backorder costs, and potential damage to customer relationships. Maintaining adequate inventory levels and safety stock, businesses can avoid these costs.

1.4.3 Balancing Between Overstocking and Stockouts

Achieving a balance between overstocking and stockouts is a critical objective of inventory management. Both scenarios have significant implications for business operations and profitability:

- **Overstocking:** Overstocking occurs when a company holds more inventory than necessary to meet demand. This situation can lead to increased holding costs, reduced liquidity, and potential obsolescence or spoilage of inventory. Overstocking also ties up capital that could be used more effectively elsewhere in the business.

- **Stockouts:** Stockouts happen when inventory levels fall below the required amount to meet demand. This can result in lost sales, customer dissatisfaction, and potential damage to the company's reputation. Frequent stockouts can also lead to increased expediting costs and production disruptions.

Balancing these two extremes involves:

- **Inventory Optimization:** Utilizing inventory optimization techniques, such as JIT and lean inventory practices, helps in maintaining optimal inventory levels. These methods focus on reducing excess inventory while ensuring sufficient supply to meet demand.
- **Supply Chain Coordination:** Effective coordination and communication with suppliers and other supply chain partners are crucial for timely replenishment and minimizing lead times. Strong supplier relationships can enhance reliability and flexibility in inventory management.
- **Continuous Monitoring and Adjustment:** Regular monitoring of inventory levels, demand patterns, and market conditions allows businesses to adjust their inventory strategies dynamically. Implementing IMS and using real-time data analytics can support this continuous adjustment process.

The objectives of inventory management are multifaceted and interrelated. Ensuring sufficient supply to meet demand, minimizing costs, and balancing overstocking and stockouts are all essential for maintaining an efficient and responsive supply chain. Achieving these objectives, businesses can enhance their operational efficiency, reduce costs, and improve customer satisfaction. Effective inventory management thus plays a pivotal role in supporting overall business success and competitiveness in the market.

1.5 Costs Associated with Inventory Management

Inventory management involves various costs that can significantly impact a company's profitability and operational efficiency. Understanding these costs is essential for effective inventory management and decision-making. The primary costs associated with inventory management include holding costs, ordering costs, and stockout costs. This section details these costs and their implications.

1.5.1 Holding Costs

Holding costs, also known as carrying costs, are the expenses incurred for storing and maintaining inventory over a period of time. These costs can be substantial and include several components:

- **Storage Costs:** These are the expenses related to warehousing inventory. They include rent or mortgage payments for warehouse space, utilities, and costs associated with maintaining the storage facility.

- **Insurance:** Inventory must be insured against risks such as theft, fire, and natural disasters. Insurance premiums contribute to the holding costs.
- **Depreciation and Obsolescence:** Inventory items may lose value over time due to depreciation or become obsolete as new products are introduced. This can lead to a loss in inventory value, contributing to holding costs.
- **Spoilage and Shrinkage:** Perishable goods can spoil if not sold in time, and inventory shrinkage can occur due to theft, damage, or administrative errors. Both spoilage and shrinkage add to holding costs.
- **Opportunity Costs:** Capital tied up in inventory represents money that could be invested elsewhere in the business. The opportunity cost of this capital is an important component of holding costs.

The implications of high holding costs are significant. They can strain a company's financial resources, reduce profitability, and lead to inefficiencies in the supply chain. Effective inventory management strategies aim to minimize holding costs by optimizing inventory levels and improving turnover rates.

1.5.2 Ordering Costs

Ordering costs are the expenses associated with placing and receiving orders for inventory replenishment. These costs are incurred each time an order is placed and include:

- **Procurement Costs:** These are the administrative expenses involved in preparing purchase orders, processing payments, and managing supplier relationships. They include labor costs for employees who handle procurement tasks.
- **Transportation Costs:** The costs of shipping and handling goods from suppliers to the warehouse or point of sale are part of ordering costs. This includes freight charges, handling fees, and any special delivery requirements.
- **Receiving Costs:** Expenses related to receiving and inspecting incoming inventory, such as labor for unloading and quality control checks, are also part of ordering costs.
- **Setup Costs:** In manufacturing environments, setup costs include the expenses incurred to prepare equipment and processes for production runs. These costs are relevant when inventory orders involve production changes.

High ordering costs can deter frequent ordering, leading to larger, less frequent orders that may increase holding costs. Balancing ordering costs with holding costs is essential for inventory optimization. Techniques such as EOQ help determine the optimal order size to minimize total inventory costs.

1.5.3 Stockout Costs

Stockout costs arise when inventory levels are insufficient to meet customer demand. These costs can have direct and indirect implications, including:

- **Lost Sales:** The most immediate impact of a stockout is the loss of sales revenue. Customers who cannot find the products they need may take their business elsewhere, leading to lost sales opportunities.
- **Backorder Costs:** When stockouts occur, companies may place back orders to fulfill customer demand once inventory is replenished. This process can incur additional administrative and shipping costs.
- **Customer Dissatisfaction:** Frequent stockouts can damage customer relationships and erode brand loyalty. Dissatisfied customers may switch to competitors, leading to long-term revenue loss.
- **Expediting Costs:** To mitigate the impact of stockouts, companies may expedite orders, incurring higher shipping and handling costs to replenish inventory quickly.
- **Production Disruptions:** In manufacturing, stockouts of raw materials or components can halt production lines, leading to downtime and increased production costs.

The implications of stockout costs highlight the importance of maintaining adequate inventory levels. Companies must balance the risk of stockouts with the costs of holding excess inventory to ensure customer satisfaction and operational efficiency. The costs associated with inventory management holding costs, ordering costs, and stockout costs are critical factors in determining the efficiency and profitability of supply chain operations. Effective inventory management strategies seek to balance these costs by optimizing inventory levels, improving turnover rates, and ensuring timely replenishment. Understanding and managing these costs, businesses can enhance their overall performance and competitiveness in the market.

1.6 Problems and Challenges in Inventory Management

Inventory management is fraught with numerous challenges that can impact a company's ability to maintain optimal inventory levels and ensure smooth operations. This section addresses some of the most common issues faced in managing inventory, including demand variability, lead time variability, and other significant challenges, and discusses strategies to overcome these challenges.

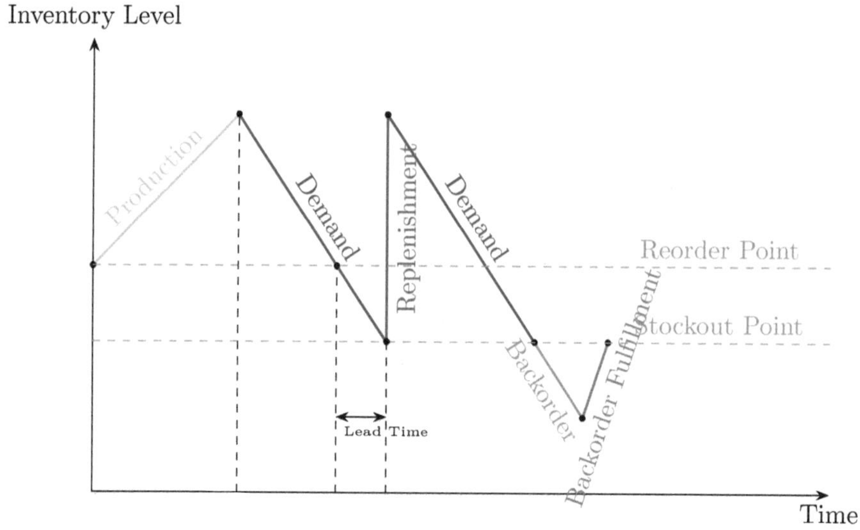

FIGURE 1.1
Inventory management stages.

1.6.1 Demand Variability

Demand variability refers to the fluctuations in customer demand for products over time. This can be due to seasonal trends, market dynamics, consumer preferences, economic conditions, and other factors. Demand variability poses several challenges for inventory management:

- **Forecasting Inaccuracies:** Predicting future demand accurately is difficult, and inaccuracies in demand forecasting can lead to either overstocking or stockouts.
- **Safety Stock Levels:** Determining the appropriate level of safety stock to buffer against demand variability without incurring excessive holding costs is challenging.
- **Promotions and Marketing Campaigns:** Sales promotions and marketing campaigns can cause sudden spikes in demand that are hard to predict and manage.

1.6.2 Strategies to Overcome Demand Variability

The strategies to overcome demand variability include:

- **Improved Forecasting Techniques:** Utilize advanced forecasting methods, such as machine learning and statistical models, to enhance accuracy. Incorporate real-time data and market intelligence to adjust forecasts dynamically.

- **Demand Sensing:** Implement demand sensing techniques to capture and analyze short-term demand signals from various sources, enabling quicker response to demand changes.
- **Collaborative Planning:** Engage in collaborative planning with key customers and suppliers to share insights and align forecasts. Collaborative planning, forecasting, and replenishment (CPFR) can improve forecast accuracy and inventory management.
- **Flexible Inventory Policies:** Develop flexible inventory policies that can adapt to changing demand patterns. Use dynamic safety stock calculations that consider demand variability and lead times.

1.6.3 Lead Time Variability

Lead time variability refers to the fluctuations in the time it takes for an order to be fulfilled from the moment it is placed until it is received. This variability can result from supplier delays, transportation issues, production bottlenecks, and other factors. Lead time variability creates several challenges:

- **Inventory Planning:** Variability in lead times complicates inventory planning and makes it difficult to determine the optimal reorder point.
- **Safety Stock Requirements:** To buffer against lead time variability, higher levels of safety stock may be required, increasing holding costs.
- **Service Levels:** Maintaining high service levels becomes challenging when lead times are unpredictable, potentially leading to stockouts and customer dissatisfaction.

1.6.4 Strategies to Overcome Lead Time Variability

The strategies to overcome lead time variability include:

- **Supplier Relationship Management:** Develop strong relationships with suppliers to enhance communication, reliability, and collaboration. Work with suppliers to reduce lead time variability and improve consistency.
- **Multiple Sourcing:** Diversify the supplier base to reduce dependency on a single source. Multiple sourcing can mitigate the impact of delays from any one supplier.
- **Buffer Inventory:** Maintain buffer inventory or safety stock to protect against lead time variability. This stock acts as a cushion to cover unexpected delays.
- **Lead Time Analysis:** Regularly analyze and monitor lead times to identify patterns and potential issues. Use this analysis to adjust inventory policies and reorder points accordingly.

1.6.5 Other Common Challenges

Apart from demand and lead time variability, several other challenges can impact inventory management:

- **Inventory Accuracy:** Inaccurate inventory records can lead to stockouts, overstocking, and operational inefficiencies. Causes of inaccuracies include data entry errors, theft, and misplacement of items.
- **Product Life Cycle Management:** Managing inventory for products with varying life cycles, from introduction to decline, requires different strategies and can be complex.
- **Supply Chain Disruptions:** Natural disasters, geopolitical events, and other disruptions can impact the supply chain, causing delays and affecting inventory levels.
- **Cost Management:** Balancing the costs associated with ordering, holding, and stockouts to achieve overall cost efficiency is a perpetual challenge.

1.6.6 Strategies to Overcome Other Common Challenges

The strategies to overcome other common challenges include:

- **Inventory Audits and Cycle Counting:** Conduct regular inventory audits and implement cycle counting to ensure accuracy of inventory records and promptly address discrepancies.
- **Product Segmentation:** Use product segmentation to categorize inventory based on demand patterns, life cycle stages, and profitability. Apply tailored inventory strategies for each segment.
- **Risk Management and Contingency Planning:** Develop risk management strategies and contingency plans to address potential supply chain disruptions. This includes identifying alternative suppliers and transportation routes.
- **Integrated Inventory Management Systems:** Utilize integrated inventory management systems that provide real-time visibility, data analytics, and automation. These systems help in optimizing inventory levels and enhancing decision-making.

Inventory management is a complex and challenging aspect of supply chain management. Addressing demand variability, lead time variability, and other common challenges requires a combination of advanced techniques, strategic planning, and continuous improvement. Implementing effective strategies and leveraging technology, businesses can overcome these challenges and achieve optimal inventory management, ensuring operational efficiency and customer satisfaction.

1.7 Types of Inventories

Effective inventory management requires a clear understanding of the different types of inventories that a business might hold. Each type of inventory serves a distinct purpose in the production and distribution process, and managing them appropriately is crucial for operational efficiency and customer satisfaction. This section explores the different types of inventories, including raw materials; WIP; finished goods; and maintenance, repair, and operations (MRO) supplies.

1.7.1 Raw Materials

Raw materials are the basic inputs that are used in the production process to create finished products. These materials are purchased from suppliers and stored until they are needed for manufacturing. Key characteristics and considerations for managing raw materials include:

- **Variety:** Raw materials can range from simple components like screws and bolts to complex materials like specialized chemicals or metals.
- **Storage:** Proper storage conditions are essential to maintain the quality and usability of raw materials. This may involve temperature control, humidity management, and protection from contamination.
- **Inventory Levels:** Maintaining optimal inventory levels of raw materials is critical to avoid production delays while minimizing holding costs. Techniques like JIT can help in reducing excess inventory.
- **Lead Times:** Understanding and managing lead times for raw material procurement is crucial to ensure that production schedules are not disrupted.

1.7.2 Work-in-Progress

WIP inventory refers to partially completed products that are still undergoing the production process. WIP inventory is a critical component of manufacturing and involves several considerations:

- **Production Stages:** WIP inventory can be found at various stages of the production process, from initial assembly to near completion. Effective tracking of WIP is essential to manage production flow and identify bottlenecks.
- **Value Addition:** As products move through different stages of production, their value increases. Managing WIP inventory involves ensuring that

resources are efficiently utilized and that production timelines are adhered to.

- **Cycle Time:** Reducing cycle time, or the time it takes for a product to move from one stage of production to the next, is key to optimizing WIP inventory. Techniques such as lean manufacturing and continuous improvement can help in minimizing cycle time.

- **Inventory Control:** Effective control of WIP inventory involves real-time monitoring and tracking systems to ensure accurate visibility of production progress and inventory levels.

1.7.3 Finished Goods

Finished goods are the final products that have completed the production process and are ready for sale to customers. Managing finished goods inventory is crucial for meeting customer demand and ensuring timely delivery. Key considerations include:

- **Storage and Handling:** Finished goods must be stored in conditions that preserve their quality and prevent damage. This may involve appropriate packaging, temperature control, and secure warehousing.

- **Inventory Levels:** Balancing inventory levels of finished goods is essential to avoid overstocking, which increases holding costs, and stockouts, which can lead to lost sales and customer dissatisfaction. Inventory optimization techniques, such as EOQ, can help in maintaining optimal levels.

- **Demand Forecasting:** Accurate demand forecasting is critical for managing finished goods inventory. It helps in anticipating customer needs and planning production and inventory accordingly.

- **Distribution:** Efficient distribution systems are necessary to ensure that finished goods reach customers promptly. This involves coordinating with logistics providers and managing transportation and delivery schedules.

1.7.4 Maintenance, Repair, and Operations Supplies

MRO supplies refer to the items that are used in the maintenance, repair, and operation of manufacturing facilities and equipment. While MRO supplies are not part of the final product, they are essential for ensuring smooth and uninterrupted production processes. Considerations for managing MRO supplies include:

- **Variety and Usage:** MRO supplies encompass a wide range of items, including tools, spare parts, lubricants, cleaning agents, and safety equipment. Understanding the usage patterns and criticality of these items is important for effective inventory management.

- **Stock Levels:** Maintaining appropriate stock levels of MRO supplies is crucial to avoid production downtime due to equipment failure or maintenance needs. Stockouts of critical MRO items can halt production and lead to significant losses.
- **Supplier Relationships:** Building strong relationships with suppliers of MRO supplies can ensure reliable and timely replenishment. This helps in maintaining consistent inventory levels and reducing lead times.
- **Inventory Control Systems:** Implementing inventory control systems for MRO supplies, such as automated replenishment and real-time tracking, can enhance efficiency and reduce administrative burdens.

Understanding the different types of inventories such as raw materials, WIP, finished goods, and MRO supplies is fundamental to effective inventory management. Each type of inventory has its own unique characteristics and management requirements. Addressing these specific needs and employing appropriate strategies, businesses can optimize inventory levels, reduce costs, and ensure smooth production and distribution processes. Effective management of all inventory types contributes to overall operational efficiency and business success.

1.8 Inventory Control and Optimization

Effective inventory control and optimization are crucial for maintaining optimal inventory levels, minimizing costs, and ensuring smooth operations. This section explains various techniques for monitoring and controlling inventory levels, such as periodic review systems and continuous review systems, and explores methods to optimize inventory, including EOQ, JIT, and safety stock calculations.

1.8.1 Definition of Inventory Optimization

Inventory optimization is a systematic approach to managing an organization's inventory levels to achieve a balance between supply and demand while minimizing costs and maximizing service levels. It involves the use of mathematical models, algorithms, and software tools to determine the most efficient inventory policies.

1.8.2 What Is Inventory Optimization

Inventory optimization refers to the process of determining the optimal quantities and timing of inventory to be held by an organization. The aim is to ensure that the right amount of inventory is available at the right time to meet customer demand, while minimizing the costs associated with holding and ordering inventory. This involves balancing various factors such as lead times, demand variability, holding costs, and service level requirements.

1.8.3 Importance of Inventory Optimization

Inventory optimization is crucial for several reasons:

- **Cost Efficiency:** Proper inventory optimization reduces the costs associated with holding inventory, such as storage costs, insurance, and taxes. It also minimizes ordering costs by determining the most cost-effective order quantities and timing.

- **Service Level Improvement:** Maintaining optimal inventory levels, companies can better meet customer demand, reducing the likelihood of stockouts and backorders. This leads to higher customer satisfaction and loyalty.

- **Operational Efficiency:** Efficient inventory management ensures smooth operations by preventing overstocking and understocking, which can disrupt production and sales processes.

- **Capital Allocation:** Optimized inventory management frees up capital that can be used for other strategic initiatives, enhancing the overall financial health of the organization.

1.8.4 Goals of Inventory Optimization

The primary goals of inventory optimization include:

- **Minimizing Costs:** Reducing holding, ordering, and shortage costs while maintaining sufficient inventory to meet demand.

- **Maximizing Service Levels:** Ensuring high product availability to meet customer demand promptly, thereby improving customer satisfaction and loyalty.

- **Balancing Supply and Demand:** Maintaining a balance between supply and demand to avoid overstocking and stockouts, which can lead to lost sales and increased costs.

- **Enhancing Forecast Accuracy:** Using advanced forecasting techniques to predict future demand accurately and adjust inventory levels accordingly.

1.9 Relevance to Organizations

Inventory optimization has significant implications for organizations, impacting various aspects of their operations and strategic goals.

1.9.1 Impact on Customer Satisfaction and Profitability

- **Customer Satisfaction:** Optimized inventory levels ensure that products are available when customers need them, leading to higher satisfaction and repeat business. Avoiding stockouts and backorders is crucial for maintaining a positive customer experience.

- **Profitability:** Effective inventory optimization reduces costs associated with holding and ordering inventory, directly impacting the bottom line. By minimizing excess inventory and avoiding stockouts, companies can improve their profitability.

1.9.2 Importance of Maintaining Optimal Inventory Levels

Maintaining optimal inventory levels is critical for several reasons:

- **Operational Efficiency:** Proper inventory levels prevent production delays and disruptions, ensuring smooth and efficient operations. This is particularly important in manufacturing and retail sectors where timely availability of inventory is crucial.

- **Cost Management:** Optimal inventory levels help in managing and reducing costs related to storage, insurance, and obsolescence. Overstocking leads to higher holding costs, while understocking can result in lost sales and customer dissatisfaction.

- **Competitive Advantage:** Companies that effectively manage their inventory can respond more quickly to market changes and customer needs, providing a competitive edge over those with less efficient inventory systems. This agility can be crucial in industries with rapidly changing demand patterns.

- **Sustainability:** Reducing excess inventory helps in minimizing waste and contributes to more sustainable business practices. This aligns with the growing emphasis on sustainability and corporate social responsibility.

- **Financial Health:** Optimized inventory management improves cash flow by reducing the amount of capital tied up in inventory. This allows companies to invest in other strategic areas and enhance their overall financial health.

1.9.3 Inventory Control Systems

Inventory control systems are methodologies used to track and manage inventory levels, orders, sales, and deliveries. They ensure that the right amount of inventory is available to meet demand while minimizing costs. Two primary inventory control systems are periodic review systems and continuous review systems.

1.9.3.1 Periodic Review Systems

In periodic review systems, inventory levels are checked at regular intervals, and orders are placed to replenish inventory to a predetermined level. Key characteristics and considerations include:

- **Review Period:** The frequency of inventory reviews is a critical factor. Common intervals include weekly, biweekly, or monthly reviews, depending on the product and demand variability.
- **Order Quantity:** During each review period, the order quantity is determined based on the difference between the current inventory level and the target inventory level.
- **Advantages:** Periodic review systems are simple to implement and manage, making them suitable for businesses with relatively stable demand patterns and lower-value items.
- **Disadvantages:** These systems may lead to higher safety stock levels to buffer against demand variability within the review period, potentially increasing holding costs.

1.9.3.2 Continuous Review Systems

Continuous review systems, also known as perpetual inventory systems, involve constant monitoring of inventory levels. Orders are placed as soon as inventory drops to a predefined reorder point. Key characteristics and considerations include:

- **Real-Time Monitoring:** Inventory levels are tracked in real time using inventory management software, allowing for immediate replenishment actions.
- **Reorder Point:** A specific reorder point is set based on lead time and demand patterns. When inventory reaches this point, an order is triggered.
- **Advantages:** Continuous review systems provide better control over inventory levels, reducing the likelihood of stockouts and minimizing excess inventory.
- **Disadvantages:** These systems can be more complex and costly to implement due to the need for advanced technology and continuous monitoring.

1.9.4 Inventory Optimization Methods

Optimizing inventory involves determining the optimal order quantities and timing to balance costs and meet demand efficiently. Several methods are commonly used for inventory optimization:

1.9.4.1 Economic Order Quantity

The EOQ model is a fundamental inventory optimization technique that determines the optimal order size to minimize total inventory costs, which include ordering and holding costs. The EOQ formula is given by:

$$EOQ = \sqrt{\frac{2DS}{H}}$$

where:

D = Annual demand for the product

S = Ordering cost per order

H = Holding cost per unit per year

Key considerations for EOQ include:

- **Balancing Costs:** EOQ balances ordering costs (which decrease with larger orders) and holding costs (which increase with larger orders) to find the most cost-effective order quantity.
- **Assumptions:** The EOQ model assumes constant demand and lead time, which may not always hold true in practice.
- **Applications:** EOQ is widely used for products with stable demand and predictable lead times, providing a straightforward approach to inventory optimization.

1.9.4.2 Just-In-Time

JIT is an inventory management philosophy aimed at minimizing inventory levels by receiving goods only as they are needed in the production process. Key principles and benefits of JIT include:

- **Reduced Inventory:** JIT reduces the need for holding large amounts of inventory, leading to lower holding costs and less waste.
- **Improved Efficiency:** Synchronizing production schedules with demand, JIT improves overall operational efficiency and reduces lead times.
- **Supplier Relationships:** JIT requires strong relationships with reliable suppliers who can deliver goods on short notice and with high precision.

- **Challenges:** Implementing JIT can be challenging due to the need for precise demand forecasting, flexible production processes, and dependable supply chains.

1.9.4.3 Safety Stock Calculations

Safety stock is additional inventory held to protect against uncertainties in demand and lead time. Calculating the appropriate level of safety stock is crucial for preventing stockouts while minimizing excess inventory. The basic formula for safety stock is:

$$\textit{Safety Stock} = Z \times \sigma_{LT}$$

where:

Z = Service level factor (based on desired service level)

σ_{LT} = Standard deviation of demand during lead time

Key considerations for safety stock calculations include:

- **Service Levels:** The desired service level reflects the probability of not facing a stockout. Higher service levels require more safety stock.
- **Demand and Lead Time Variability:** Higher variability in demand and lead time necessitates higher safety stock levels to buffer against uncertainties.
- **Cost-Benefit Analysis:** Balancing the cost of holding safety stock against the potential cost of stockouts is essential for determining optimal safety stock levels.

Example 1.1. *A company sells a product with a highly variable demand. The lead time for replenishment is also uncertain. The company aims for a 95% service level to ensure a high probability of meeting customer demand without stockouts. The average demand during lead time is 100 units, and the standard deviation of demand during lead time is 20 units. What should the company maintain as safety stock to avoid stockouts?*

Solution 1.1. *Given:*

- *The average demand during lead time (μ_{LT}) is 100 units.*
- *The standard deviation of demand during lead time (σ_{LT}) is 20 units.*
- *The desired service level is 95%, corresponding to a Z-value (service level factor) of approximately 1.645 (from standard normal distribution tables).*

The safety stock (SS) can be calculated using the formula:

$$SS = Z \cdot \sigma_{LT}$$

Substituting the given values:

$$SS = 1.645 \cdot 20 = 32.9$$

Therefore, the company should maintain a safety stock of approximately 33 units to achieve a 95% service level.

1.10 Fundamental Concepts and Practices

Effective inventory management relies on a thorough understanding of the fundamental concepts and practices that provide the foundation for making informed decisions. This section provides an overview of inventory analytics and its role in decision-making. We also introduce key inventory models, including deterministic models and probabilistic models.

1.10.1 Inventory Analytics

Inventory analytics involves the use of data analysis, statistical methods, and optimization techniques to make informed decisions about inventory management. It plays a crucial role in understanding inventory performance, identifying trends, and optimizing inventory levels. Key aspects of inventory analytics include:

1.10.1.1 Data Collection and Management

Accurate and comprehensive data collection is the cornerstone of effective inventory analytics. This involves gathering data on:

- **Sales and Demand:** Historical sales data and demand forecasts.
- **Inventory Levels:** Current stock levels, including raw materials, WIP, and finished goods.
- **Lead Times:** Time taken for replenishment orders to be delivered.
- **Costs:** Ordering costs, holding costs, and stockout costs.

1.10.1.2 Data Analysis and Visualization

Analyzing and visualizing inventory data helps in identifying patterns, trends, and anomalies. Key techniques include:

- **Descriptive Analytics:** Summarizes historical data to understand past inventory performance.
- **Diagnostic Analytics:** Investigates the reasons behind inventory issues, such as stockouts or excess inventory.
- **Predictive Analytics:** Uses statistical models and machine learning to forecast future demand and inventory needs.
- **Prescriptive Analytics:** Recommends optimal inventory decisions based on predictive models and optimization algorithms.

1.10.1.3 Role in Decision-Making

Inventory analytics supports decision-making by providing actionable insights and optimizing inventory policies. Key applications include:

- **Demand Forecasting:** Predicting future demand to plan inventory levels and replenishment schedules.
- **Inventory Optimization:** Determining optimal order quantities, reorder points, and safety stock levels.
- **Performance Monitoring:** Tracking KPIs such as inventory turnover, service levels, and carrying costs.

1.10.2 Inventory Models

Inventory models are mathematical frameworks used to determine optimal inventory policies and decision rules. They can be broadly classified into deterministic models and probabilistic models.

1.10.2.1 Deterministic Models

Deterministic models assume that all parameters, such as demand and lead time, are known with certainty. These models provide precise solutions for inventory management. Key deterministic models include:

- **Economic Order Quantity:** The EOQ model determines the optimal order quantity that minimizes the total cost, considering ordering and holding costs. The EOQ formula is:

$$EOQ = \sqrt{\frac{2DS}{H}}$$

  where D is the annual demand, S is the ordering cost per order, and H is the holding cost per unit per year.

Example 1.2. *A small business sells 10,000 units of a product annually. The ordering cost is \$50 per order, and the holding cost is \$2 per unit per year. Calculate the EOQ of the business.*

Solution 1.2.

$$EOQ = \sqrt{\frac{2 \times 10{,}000 \times 50}{2}} = \sqrt{500{,}000} \approx 707 \ units$$

The business should order 707 units each time to minimize total costs.

- **Reorder Point:** The reorder point (ROP) model specifies the inventory level at which a new order should be placed. It is calculated as:

$$ROP = d \times L$$

where d is the average daily demand and L is the lead time in days.

Example 1.3. *A retailer has an average daily demand of 50 units and a lead time of seven days. Calculate the ROP of the retailer.*

Solution 1.3.

$$ROP = 50 \times 7 = 350 \ units$$

The retailer should place a new order when the inventory level drops to 350 units.

- **Production Order Quantity:** Similar to EOQ, the production order quantity (POQ) model is used when inventory is replenished gradually as items are produced. The POQ formula is:

$$POQ = \sqrt{\frac{2DS}{H \times (1 - \frac{d}{p})}}$$

where p is the production rate and d is the demand rate.

Example 1.4. *A factory has an annual demand of 20,000 units, an ordering cost of \$100, a holding cost of \$5 per unit per year, a production rate of 5,000 units per month, and a monthly demand of 1,500 units. Calculate the POQ of the factory.*

Solution 1.4.

$$POQ = \sqrt{\frac{2 \times 20,000 \times 100}{5 \times (1 - \frac{1{,}500}{5{,}000})}}$$

$$= \sqrt{\frac{4,000,000}{5 \times 0.7}} = \sqrt{\frac{4,000,000}{3.5}} \approx 1,070 \ units$$

The factory should produce 1,073 units per production run to minimize total costs.

- **Lot-for-Lot:** The lot-for-lot (L4L) model orders exactly what is needed for production based on the demand for each period, minimizing inventory holding costs but potentially increasing ordering costs.

  Example 1.5. *XYZ Electronics manufactures custom circuit boards. They have received orders for 200 units in January, 300 units in February, and 250 units in March. To meet these demands without carrying excess inventory, XYZ Electronics decides to use the L4L ordering approach.*

  Solution 1.5. *Using the L4L approach, XYZ Electronics orders exactly 200 units in January, 300 units in February, and 250 units in March. This aligns their orders with the monthly demand, thereby minimizing inventory holding costs. However, this approach may increase ordering costs due to the higher frequency of orders.*

- **Economic Production Quantity:** The economic production quantity (EPQ) model determines the optimal production quantity that minimizes total costs, accounting for the production rate. The EPQ formula is:

$$EPQ = \sqrt{\frac{2DS}{H} \times \frac{p}{p-d}}$$

  where D is the annual demand, S is the setup cost per production run, H is the holding cost per unit per year, p is the production rate, and d is the demand rate.

  Example 1.6. *A company has an annual demand of 15,000 units, a setup cost of \$80, a holding cost of \$4 per unit per year, a production rate of 6,000 units per month, and a monthly demand of 1,250 units. Calculate the EPQ of the company.*

  Solution 1.6.

$$EPQ = \sqrt{\frac{2 \times 15{,}000 \times 80}{4} \times \frac{6{,}000}{6{,}000 - 1{,}250}} = \sqrt{\frac{2{,}400{,}000}{4} \times \frac{6{,}000}{4{,}750}}$$

$$= \sqrt{600{,}000 \times 1.263} \approx 871 \textit{ units}$$

  The company should produce 871 units per production run to minimize total costs.

1.10.2.2 Probabilistic Models

Probabilistic models account for uncertainty in demand and lead time by incorporating variability into the calculations. These models provide more realistic solutions for inventory management. Key probabilistic models include:

- **Safety Stock Calculation:** Safety stock protects against demand and lead time variability. The basic formula for safety stock is:

$$\textit{Safety Stock} = Z \times \sigma_{LT}$$

where Z is the service level factor, and σ_{LT} is the standard deviation of demand during lead time.

Example 1.7. *A company aims for a 95% service level ($Z = 1.65$) and has a standard deviation of demand during lead time (σ_{LT}) of 30 units. Calculate the safety stock of the company.*

Solution 1.7.

$$\textit{Safety Stock} = 1.65 \times 30 = 49.5 \approx 50 \textit{ units}$$

The company should hold 50 units of safety stock to achieve the desired service level.

- **Continuous Review Model:** The continuous review (Q,R) model determines the optimal order quantity (Q) and reorder point (R) considering demand variability. The reorder point is calculated as:

$$R = d \times L + Z \times \sigma_{LT}$$

where d is the average demand, L is the lead time, Z is the service level factor, and σ_{LT} is the standard deviation of demand during lead time.

Example 1.8. *A retailer has an average demand of 40 units per day, a lead time of five days, a service level factor (Z) of 1.96, and a standard deviation of demand during lead time (σ_{LT}) of 20 units. Calculate the reorder point of the retailer.*

Solution 1.8.

$$R = 40 \times 5 + 1.96 \times 20 = 200 + 39.2 = 239.2 \approx 240 \textit{ units}$$

The retailer should place a new order when the inventory level drops to 240 units.

- **Periodic Review Model:** In periodic review (P,S) model, inventory is reviewed at regular intervals (P), and orders are placed to raise the inventory level to a target level (S). The order quantity is:

$$Q = S - I$$

where I is the current inventory level at the review time.

Example 1.9. *A store reviews its inventory every month (P). The target inventory level (S) is 500 units, and the current inventory level (I) at the review time is 150 units. Calculate the order quantity of the store.*

Solution 1.9.

$$Q = 500 - 150 = 350 \ units$$

The store should order 350 units to reach the target inventory level.

- **Newsvendor Model:** This model determines the optimal order quantity for perishable or single-period products. The optimal order quantity Q^* is given by:

$$Q^* = F^{-1}\left(\frac{C_u}{C_o + C_u}\right)$$

where F^{-1} is the inverse cumulative distribution function of demand, C_u is the cost of understocking, and C_o is the cost of overstocking.

Proof

Let $f(D)$ be the ***probability density function (PDF)*** *of demand and $F(D)$ the* ***cumulative distribution function (CDF)****.*

The total expected cost $C(Q)$ is given by:

$$C(Q) = C_u \cdot \mathbb{E}[(D - Q)^+] + C_o \cdot \mathbb{E}[(Q - D)^+]$$

where:

$$(x)^+ = \max(x, 0)$$

Understocking Cost ($D > Q$):

$$\mathbb{E}[(D - Q)^+] = \int_Q^\infty (D - Q) f(D) \, \mathrm{d}D$$

Overstocking Cost ($D < Q$):

$$\mathbb{E}[(Q - D)^+] = \int_{-\infty}^Q (Q - D) f(D) \, \mathrm{d}D$$

Substituting these into $C(Q)$, we get:

$$C(Q) = C_u \int_Q^\infty (D - Q) f(D) \, \mathrm{d}D + C_o \int_{-\infty}^Q (Q - D) f(D) \, \mathrm{d}D$$

To find the optimal Q^*, we minimize $C(Q)$. This requires computing $\frac{\mathrm{d}C(Q)}{\mathrm{d}Q}$.

Leibniz's Rule for Differentiation:

For an integral of the form:

$$I(a) = \int_{g(a)}^{h(a)} f(x, a) \, \mathrm{d}x$$

the derivative is:

$$\frac{\mathrm{d}I}{\mathrm{d}a} = f(h(a), a) \cdot \frac{\mathrm{d}h(a)}{\mathrm{d}a} - f(g(a), a) \cdot \frac{\mathrm{d}g(a)}{\mathrm{d}a} + \int_{g(a)}^{h(a)} \frac{\partial f(x, a)}{\partial a} \, \mathrm{d}x$$

Applying Leibniz's Rule to $C(Q)$:

The first term in $C(Q)$ is:

$$C_u \int_Q^{\infty} (D - Q) f(D) \, \mathrm{d}D$$

(a) ***Upper limit:*** ∞ *(constant), so* $\frac{\mathrm{d}(\infty)}{\mathrm{d}Q} = 0$

(b) ***Lower limit:*** Q, *so* $\frac{\mathrm{d}(Q)}{\mathrm{d}Q} = 1$

(c) ***Partial derivative of the integrand:***

$$\frac{\partial}{\partial Q} [(D - Q) f(D)] = -f(D)$$

Applying Leibniz's Rule:

$$\frac{\mathrm{d}}{\mathrm{d}Q} \left[C_u \int_Q^{\infty} (D - Q) f(D) \, \mathrm{d}D \right] = -C_u \cdot (Q - Q) f(Q) + C_u \cdot \int_Q^{\infty} (-f(D)) \, \mathrm{d}D$$

The second term evaluates to zero because it integrates over $f(D)$, leaving:

$$\frac{\mathrm{d}}{\mathrm{d}Q} \left[C_u \int_Q^{\infty} (D - Q) f(D) \, \mathrm{d}D \right] = -C_u (1 - F(Q))$$

The second term in $C(Q)$ is:

$$C_o \int_{-\infty}^{Q} (Q - D) f(D) \, \mathrm{d}D$$

(a) ***Upper limit:*** Q, *so* $\frac{\mathrm{d}(Q)}{\mathrm{d}Q} = 1$

(b) ***Lower limit:*** $-\infty$ *(constant), so* $\frac{\mathrm{d}(-\infty)}{\mathrm{d}Q} = 0$

(c) ***Partial derivative of the integrand:***

$$\frac{\partial}{\partial Q}[(Q-D)f(D)] = f(D)$$

Applying Leibniz's Rule:

$$\frac{\mathrm{d}}{\mathrm{d}Q}\left[C_o\int_{-\infty}^{Q}(Q-D)f(d)\,\mathrm{d}d\right] = C_o\cdot(Q-Q)f(Q) - C_o\cdot\int_{-\infty}^{Q}f(D)\,\mathrm{d}D$$

Simplifying:

$$\frac{\mathrm{d}}{\mathrm{d}Q}\left[C_o\int_{-\infty}^{Q}(Q-D)f(D)\,\mathrm{d}D\right] = C_oF(Q)$$

Combining the results:

$$\frac{\mathrm{d}C(Q)}{\mathrm{d}Q} = -C_u(1-F(Q)) + C_oF(Q)$$

Setting $\frac{\mathrm{d}C(Q)}{\mathrm{d}Q} = 0$*:*

$$-C_u(1-F(Q)) + C_oF(Q) = 0$$

Rearranging:

$$C_u(1-F(Q)) = C_oF(Q)$$

Simplify:

$$F(Q) = \frac{C_u}{C_u+C_o}$$

The optimal order quantity is:

$$Q^* = F^{-1}\left(\frac{C_u}{C_u+C_o}\right)$$

where F^{-1} *is the inverse CDF of the demand distribution.*

Example 1.10. *A bakery sells fresh pastries daily. The manager must determine the optimal number of pastries to prepare each day to minimize costs associated with understocking and overstocking. The cost of understocking* (C_u) *is \$2 per unit (lost profit), and the cost of overstocking* (C_o) *is \$1 per unit (waste). If the demand follows a normal distribution with a mean of 100 units and a standard deviation of 20 units, what is the optimal number of pastries to prepare each day?*

Solution 1.10. *To find the optimal number of pastries to prepare each day, we need to use the newsvendor model, which balances the costs of understocking and overstocking. The formula for the optimal order quantity Q^* is:*

$$Q^* = F^{-1}\left(\frac{C_u}{C_o + C_u}\right)$$

Given:

1. *Cost of understocking (C_u) = \$2 per unit*
2. *Cost of overstocking (C_o) = \$1 per unit*
3. *Mean demand (μ) = 100 units*
4. *Standard deviation of demand (σ) = 20 units*

First, we calculate the critical ratio:

$$CR = \frac{C_u}{C_o + C_u} = \frac{2}{1+2} = \frac{2}{3} \approx 0.667$$

Next, we find the 66.7th percentile of the normal distribution. This percentile corresponds to the z-score from the standard normal distribution. Using a standard normal distribution table or a statistical calculator, the z-score for 0.667 is approximately 0.43.

Now, we use the z-score to find the optimal order quantity Q^*:

$$Q^* = \mu + z\sigma = 100 + 0.43 \times 20 = 100 + 8.6 = 108.6$$

Therefore, the bakery should prepare approximately 109 units (rounding up to the nearest whole number) to balance the costs of understocking and overstocking.

Steps:

1. *Calculate the critical ratio $CR = \frac{C_u}{C_o+C_u} = 0.667$.*
2. *Find the z-score corresponding to the 66.7th percentile, which is approximately 0.43.*
3. *Compute the optimal order quantity $Q^* = 100 + 0.43 \times 20 = 108.6$.*
4. *Round to the nearest whole number: 109 units.*

1.11 Impact of Inventory Management on Organizational Operations

Effective inventory management plays a pivotal role in the overall operational efficiency and performance of organizations across various industries. This section analyzes the real-world impact of inventory management practices, supported by case studies or examples.

1.11.1 Optimized Inventory Levels

Maintaining optimal inventory levels has a direct impact on operational efficiency. Avoiding stockouts and overstocking, organizations can streamline production processes, reduce lead times, and improve customer satisfaction. For example, a manufacturing company that implements JIT inventory management can minimize inventory holding costs while ensuring timely production to meet customer demand. As a result, the organization achieves higher throughput, reduced storage costs, and improved cash flow.

1.11.2 Reduced Costs

Effective inventory management leads to cost savings across various aspects of operations. Optimizing inventory levels, organizations can reduce holding costs associated with storage, insurance, and obsolescence. Additionally, efficient inventory practices minimize ordering costs by consolidating orders, negotiating better pricing with suppliers, and reducing the need for rush orders. For instance, a retail chain that implements advanced demand forecasting techniques can reduce excess inventory and markdown costs, leading to improved profitability and financial performance.

1.11.3 Improved Supply Chain Collaboration

Inventory management practices often involve collaboration with suppliers, distributors, and other stakeholders in the supply chain. Sharing inventory data and coordinating replenishment activities, organizations can enhance supply chain visibility, reduce lead times, and mitigate supply chain risks. For example, a food manufacturer that implements collaborative planning and forecasting with its suppliers can ensure timely delivery of raw materials, minimize production disruptions, and maintain product quality standards. This collaborative approach fosters stronger partnerships and enables organizations to respond effectively to market changes and customer demands.

1.11.4 Enhanced Customer Service

Efficient inventory management directly impacts customer service levels and satisfaction. Ensuring product availability and timely delivery, organizations can meet customer expectations and build loyalty. For instance, an e-commerce company that invests in inventory optimization technologies can provide accurate stock information, offer fast order fulfillment, and deliver superior customer experiences. As a result, the organization gains a competitive edge in the market, attracts repeat business, and earns positive reviews and referrals from satisfied customers.

1.11.5 Toyota Production System

The Toyota Production System (TPS) is a renowned example of effective inventory management practices. Implementing JIT principles, Toyota minimized inventory levels, reduced waste, and improved production efficiency. As a result, Toyota achieved significant cost savings, shortened lead times, and maintained high product quality standards. TPS became a benchmark for lean manufacturing practices adopted by organizations worldwide [1,2].

1.11.6 Amazon's Fulfillment Centers

Amazon revolutionized e-commerce through its advanced inventory management and fulfillment strategies. Leveraging sophisticated algorithms and robotics technology, Amazon optimizes inventory storage, picking, and packing processes in its fulfillment centers. This enables Amazon to offer fast and reliable order fulfillment, expand product offerings, and maintain high service levels for millions of customers globally. Effective inventory management has a profound impact on organizational operations, driving improvements in efficiency, cost savings, supply chain collaboration, and customer service. Optimizing inventory levels, reducing costs, enhancing supply chain visibility, and delivering superior customer experiences, organizations can gain a competitive advantage and achieve sustainable growth. Through case studies and examples like Toyota's TPS and Amazon's fulfillment centers, we see how innovative inventory management practices can transform businesses and shape industry standards [1,3].

1.12 Practical Applications

In this section, we delve into practical tips for practitioners in various industries and discuss the tools and technologies used in inventory management today. Effective inventory management requires a combination of strategic planning, operational excellence, and leveraging the right tools and technologies to streamline processes and enhance efficiency.

1.12.1 Practical Tips for Practitioners

- **Demand Forecasting:** Invest in robust demand forecasting techniques to anticipate future demand patterns accurately. Utilize historical sales data, market trends, and predictive analytics to develop reliable demand forecasts [4].
- **Supplier Collaboration:** Foster strong relationships with suppliers through collaborative planning and forecasting. Share demand forecasts, inventory plans, and production schedules to ensure timely delivery of raw materials and components [5].
- **Inventory Segmentation:** Classify inventory items based on their demand patterns, value, and criticality. Implement different inventory management strategies for high-value, fast-moving items compared to low-value, slow-moving items [6].
- **Safety Stock Optimization:** Calculate safety stock levels dynamically based on demand variability and lead time uncertainty. Use statistical methods and inventory optimization models to determine optimal safety stock levels for different products [7].
- **Continuous Improvement:** Embrace a culture of continuous improvement and optimization. Regularly review inventory performance metrics, identify areas for improvement, and implement corrective actions to enhance efficiency and reduce costs [8].

1.12.2 Tools and Technologies

- **Inventory Management Software:** Utilize advanced inventory management software systems to streamline inventory tracking, order management, and replenishment processes. Look for features such as real-time inventory visibility, automated reorder notifications, and integration with other business systems [9].
- **Barcode and RFID Technology:** Implement barcode and RFID technology for accurate inventory tracking and management. Barcode scanners and RFID tags enable rapid identification and location of inventory items, reducing manual errors and improving efficiency [10].
- **Warehouse Automation:** Invest in warehouse automation technologies such as AS/RS, conveyor systems, and robotic picking systems. Automation improves warehouse efficiency, reduces labor costs, and speeds up order fulfillment [11].

- **Supply Chain Analytics:** Leverage supply chain analytics tools to gain insights into inventory performance, demand patterns, and supply chain risks. Advanced analytics platforms use data visualization, predictive modeling, and machine learning algorithms to optimize inventory decisions and improve operational efficiency [12].
- **Cloud-Based Solutions:** Consider adopting cloud-based inventory management solutions for scalability, flexibility, and accessibility. Cloud-based systems offer real-time data synchronization, remote access, and seamless integration with other business applications, enabling agile and efficient inventory management operations [13].

1.13 Conclusion

Implementing robust demand forecasting techniques, fostering collaboration with suppliers, segmenting inventory effectively, and leveraging advanced tools such as inventory management software and warehouse automation, practitioners can optimize inventory levels, reduce costs, and enhance operational efficiency. Inventory management is a dynamic and multifaceted discipline that plays a critical role in the success of organizations. Mastering the principles, techniques, and strategies discussed in this chapter and pursuing further learning opportunities, individuals can become adept practitioners and leaders in the field of inventory management. We hope this chapter has provided valuable insights and inspiration for your journey in mastering inventory management. In conclusion, this chapter has provided a comprehensive overview of inventory management, covering key concepts, principles, techniques, and their real-world impact on organizational operations. *Happy learning and happy managing!*

Exercises

Exercise 1.1. *Consider a company with a lead time demand of 500 units and a standard deviation of demand during lead time of 60 units. If the company desires a service level of 95%, calculate the safety stock required.*

Exercise 1.2. *A retailer expects a lead time demand of 200 units, and the standard deviation of demand during lead time is 30 units. The desired service level is 90%. What is the safety stock needed for this service level?*

Exercise 1.3. *An electronics store wants to maintain a service level of 99% for a product with a lead time demand of 1,000 units and a standard deviation of 80 units. Calculate the required safety stock.*

Exercise 1.4. *A wholesaler calculates the lead time demand for a product to be 750 units with a standard deviation of 50 units. If the company targets a service level of 85%, determine the safety stock required.*

Exercise 1.5. *A company aims for a 98% service level for its products. If the lead time demand is 400 units and the standard deviation of demand during lead time is 25 units, how much safety stock should be maintained?*

Exercise 1.6. *A company has an annual demand of 10,000 units for a product. The ordering cost is $50 per order, and the holding cost per unit per year is $2. Calculate the EOQ.*

Exercise 1.7. *A retailer needs to order 15,000 units of a product annually. The ordering cost is $75 per order, and the holding cost is $1.50 per unit per year. Determine the EOQ and the total annual cost of inventory if the purchase cost per unit is $10.*

Exercise 1.8. *An electronics company has an annual demand of 25,000 units for a component. The cost to place an order is $100, and the annual holding cost per unit is $3. Additionally, the company faces a storage constraint and can only hold a maximum of 1,200 units at a time. Determine if the EOQ is feasible given the storage constraint.*

Exercise 1.9. *A manufacturer has a yearly requirement of 8,000 units. The ordering cost is $60 per order, and the holding cost is $2.50 per unit per year. If the lead time for each order is two weeks and the company operates 50 weeks per year, calculate the reorder point in addition to the EOQ.*

Exercise 1.10. *A supplier needs to order 12,000 units of an item each year. The ordering cost is \$40 per order, and the holding cost per unit per year is \$2. However, the supplier offers a discount that reduces the ordering cost to \$30 if the order quantity exceeds 1,000 units. Determine the optimal order quantity and the total annual cost of inventory.*

Exercise 1.11. *A retailer sells a seasonal product. The demand for the product is normally distributed with a mean of 500 units and a standard deviation of 100 units. The retailer purchases each unit for \$10 and sells it for \$20. Unsold units have a salvage value of \$5 each. Determine the optimal order quantity using the newsvendor model.*

Exercise 1.12. *A bakery needs to decide the quantity of a special type of bread to bake daily. The daily demand follows a normal distribution with a mean of 200 loaves and a standard deviation of 40 loaves. Each loaf costs \$1 to produce and is sold for \$3. Any unsold loaf can be donated with a salvage value of \$0.50 per loaf. Calculate the optimal daily production quantity.*

Exercise 1.13. *An electronics store sells a high-tech gadget with uncertain demand. The demand is normally distributed with a mean of 1,000 units and a standard deviation of 200 units. The store buys each gadget for \$150 and sells it for \$300. The salvage value for any unsold gadget is \$50. Determine the optimal order quantity to maximize the expected profit.*

Exercise 1.14. *A fashion retailer needs to decide the order quantity for a new line of designer jackets. The demand for the jackets is normally distributed with a mean of 600 units and a standard deviation of 150 units. Each jacket costs \$80 to produce and sells for \$200. The salvage value for unsold jackets is \$40 each. Determine the optimal order quantity.*

Exercise 1.15. *A newspaper stand needs to determine the number of newspapers to stock daily. The daily demand is normally distributed with a mean of 300 newspapers and a standard deviation of 50 newspapers. Each newspaper costs \$0.50 and sells for \$1. Unsold newspapers can be recycled with a salvage value of \$0.10 per newspaper. Calculate the optimal daily order quantity.*

2

Inventory Control Techniques

This chapter examines a range of inventory control techniques that are vital for efficient inventory management. These techniques, including ABC analysis, the economic order quantity (EOQ) model, safety stock, and reorder point, are tools designed to optimize inventory levels and reduce costs. The chapter compares these methods, demonstrating their practical applications and their impact on organizational performance. It highlights the significance of inventory control in enhancing customer satisfaction and profitability. Offering a thorough exploration of inventory control concepts and methodologies, this chapter provides valuable insights into how inventory management systems function within organizations, emphasizing the importance of coordination and cost-effective material management.

2.1 Introduction to Inventory Control Techniques

Inventory control techniques are systematic methods and tools used to manage inventory levels, ensure the availability of products, and minimize costs associated with holding and ordering inventory. These techniques are essential for maintaining a balance between having enough inventory to meet customer demand and avoiding excess stock that ties up capital and incurs additional costs. Optimization techniques and models play a crucial role in inventory management, offering strategies to enhance efficiency, minimize costs, and optimize resource allocation. In the realm of inventory management, businesses seek to strike a balance between having sufficient stock to meet demand and minimizing holding costs. The choice of a specific optimization approach depends on factors such as demand patterns, supply chain complexity, and the desired balance between costs and service levels. As technology and analytical methods continue to advance, the field of inventory management optimization evolves, offering increasingly sophisticated solutions to meet the challenges of modern supply chains. This introduction provides an overview of key optimization techniques and models relevant to inventory management.

DOI: 10.1201/9781003534860-2

2.1.1 Definition of Inventory Control Techniques

Inventory control techniques encompass a variety of strategies and methodologies aimed at optimizing the management of inventory. These techniques help organizations determine the appropriate quantity of stock to hold, the timing for reordering, and the methods for monitoring and adjusting inventory levels. Implementing these techniques, companies can achieve greater efficiency, reduce waste, and improve overall operational performance.

2.1.2 Importance of Inventory Control Techniques

Effective inventory control is critical for several reasons:

- **Cost Reduction:** Proper inventory control minimizes costs associated with overstocking, such as storage, insurance, and obsolescence costs. It also reduces costs related to stockouts, such as expedited shipping fees and lost sales.
- **Customer Satisfaction:** Maintaining optimal inventory levels ensures that products are available when customers need them, leading to higher customer satisfaction and loyalty.
- **Efficient Resource Utilization:** Inventory control techniques help in better allocation of resources, ensuring that capital is not unnecessarily tied up in excess inventory. This allows for more investment in other critical areas of the business.
- **Improved Cash Flow:** Optimizing inventory levels, companies can improve their cash flow, as less money is tied up in unsold inventory. This financial flexibility can be crucial for funding operations and growth.
- **Operational Efficiency:** Effective inventory management reduces the complexity of managing stock, leading to more streamlined operations and reduced administrative workload.

2.1.3 Contribution to Effective Inventory Management

Inventory control techniques contribute significantly to effective inventory management by providing a structured approach to managing stock levels. Here's how these techniques facilitate better inventory management:

- **Demand Forecasting:** Techniques such as ABC analysis help in understanding demand patterns and prioritizing inventory management efforts on high-value and high-demand items. This aids in more accurate demand forecasting and planning.
- **Order Optimization:** Models like EOQ assist in determining the optimal order size that minimizes the total cost of inventory, including ordering and holding costs.

- **Risk Mitigation:** Implementing safety stock strategies helps mitigate risks associated with demand variability and supply chain disruptions, ensuring that stockouts are minimized.
- **Replenishment Planning:** Reorder point calculations provide a systematic way to determine when to replenish inventory, ensuring that new stock is ordered before existing stock runs out.
- **Performance Measurement:** Inventory control techniques provide metrics and benchmarks for evaluating inventory performance. This helps in identifying inefficiencies and areas for improvement.

Inventory control techniques are vital for achieving a well-balanced inventory system that supports operational efficiency, cost-effectiveness, and high customer satisfaction. Integrating these techniques into their inventory management practices, organizations can better align their inventory levels with business objectives and market demands, leading to enhanced overall performance.

2.2 Just-In-Time

Just-in-time (JIT) is a manufacturing, inventory management philosophy and strategy that emphasizes producing goods or providing services exactly when they are needed in the production process or by the customer. The core principle of JIT is to minimize or eliminate the amount of inventory held in the production process, which includes raw materials, work-in-progress, and finished goods. JIT aims to synchronize production with customer demand, ensuring that items are produced or delivered just in time to meet immediate needs. The principles of JIT were implemented at Toyota as part of the company's efforts to improve efficiency, reduce waste, and enhance overall production processes. Taiichi Ohno is often credited as the key architect of the Toyota Production System, which has since become a foundational model for lean manufacturing and production systems worldwide. The specific development of JIT principles at Toyota occurred in the post–World War II era, with significant advancements made in the 1950s and beyond. However, the principles of JIT have been applied across various industries to improve operational efficiency, reduce costs, and enhance overall competitiveness. While JIT offers significant benefits, its successful implementation requires careful planning, coordination, and a commitment to quality and continuous improvement. The key characteristics and principles of JIT include:

2.2.1 Key Principles of JIT

- **Waste Reduction:** JIT focuses on eliminating waste in all forms, including overproduction, excess inventory, waiting time, unnecessary transportation, inefficient processes, excess motion, and defects. Minimizing waste, JIT enhances productivity and cost-effectiveness.
- **Continuous Improvement (Kaizen):** JIT encourages a culture of continuous improvement where employees at all levels are involved in identifying and implementing enhancements to processes, systems, and products. Kaizen, the Japanese term for continuous improvement, is fundamental to JIT implementation.
- **Pull System:** Instead of relying on forecasts and pushing products through the production process, JIT employs a pull system where goods and services are produced based on actual customer demand. This minimizes the risk of overproduction and excess inventory.
- **Takt Time:** Takt time refers to the rate at which products must be produced to meet customer demand. JIT aligns production processes with takt time to ensure a smooth flow of materials and resources, thereby avoiding bottlenecks and delays.
- **Flexible Workforce:** JIT emphasizes cross-training and flexibility among employees to respond quickly to changing demand and production requirements. A multi-skilled workforce enables organizations to adapt more effectively to fluctuations in customer needs and market conditions.
- **Supplier Partnerships:** JIT fosters close relationships with suppliers based on trust, collaboration, and mutual benefit. Working closely with suppliers, organizations can minimize lead times, reduce costs, and improve the quality of inputs, thus enhancing overall efficiency.
- **Jidoka (Autonomation):** Jidoka is the principle of autonomation, which involves designing processes to automatically stop when abnormalities or defects occur. This enables timely detection and resolution of issues, preventing them from propagating downstream and ensuring higher quality standards.

2.2.2 Benefits of JIT

Implementing JIT offers several benefits to organizations, including:

- **Reduced Inventory Costs:** Producing goods in response to actual demand, organizations can minimize the need for excess inventory, thereby reducing carrying costs and the risk of obsolescence.

- **Improved Quality:** JIT encourages a focus on quality at every stage of the production process, leading to fewer defects, lower rework costs, and higher customer satisfaction.
- **Increased Efficiency:** JIT streamlines processes, reduces lead times, and eliminates non-value-added activities, resulting in higher productivity and efficiency.
- **Lower Operating Costs:** Through waste reduction, improved resource utilization, and enhanced process efficiency, organizations can lower their operating costs and improve profitability.
- **Enhanced Flexibility:** JIT enables organizations to respond more quickly to changes in customer demand, market conditions, and production requirements, thereby improving adaptability and competitiveness.
- **Stronger Supplier Relationships:** Fostering closer partnerships with suppliers, organizations can benefit from improved reliability, shorter lead times, and better coordination throughout the supply chain.

2.2.3 Challenges of JIT

Despite its many advantages, JIT implementation can pose several challenges, including:

- **Dependency on Suppliers:** JIT relies on timely and reliable supply of materials and components from suppliers. Any disruptions in the supply chain can have significant ramifications for production schedules and customer satisfaction.
- **Risk of Stockouts:** Since JIT maintains minimal inventory levels, there is a heightened risk of stockouts if demand exceeds expectations or if there are delays in the supply chain.
- **Sensitivity to Variability:** JIT systems are highly sensitive to fluctuations in demand, production, and supply chain performance. Variability can disrupt production schedules and increase the likelihood of bottlenecks and inefficiencies.
- **Cost of Implementation:** Implementing JIT requires significant upfront investment in terms of process redesign, training, technology, and organizational change. The initial costs and complexities involved can be daunting for some organizations.
- **Cultural Resistance:** JIT implementation often necessitates a cultural shift within organizations, with employees needing to embrace new ways of working, greater accountability, and a focus on continuous improvement. Resistance to change can impede successful implementation.

- **Risk of Overemphasis on Efficiency:** In pursuit of efficiency gains, organizations may overlook other important factors such as innovation, product diversity, and long-term strategic considerations. Overemphasis on efficiency at the expense of other priorities can limit organizational resilience and competitiveness.

2.2.4 Applications of JIT

While JIT was originally developed in the context of manufacturing, its principles have been widely adopted across various industries and functional areas, including:

- **Manufacturing:** JIT is extensively used in manufacturing industries to streamline production processes, minimize inventory, and improve quality and efficiency.
- **Supply Chain Management:** JIT principles are applied in supply chain management to optimize inventory levels, reduce lead times, and enhance coordination among suppliers, manufacturers, and distributors.
- **Retail:** Retailers leverage JIT to manage inventory levels, align product availability with customer demand, and reduce carrying costs associated with excess stock.
- **Healthcare:** In healthcare, JIT principles are employed to improve patient flow, minimize wait times, optimize resource utilization, and enhance the overall quality and efficiency of care delivery.
- **Software Development:** Agile methodologies in software development share similarities with JIT, emphasizing iterative development, continuous delivery, and responsiveness to changing customer requirements.
- **Service Industries:** JIT concepts are increasingly applied in service industries such as hospitality, transportation, and banking to enhance operational efficiency, improve service quality, and optimize resource utilization.

Example 2.1. *TechGadgets Inc. is a company that manufactures high-end electronic devices. Despite its success, the company faces several issues related to inventory management, which include:*

- *High Inventory Costs: TechGadgets Inc. is incurring significant costs due to large amounts of raw materials and finished goods inventory.*
- *Storage Space Limitation: The warehouse is overcrowded, leading to inefficient storage and retrieval processes.*
- *Obsolete Inventory: Rapid technological advancements result in a high rate of inventory obsolescence, leading to financial losses.*

How can the company address these problems?

Solution 2.1. *To address these problems, TechGadgets Inc. needs to implement JIT inventory management. Here are the steps to take:*

- *Supplier Collaboration: TechGadgets Inc. has to collaborate with its key suppliers ensuring timely and frequent delivery of raw materials. They have to establish agreements that will allow for smaller, more frequent shipments, reducing the need to hold large inventories.*
- *Demand Forecasting: The company has to invest in advanced forecasting tools to predict customer demand accurately. This will allow them to synchronize their production schedules with market demand, minimizing the risk of overproduction and stockouts.*
- *Lean Manufacturing Practices: TechGadgets Inc. has to adopt lean manufacturing practices to streamline their production process. This will include cross-training employees to handle multiple tasks, reducing setup times, and eliminating waste.*
- *Inventory Tracking Systems: The company has to implement robust inventory tracking systems that provide real-time data on inventory levels and production needs. This will ensure materials are ordered and received just in time for production.*

2.3 ABC Analysis

ABC analysis is an inventory categorization technique often used in inventory management. It is based on the principle that not all inventory items hold the same value or importance for a business. Categorizing inventory items into three groups, namely A, B, and C, organizations can prioritize their resources and management efforts more effectively. Here is a comprehensive explanation of each category and the optimization strategies involved:

2.3.1 Categories in ABC Analysis

ABC analysis is a method of categorizing inventory items based on their importance to the business. The importance is typically measured by the items' consumption values, which is a product of an item's annual demand and its cost per unit. ABC analysis divides inventory into three categories:

- **Category A:** These are high-value items that typically constitute a small percentage of the total inventory but a large percentage of the total inventory value. Tight control and accurate record-keeping are essential for these items to avoid stockouts or excess inventory. Often, these items are critical to the production process or customer satisfaction. Typically, Category A items represent around 70–80% of the inventory value but only 10–20% of the total items.

- **Category B:** These items are of moderate value and usually fall between A and C items in terms of both quantity and value. They require moderate control and monitoring. B items may not need as rigorous oversight as A items, but they still require regular review and management. Category B items usually make up about 15–25% of the inventory value and 30% of the total items.

- **Category C:** These are low-value items that generally account for a large percentage of the total inventory in terms of quantity but only a small percentage of the total inventory value. Less stringent control is needed for C items, as they are relatively inexpensive and not critical to the core business operations. However, managing them efficiently can still contribute to overall cost savings. Category C items generally account for about 5–10% of the inventory value and 50–60% of the total items.

2.3.2 Methodology for Categorizing Inventory Items

The process of conducting an ABC analysis involves several steps:

1. **Data Collection:** Gather data on the annual consumption and cost per unit for each inventory item.
2. **Calculation of Annual Consumption Value:** Calculate the annual consumption value for each item by multiplying its annual demand by its cost per unit.
3. **Ranking Items:** Rank the inventory items in descending order based on their annual consumption value.
4. **Cumulative Calculation:** Calculate the cumulative annual consumption value and the cumulative percentage of each item.
5. **Categorization:** Categorize the items into A, B, and C categories based on the cumulative percentage values, typically using the following thresholds:
    - **Category A:** Top 70–80% of the cumulative consumption value.
    - **Category B:** Next 15–25% of the cumulative consumption value.
    - **Category C:** Remaining 5–10% of the cumulative consumption value.

2.3.3 Optimization Strategies

Optimizing inventory management using ABC analysis involves several key strategies tailored to the criticality and value of each category:

2.3.3.1 Category A Optimization

- Frequent Review: Perform regular and frequent reviews of A items to ensure they are in stock and meet demand without overstocking.
- Accurate Forecasting: Use sophisticated forecasting techniques to predict demand accurately for A items. This helps in maintaining the right balance between stock availability and minimizing holding costs.
- Supplier Management: Develop strong relationships with reliable suppliers to ensure timely delivery and potentially negotiate better terms for high-value items.
- Safety Stock: Maintain an optimal level of safety stock to cushion against variability in demand or supply delays, ensuring that stockouts are minimized.

2.3.3.2 Category B Optimization

- Periodic Review: Conduct periodic reviews of B items, which may not need as frequent attention as A items but still require regular oversight.
- Economic Order Quantity (EOQ): Apply EOQ models to determine the optimal order quantity that balances ordering costs with holding costs.
- Inventory Turnover: Monitor inventory turnover rates to ensure that B items are moving efficiently through the supply chain and are not becoming obsolete or in excess.

2.3.3.3 Category C Optimization

- Bulk Ordering: Consider bulk ordering C items to take advantage of economies of scale, as these items are generally low-cost and less critical.
- Minimal Monitoring: Implement a less intensive monitoring system for C items, such as periodic reviews instead of continuous monitoring.
- Automated Replenishment: Use automated inventory systems to manage reordering of C items, reducing the administrative burden.
- Space Optimization: Allocate storage space efficiently, ensuring that C items do not take up valuable space needed for more critical A and B items.

2.3.4 Apply Control Policies

Implement the appropriate control policies and optimization strategies for each category.

2.3.4.1 Category A Control Policies

- Regular Audits: Conduct regular and detailed audits of A items to ensure inventory records are accurate. Discrepancies in high-value items can lead to significant financial losses.
- Just-In-Time (JIT) Inventory: Implement JIT inventory management for A items to reduce holding costs and minimize waste. JIT involves receiving goods only as they are needed in the production process.
- Supplier Relationships: Establish strong partnerships with key suppliers to ensure a reliable supply chain. Negotiate better terms, such as bulk discounts or priority delivery, to secure the best possible deals.
- Security Measures: Implement stringent security measures to prevent theft or loss of high-value items. This can include surveillance systems, restricted access areas, and regular inventory checks.

2.3.4.2 Category B Control Policies

- Cycle Counting: Use cycle counting as a method to ensure inventory accuracy for B items. This involves counting a subset of inventory items on a rotating schedule throughout the year.
- Automated Systems: Implement automated inventory management systems to track and reorder B items. These systems can help maintain optimal inventory levels without manual intervention.
- Vendor Managed Inventory (VMI): Consider VMI agreements where suppliers manage the inventory levels of B items. This can reduce the administrative burden on the company and improve efficiency.
- Performance Monitoring: Regularly monitor the performance and turnover rates of B items to identify any inefficiencies or areas for improvement. This ensures that B items are not overstocked or underutilized.

2.3.4.3 Category C Control Policies

- Bulk Purchasing: Leverage bulk purchasing to reduce the cost per unit of C items. Since these items are low-cost, buying in bulk can provide significant savings.
- Simplified Replenishment: Use simplified and automated replenishment processes for C items to ensure they are always in stock without extensive

management. Reorder points and automatic triggers can be set in inventory management software.

- Stock Rotation: Implement stock rotation practices, such as first-in, first-out (FIFO), to ensure that older stock is used first, reducing the risk of obsolescence.
- Minimize Handling Costs: Reduce handling and storage costs for C items by optimizing storage space. Use bin shelving or other efficient storage methods to keep C items organized and easily accessible.

2.3.5 Benefits

- **Focused Management:** ABC analysis allows organizations to focus their efforts on the most valuable items, improving inventory turnover and reducing holding costs.
- **Improved Forecasting:** Concentrating on high-value items, organizations can achieve more accurate demand forecasting and better inventory planning.
- **Efficient Resource Utilization:** Resources such as time, effort, and capital can be allocated more effectively, ensuring that critical items are adequately managed.
- **Enhanced Supplier Relationships:** Better management of Category A items often leads to improved supplier relationships and more favorable terms.

2.3.6 Challenges of ABC Analysis

While ABC analysis provides valuable inventory categorization insights, its implementation faces several systemic and operational challenges.

2.3.6.1 Dynamic Nature of Inventory

- Changing Demand Patterns: Demand for items can change over time, necessitating frequent reclassification and updates to the ABC categories.
- Market Trends: Rapid changes in market trends can make the analysis outdated quickly.

2.3.6.2 Complexity in Categorization

- Criteria Selection: Determining the appropriate criteria for categorization (e.g., sales volume, revenue, consumption value) can be complex and may vary between industries.

- Multifactor Influence: Items might need to be evaluated based on multiple factors (e.g., lead time, criticality), complicating the categorization process.

2.3.6.3 Data Accuracy and Availability

- Data Quality: Reliable and accurate data is essential for effective analysis. Inaccurate or incomplete data can lead to incorrect categorization and poor decision-making.
- Data Integration: Integrating data from various sources (e.g., sales, procurement, inventory systems) can be challenging.

2.3.6.4 Resource Constraints

- Implementation Effort: The process of conducting ABC analysis requires time and effort, which might be significant for companies with a large number of inventory items.
- Continuous Monitoring: Continuous monitoring and updating of the ABC categories require dedicated resources, which may be a constraint for some organizations.

2.3.6.5 Resistance to Change

- Organizational Buy-in: Gaining acceptance and support from all stakeholders (e.g., procurement, sales, finance) can be difficult.
- Change Management: Implementing changes based on ABC analysis may encounter resistance due to entrenched practices and reluctance to adopt new methods.

2.3.6.6 Implementation Costs

- Technology Investment: Effective ABC analysis might require investment in technology and software for data collection, analysis, and reporting.
- Training: Employees need to be trained on the new processes and tools, which incurs additional costs and time.

2.3.7 Limitations

- **Oversimplification:** ABC analysis may oversimplify inventory management by focusing only on consumption value and ignoring other factors such as lead time variability and supply risk.
- **Static Classification:** The classification may become outdated if not reviewed regularly, as demand patterns and item importance can change over time.

- **Neglect of Category C Items:** Category C items, though low in value, might still be critical for certain operations. Overlooking these items can lead to stockouts and operational disruptions.
- **Limited Scope:** ABC analysis is primarily based on historical data and may not account for future trends or market changes that could impact inventory needs.

Example 2.2. *TechGear Inc. is a company specializing in the manufacturing of high-end electronic gadgets. To enhance its operational efficiency and reduce costs, the company aims to optimize its inventory management system. The inventory consists of various components and raw materials, each with unique usage rates and costs. Effective inventory management is crucial for maintaining production schedules and minimizing expenses. Here is a snapshot of TechGear Inc.'s inventory data:*

- *Microchip A: Annual usage = 1,000 units, Unit cost = $100*
- *Display Screen: Annual usage = 500 units, Unit cost = $200*
- *Battery B: Annual usage = 2,000 units, Unit cost = $50*
- *Plastic Housing: Annual usage = 5,000 units, Unit cost = $10*
- *Screws and Fasteners: Annual usage = 10,000 units, Unit cost = $0.10*

Solution 2.2. *To optimize inventory management using ABC Analysis, TechGear Inc. follows these steps:*

Calculate Annual Consumption Value

The annual consumption value is calculated by multiplying the annual usage by the unit cost:

$$\begin{aligned}
\textit{Microchip A} &: 1{,}000 \times 100 = \$100{,}000 \\
\textit{Display Screen} &: 500 \times 200 = \$100{,}000 \\
\textit{Battery B} &: 2{,}000 \times 50 = \$100{,}000 \\
\textit{Plastic Housing} &: 5{,}000 \times 10 = \$50{,}000 \\
\textit{Screws and Fasteners} &: 10{,}000 \times 0.10 = \$1{,}000
\end{aligned}$$

Rank Items by Value

Rank the items in descending order based on their annual consumption value:

- *Microchip A: $100,000*
- *Display Screen: $100,000*
- *Battery B: $100,000*

- *Plastic Housing: $50,000*
- *Screws and Fasteners: $1,000*

Categorize Items

Categorize the items into A, B, and C based on their contribution to the total inventory value:

- *Category A: Microchip A, Display Screen, Battery B (top 70–80% of total value)*
- *Category B: Plastic Housing (next 15–25% of total value)*
- *Category C: Screws and Fasteners (remaining 5–10% of total value)*

Apply Control Policies

Implement the following control policies for each category:

Category A

- *Frequent Review: Weekly inventory reviews.*
- *Accurate Forecasting: Use advanced forecasting methods.*
- *Supplier Management: Establish strong relationships with suppliers.*
- *Safety Stock: Maintain safety stock levels.*

Category B

- *Periodic Review: Monthly inventory reviews.*
- *EOQ Models: Determine optimal order quantities.*
- *Inventory Turnover: Monitor turnover rates.*

Category C

- *Bulk Ordering: Purchase in bulk.*
- *Minimal Monitoring: Quarterly reviews.*
- *Automated Replenishment: Use automated systems.*
- *Space Optimization: Efficient storage.*

ABC analysis is a powerful tool for prioritizing inventory management efforts and optimizing resource allocation. Categorizing inventory items based on their value and impact, organizations can implement more effective control measures, improve forecasting accuracy, and enhance overall operational efficiency. However, it is essential to recognize the limitations of ABC analysis and complement it with other inventory management techniques to achieve a comprehensive and adaptive inventory strategy. Following is an example of how to use R Programming Language to carry out ABC analysis by randomly generating 50 values and display the result in Table 2.1 and Figure 2.1.

```
# Set items' SKUs
SKU <- c(1:50)

# Set items' demands
demand <- ceiling(runif(50, 0, 100))

# Set items' prices
price <- ceiling(runif(50, 1, 100))

# Get items' sales
sales <- demand * price

# Construct inventory data frame
inventory <- data.frame(SKU, sales)

# Sort sales column in descending order
inventory <- inventory[order(-inventory$sales), ]

# Get the total sales
total_sales <- sum(inventory$sales)

# Create an empty vector for sales percentages
percentage <- c()

# Get the sales percentage for each item
for (i in inventory$sales) {
  percentage <- append(percentage, i / total_sales * 100)
}

# Add the percentage column to inventory data frame
inventory <- mutate(inventory, percentage)

# Get the cumulative sum of the products sales' percentages
cumulative <- cumsum(inventory$percentage)

# Add the cumulative column to inventory data frame
inventory <- mutate(inventory, cumulative)

# Create an empty vector for items classification
classification <- c()

# Classify the items by A, B, and C
for (i in inventory$cumulative) {
```

```
  if (i < 70) {
    classification <- append(classification, "A")
  } else if (i >= 70 & i < 90) {
    classification <- append(classification, "B")
  } else {
    classification <- append(classification, "C")
  }
}

# Add the classification column to inventory data frame
inventory <- mutate(inventory, classification)

# Build inventory classification graph
plot <- ggplot(data = inventory, aes(x = classification)) +
  geom_bar(aes(fill = classification)) +
  labs(
    title = "Inventory ABC Distribution",
    subtitle = "Inventory Planning and Managing",
    x = "Items Classification",
    y = "Count"
  )

# Save plot with 300 DPI
ggsave(filename = "inventory_classification.png", plot =
    plot, dpi = 300, width = 8, height = 6)

# Print results
summary <- count(inventory, classification)
summary$percentage <- summary$n / sum(summary$n) * 100
print(summary)

# Create LaTeX table and display it
latex_table <- kable(summary, format = "latex", booktabs =
    TRUE,
                      col.names = c("Classification",
                          "Count", "Percentage"),
                      caption = "Summary of Inventory
                          Classification") %>%
  kable_styling(latex_options = c("striped",
      "hold_position"))

cat(latex_table)
```

TABLE 2.1
Summary of Inventory Classification

Classification	Count	Percentage
A	17	34
B	11	22
C	22	44

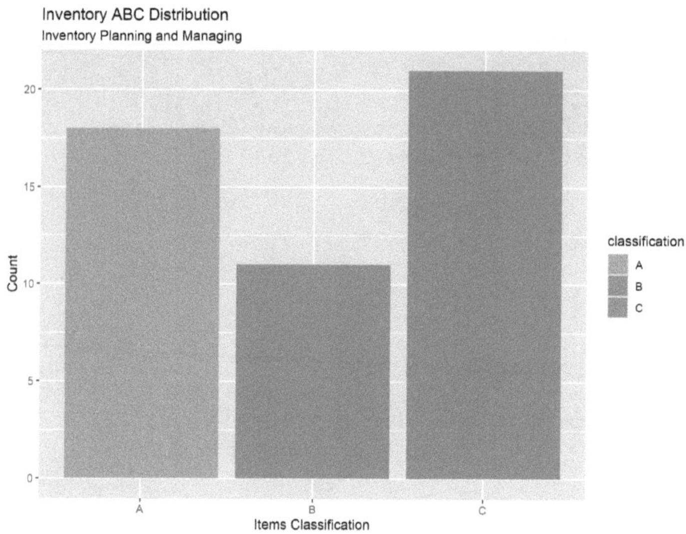

FIGURE 2.1
Inventory classification.

2.4 VED Analysis

VED analysis is a classification method used in inventory management to categorize items based on their criticality and importance for the business. The acronym "VED" stands for Vital, Essential, and Desirable, and each category represents a different level of criticality in terms of the impact of an item's scarcity or unavailability on business operations. VED analysis is often used in conjunction with ABC analysis, where items are categorized based on their value. The combination of VED and ABC analyses helps businesses prioritize their focus and resources on managing and controlling inventory items effectively. Understanding the criticality of items, businesses can make informed decisions about inventory levels, storage, and procurement strategies, ensuring that essential and vital items receive the necessary attention and resources. Here's a breakdown of each category:

2.4.1 Vital (V)

Items classified as Vital are those that are critical to the business operations. The scarcity or unavailability of these items can cause significant disruptions, halt production, or lead to severe financial losses. These items are typically high priority and require close monitoring and control. Characteristics of Vital items include:

- **High Impact on Operations:** The absence of Vital items can severely affect production processes, service delivery, or overall business continuity.
- **Critical to Customer Satisfaction:** These items are essential for meeting customer requirements and maintaining service levels.
- **Strict Inventory Control:** Due to their critical nature, Vital items often necessitate stringent inventory control measures, such as regular audits, safety stock levels, and priority in procurement.
- **Higher Safety Stock Levels:** To mitigate the risk of stockouts, higher safety stock levels are maintained for Vital items.

2.4.2 Essential (E)

Essential items are important for business operations, but their scarcity does not lead to immediate and severe disruptions. However, a prolonged shortage can cause inefficiencies, increased costs, or reduced service quality. These items require significant attention but do not need as stringent control measures as Vital items. Characteristics of Essential items include:

- **Moderate Impact on Operations:** The unavailability of Essential items can impact operations, but usually does not cause immediate stoppage.
- **Important for Efficiency:** These items contribute to operational efficiency and productivity.
- **Balanced Inventory Control:** Essential items are managed with balanced inventory control practices, focusing on maintaining adequate stock levels without excessive overstocking.
- **Reasonable Safety Stock Levels:** Safety stock levels are maintained to cover moderate demand fluctuations and lead time variability.

2.4.3 Desirable (D)

Desirable items are those that are nice to have but do not significantly impact business operations if they are scarce or unavailable. These items often have lower demand variability and are not critical to the immediate functioning of the business. Characteristics of Desirable items include:

- **Low Impact on Operations:** The absence of Desirable items typically does not disrupt business operations or service delivery.
- **Low Criticality:** These items are not essential for maintaining operational efficiency or customer satisfaction.
- **Minimal Inventory Control:** Inventory control for Desirable items is less rigorous, with lower safety stock levels and less frequent reviews.
- **Cost Management Focus:** The primary focus for Desirable items is on cost management, avoiding unnecessary stocking and minimizing carrying costs.

2.4.4 Integration with ABC Analysis

VED analysis can be effectively combined with ABC analysis to provide a comprehensive inventory management strategy. ABC analysis categorizes items based on their value and contribution to overall inventory costs, dividing them into:

- **A Items:** High-value items that typically account for a significant portion of total inventory value but a small percentage of items.
- **B Items:** Moderate-value items that represent a balance between value and quantity.
- **C Items:** Low-value items that make up a large portion of the inventory but contribute minimally to the total value.

When combined, VED and ABC analyses allow businesses to:

- **Prioritize Inventory Management Efforts:** Focus resources and attention on items that are both high value and high criticality.
- **Optimize Inventory Levels:** Maintain optimal stock levels based on both value and criticality, ensuring that vital and essential items are adequately stocked.
- **Improve Decision-Making:** Make informed decisions regarding procurement, storage, and resource allocation by considering both the value and criticality of inventory items.

2.4.5 Advantages of VED Analysis

- **Enhanced Focus on Critical Items:** VED analysis ensures that the most critical items receive the necessary attention and resources, reducing the risk of operational disruptions.

- **Better Resource Allocation:** Categorizing items based on their criticality, businesses can allocate resources more efficiently, focusing on items that have the greatest impact on operations.
- **Improved Inventory Management:** The combination of VED and ABC analyses provides a comprehensive approach to inventory management, optimizing stock levels and reducing excess inventory.
- **Informed Decision-Making:** VED analysis provides valuable insights into the criticality of inventory items, supporting more informed and strategic decision-making processes.
- **Risk Mitigation:** Identifying vital and essential items, businesses can implement appropriate risk mitigation strategies, such as maintaining higher safety stock levels and prioritizing procurement.

2.4.6 Disadvantages of VED Analysis

- **Complexity in Implementation:** Implementing VED analysis can be complex and time-consuming, requiring detailed evaluation and classification of inventory items.
- **Subjectivity:** The classification of items into Vital, Essential, and Desirable categories can be subjective, leading to potential inconsistencies and inaccuracies.
- **Resource Intensive:** Regularly conducting VED analysis and maintaining updated classifications can be resource intensive requiring significant effort and investment.
- **Potential Overemphasis on Vital Items:** Focusing too heavily on Vital items may lead to neglect of Essential and Desirable items, potentially causing inefficiencies in other areas of inventory management.

VED analysis is a valuable tool for categorizing inventory items based on their criticality and impact on business operations. Integrating VED analysis with ABC analysis, businesses can enhance their inventory management practices, ensuring that vital and essential items receive the necessary attention and resources. This comprehensive approach leads to improved operational efficiency, reduced costs, and higher customer satisfaction. However, it is important to recognize the potential disadvantages and implement VED analysis thoughtfully to maximize its benefits.

2.5 Safety Stock Management

Safety stock management is a crucial aspect of inventory control and supply chain management. Safety stock refers to the additional inventory held by a company beyond the expected demand during normal replenishment cycles. The primary purpose of safety stock is to act as a buffer or insurance against unexpected fluctuations in demand or supply chain disruptions. While safety stock is essential for mitigating uncertainties, effective safety stock management requires periodic review and adjustment. As supply chain dynamics change, companies must reassess and optimize their safety stock levels to ensure they remain well-prepared to meet customer demand under various conditions.

2.5.1 Key Points About Safety Stock Management

Safety stock, also known as buffer stock, is an additional quantity of inventory kept on hand to mitigate the risk of stockouts caused by uncertainties in demand and supply. It acts as a cushion against variability in customer demand, lead time, and supply chain disruptions. The primary purpose of safety stock is to ensure that a company can continue to meet customer demand even when unexpected fluctuations occur. The importance of safety stock lies in its ability to provide a safeguard against uncertainties, thereby enhancing the reliability and responsiveness of the inventory management system. Here are the major key points on safety stock:

- **Buffer Against Variability:** Safety stock serves as a cushion to absorb variability in demand, supply lead times, and other uncertainties. It acts as a safeguard against unexpected events that could disrupt the regular flow of goods.

- **Prevention of Stockouts:** The main goal of safety stock is to prevent stockouts. If actual demand exceeds forecasts or if there are delays in the supply chain, the safety stock ensures that there are sufficient goods available to meet customer demand.

- **Uncertainty in Demand and Lead Times:** Safety stock takes into account uncertainties in both demand and lead times. Fluctuations in customer orders, unexpected surges in demand, or delays in receiving new inventory can be mitigated by maintaining an appropriate level of safety stock.

- **Optimal Service Levels:** Safety stock helps companies achieve optimal service levels by ensuring that they can fulfill customer orders even under less-than-ideal circumstances. It contributes to customer satisfaction by reducing the likelihood of backorders or delayed shipments.

- **Demand Forecasting Accuracy:** Variability in demand forecasting is common. Safety stock allows companies to manage the impact of forecast errors by providing a buffer to cover potential discrepancies between predicted and actual demand.
- **Seasonal Demand and Promotions:** Safety stock becomes particularly important in industries with seasonal demand patterns or during promotional periods when demand is expected to increase. It helps companies manage sudden spikes in orders.
- **Supply Chain Disruptions:** Safety stock is valuable during supply chain disruptions, such as delays in shipments or unexpected interruptions in the production process. It allows companies to continue operations while addressing and resolving supply chain issues.
- **Calculation and Optimization:** Determining the appropriate level of safety stock involves considering factors like lead time variability, demand variability, and desired service levels. Optimization techniques help strike a balance between carrying costs and the risk of stockouts.

2.5.2 Methods for Calculating Safety Stock Levels

Several methods can be used to calculate safety stock levels, each varying in complexity and accuracy. The choice of method depends on the specific circumstances and requirements of the organization. Common methods include:

2.5.2.1 Basic Safety Stock Formula

A simple and commonly used method is based on the standard deviation of demand during the lead time and a desired service level. The formula is:

$$SS = Z \times \sigma_{LT}$$

where:

SS is the safety stock level.

Z is the Z-score corresponding to the desired service level (e.g., a Z-score of 1.645 for a 95% service level).

σ_{LT} is the standard deviation of demand during the lead time.

2.5.2.2 Safety Stock Based on Average and Maximum Demand

Another approach calculates safety stock as a function of average and maximum demand during the lead time:

$$SS = (D_{max} - D_{avg}) \times LT$$

where:

D_{max} is the maximum demand during the lead time.

D_{avg} is the average demand during the lead time.

LT is the lead time.

2.5.2.3 Safety Stock Based on Service Level and Lead Time Variability

This method considers both demand variability and lead time variability:

$$SS = Z \times \sqrt{(LT \times \sigma_D^2) + (D^2 \times \sigma_{LT}^2)}$$

where:

σ_D is the standard deviation of demand.

D is the average demand.

σ_{LT} is the standard deviation of lead time.

2.5.3 Advantages of Safety Stock

Safety stock directly impacts customer satisfaction by ensuring product availability and timely fulfillment of orders. High customer satisfaction can lead to repeat business, positive word-of-mouth, and increased customer loyalty. Preventing stockouts and meeting customer demand consistently, companies can enhance their reputation and competitive advantage in the market. Safety stock is a crucial element of effective inventory management, providing a buffer against uncertainties in demand and supply. Calculating appropriate safety stock levels and managing them effectively, organizations can prevent stockouts, maintain high service levels, and ensure smooth operations. Advantages are:

- **Prevention of Stockouts:** Safety stock acts as a buffer against variability in demand and supply, ensuring that there are enough products available to meet customer demand even when there are unexpected fluctuations. This helps in maintaining service levels and customer satisfaction.

- **Increased Customer Satisfaction:** Preventing stockouts, safety stock ensures that customer orders can be fulfilled promptly. This leads to higher levels of customer satisfaction and can enhance the company's reputation and customer loyalty.

- **Protection Against Lead Time Variability:** Safety stock mitigates the impact of longer-than-expected lead times, providing a cushion that ensures continuous production and sales operations without interruption.

- **Handling Demand Surges:** During periods of unexpected demand surges, safety stock allows the company to respond quickly without the need for emergency orders or expedited shipping, which can be costly.
- **Operational Flexibility:** Having safety stock allows for greater flexibility in operations, enabling companies to manage supply chain disruptions more effectively and maintain smooth production schedules.

2.5.4 Disadvantages of Safety Stock

Safety stock is an additional quantity of an item held in inventory to reduce the risk of stockouts caused by uncertainties in supply and demand. While safety stock can provide significant benefits, it also has certain drawbacks.

- **Increased Holding Costs:** Maintaining safety stock involves additional holding costs, including storage, insurance, and taxes. These costs can be significant, especially for perishable or high-value items.
- **Capital Tied Up in Inventory:** Safety stock requires investment in inventory, tying up capital that could otherwise be used for other purposes such as investing in new projects, marketing, or research and development.
- **Risk of Obsolescence:** For products with short life cycles or those subject to rapid technological changes, safety stock runs the risk of becoming obsolete, leading to potential losses.
- **Space Requirements:** Additional inventory requires more storage space, which can be a limitation for companies with constrained warehouse capacity. This may necessitate investment in additional storage facilities.
- **Complex Inventory Management:** Managing safety stock adds complexity to inventory management processes, requiring accurate demand forecasting and regular review of inventory levels to ensure optimal stock levels are maintained.

2.5.5 Balancing Safety Stock

The key to effective inventory management is balancing the benefits of safety stock against its costs. Companies need to carefully determine the appropriate level of safety stock based on factors such as demand variability, lead time uncertainty, service level targets, and cost considerations. Advanced inventory optimization models and techniques can help in achieving this balance, ensuring that the benefits of safety stock outweigh the associated costs. Overall, safety stock plays a critical role in ensuring supply chain resilience and customer satisfaction, but it must be managed strategically to avoid unnecessary costs and inefficiencies.

2.6 Inventory Turnover

Inventory turnover is a key performance indicator (KPI) that measures how quickly a company sells and replaces its inventory. Optimization strategies involve improving turnover ratios, reducing excess stock, and aligning stock levels with demand to improve capital efficiency. Metrics for inventory turnover measure the speed at which inventory is sold and can be calculated as:

$$Inventory\ Turnover = \frac{COGS}{Average\ Inventory}$$

where COGS is the cost of goods sold.

2.6.1 Benefits of Calculating Inventory Turnover

Understanding and calculating inventory turnover offers several benefits for businesses such as:

- **Improved Inventory Management:** Calculating inventory turnover helps businesses understand how efficiently their inventory is being managed. A high turnover rate indicates effective inventory management with minimal overstocking, while a low turnover rate may highlight issues such as overstocking or slow-moving inventory.
- **Enhanced Cash Flow:** High inventory turnover rates often correlate with better cash flow, as products are sold quickly, reducing the amount of capital tied up in inventory. This allows businesses to reinvest cash into operations, growth initiatives, or other critical areas.
- **Reduced Holding Costs:** Optimizing inventory turnover, businesses can minimize holding costs, which include storage, insurance, and obsolescence costs. Reduced holding costs lead to higher profitability and more efficient use of resources.
- **Increased Responsiveness to Market Demand:** A high turnover rate indicates that a business can quickly respond to changes in market demand. This agility allows for better alignment of inventory levels with current sales trends, reducing the risk of stockouts or excess inventory.
- **Insight into Sales Effectiveness:** Inventory turnover provides valuable insights into sales performance and product demand. Products with high turnover are typically more popular, guiding inventory purchasing decisions and helping businesses focus on high-demand items.
- **Identification of Slow-Moving Items:** Calculating inventory turnover helps identify slow-moving items that may need to be discounted, promoted,

or discontinued. This prevents inventory from becoming obsolete and reduces the financial impact of unsold goods.

- **Benchmarking and Performance Improvement:** Inventory turnover ratios can be benchmarked against industry standards or historical performance to assess how well the business is performing relative to peers. This benchmarking can drive continuous improvement efforts and strategic decision-making.
- **Better Supplier Negotiations:** Understanding inventory turnover can enhance supplier negotiations by providing data on how quickly products are sold. This information can be used to negotiate better terms, such as bulk purchasing discounts or more flexible payment terms.
- **Optimized Reordering Processes:** Inventory turnover metrics aid in optimizing reordering processes by indicating the optimal reorder points and quantities. This ensures that inventory levels are maintained efficiently without overstocking or stockouts.
- **Enhanced Financial Analysis:** Inventory turnover is a critical component of financial analysis, affecting key metrics such as return on assets (ROA) and gross margin return on inventory (GMROI). High inventory turnover can positively impact these financial ratios, indicating better asset utilization and profitability.
- **Strategic Planning and Forecasting:** Regularly calculating inventory turnover supports strategic planning and demand forecasting. It provides historical data that can be used to predict future sales trends and plan inventory levels accordingly.

Calculating inventory turnover is an essential practice for businesses seeking to optimize their inventory management processes. It provides numerous benefits, from improving cash flow and reducing holding costs to enhancing market responsiveness and supporting strategic planning. Regularly monitoring and analyzing inventory turnover, businesses can make informed decisions that drive efficiency, profitability, and overall operational success.

2.7 Reorder Point (ROP)

The Reorder Point (ROP) serves as a pivotal control mechanism in inventory systems, balancing stock availability with carrying costs. Its calculation integrates demand forecasting, supply chain reliability, and operational efficiency metrics. This section explores the critical components and methodologies involved in determining the ROP, including lead time variability, service level targets, and the interplay between safety stock and expected demand.

2.7.1 Definition of Reorder Point and Its Role in Inventory Management

The reorder point (ROP) is a critical inventory management metric that indicates the inventory level at which a new order should be placed to replenish stock before it runs out. The primary role of the reorder point is to ensure that inventory is reordered in a timely manner, allowing for lead time—the time it takes for the order to be processed and delivered—thus preventing stockouts and ensuring continuous supply to meet customer demand.

The ROP is a threshold that triggers the replenishment process and is essential for maintaining optimal inventory levels. Setting an appropriate reorder point, companies can effectively balance inventory holding costs with the risk of stockouts, leading to improved inventory control and cost management.

2.7.2 Calculation of Reorder Point Based on Lead Time and Demand

The reorder point is typically calculated based on the average demand during the lead time, along with an allowance for safety stock to account for demand variability and lead time fluctuations. The basic formula for calculating the reorder point is:

$$\text{ROP} = \text{AVDR} \times \text{LTD} + \text{SS}$$

where:

- $AVDR$ is the average demand per period (e.g., units per day or units per week).
- LT is the lead time (the time between placing an order and receiving the goods, in the same time units as D).
- SS is the safety stock level, which provides a buffer against demand and lead time variability.

Example 2.3. *A company has an average daily demand of 50 units, a lead time of five days, and maintains a safety stock of 100 units. What would is the company's reorder point?*

Solution 2.3.

$$ROP = (50 \ units/day \times 5 \ days) + 100 \ units = 350 \ units$$

In this example, the company would place a new order when the inventory level drops to 350 units.

Example 2.4. *Suppose a company sells an average of 30 units of a product per day. The lead time for an order is seven days, and the company maintains a safety stock of 100 units to cover unexpected increases in demand or delays in delivery. Find the ROP of the company.*

Solution 2.4. *Given:*

- *Average Demand Rate (AVDR) = 30 units per day*
- *Lead Time in Days (LTD) = 7 days*
- *Safety Stock (SS) = 100 units*

The formula for the Reorder Point (ROP) is:

$$ROP = AVDR \times LTD + SS$$

Substituting the given values:

$$\begin{aligned} ROP &= 30\, units/day \times 7\, days + 100\, units \\ &= 210\, units + 100\, units \\ &= 310\, units \end{aligned}$$

Therefore, the ROP is:

$$\boxed{310\, units}$$

This means that when the inventory level drops to 310 units, it is time to place a new order to replenish the stock.

Example 2.5. *Suppose a company has determined they need to place a new order when their inventory level drops to 400 units (ROP). They know their average demand rate is 50 units per day, and they keep a safety stock of 100 units to cover unexpected increases in demand or delays in delivery. What is the company's LTD?*

Solution 2.5. *Given:*

- *Reorder Point (ROP) = 400 units*
- *Average Demand Rate (AVDR) = 50 units per day*
- *Safety Stock (SS) = 100 units*

The formula for the ROP is:

$$ROP = AVDR \times LTD + SS$$

We need to solve for the LTD. Rearranging the formula to solve for LTD:

$$LTD = \frac{ROP - SS}{AVDR}$$

Substituting the given values:

$$\begin{aligned} LTD &= \frac{400\, units - 100\, units}{50\, units/day} \\ &= \frac{300\, units}{50\, units/day} \\ &= 6\, days \end{aligned}$$

Therefore, the LTD is:

$$\boxed{6 \text{ } days}$$

This means that the lead time for the company's orders is six days.

2.7.3 Impact of Reorder Points on Inventory Control and Cost Management

The implementation of reorder points has significant implications for inventory control and cost management.

2.7.3.1 Inventory Control

Reorder points enhance inventory control by providing a systematic and proactive approach to inventory replenishment. Setting ROPs, companies can:

- **Automate Replenishment:** Trigger automated reordering processes, reducing manual intervention and the risk of human error.
- **Maintain Optimal Stock Levels:** Ensure that inventory levels are maintained at optimal levels, preventing both stockouts and overstocking.
- **Improve Forecasting Accuracy:** Use historical demand data to set accurate ROPs, improving the overall forecasting and planning process.

2.7.3.2 Cost Management

Reorder points contribute to effective cost management by balancing various inventory-related costs:

- **Reduce Holding Costs:** Setting appropriate ROPs, companies can avoid overstocking, which reduces holding costs associated with storage, insurance, and obsolescence.
- **Minimize Stockout Costs:** Properly set ROPs help in avoiding stockouts, which can lead to lost sales, customer dissatisfaction, and potential expedited shipping costs to replenish stock urgently.
- **Optimize Order Frequency:** Balancing ordering and holding costs ensures that orders are placed neither too frequently (which increases ordering costs) nor too infrequently (which increases holding costs).

Reorder points are a vital tool in inventory management, enabling organizations to maintain continuous supply, prevent stockouts, and manage inventory costs effectively. Calculating ROPs based on lead time and demand, and incorporating safety stock to buffer against uncertainties, companies can ensure efficient and responsive inventory systems. The use of reorder points enhances inventory control, automates replenishment processes, and supports overall cost management, leading to improved operational efficiency and customer satisfaction.

2.8 Economic Order Quantity (EOQ) Model

The EOQ model is a fundamental inventory management tool that determines the optimal order quantity for minimizing the total costs associated with ordering and holding inventory. The purpose of the EOQ model is to identify the most cost-effective quantity to order, balancing the trade-off between ordering costs (costs incurred every time an order is placed) and holding costs (costs associated with keeping inventory in stock).

2.8.1 Mathematical Formulation and Assumptions of the EOQ Model

The EOQ model is derived based on a set of assumptions and uses a specific mathematical formula to calculate the optimal order quantity. The basic EOQ formula is:

$$EOQ = \sqrt{\frac{2DS}{H}}$$

where:

D is the annual demand for the product (units per year).

S is the ordering cost per order (fixed cost per order).

H is the holding cost per unit per year (cost to hold one unit of inventory for a year).

The model is based on the following assumptions:

1. **Constant Demand:** The demand for the product is constant and known with certainty.
2. **Constant Lead Time:** The lead time, or the time between placing and receiving an order, is constant.
3. **Fixed Ordering Cost:** The cost per order is fixed and does not vary with the order size.
4. **Constant Holding Cost:** The holding cost per unit per year is constant.
5. **Instantaneous Replenishment:** Inventory is replenished instantaneously, meaning the entire order quantity is received at once.
6. **No Stockouts:** Stockouts (shortages) are not allowed, implying that orders are placed at the right time to avoid running out of stock.

2.8.2 Economic Production Quantity (EPQ)

EPQ builds upon EOQ by introducing production constraints into the optimization process. In manufacturing environments, where production capacity is a limiting factor, EPQ aids in determining the optimal order quantity to balance production and inventory costs effectively. The EPQ formula is:

$$EPQ = \sqrt{\frac{2DS}{H}} \cdot \sqrt{\frac{p}{p-d}}$$

where:

p is the production rate.

Holding Cost: $H = C \cdot r$, where C is the unit cost and r is the holding rate.

Ordering Cost: $S = \frac{D}{Q} \cdot K$, where Q is the order quantity and K is the ordering cost per order.

Shortage/Backorder Cost: The cost of stockouts can be calculated as the product of the shortage rate and the cost per unit shortage.

Example 2.6. *AJK Manufacturing Company produces widgets. The company needs to decide the optimal order quantity for the raw materials used to produce these widgets. Here are the details:*

- *The annual demand for widgets (D) is 10,000 units.*
- *The cost to place an order (S) is \$50 per order.*
- *The holding cost per unit per year (H) is \$2.*

Calculate the optimal quantity and total inventory cost for AJK company.

Solution 2.6.

$$EOQ = \sqrt{\frac{2DS}{H}}$$

Here, $D = 10{,}000$, $S = 50$, *and* $H = 2$.

$$EOQ = \sqrt{\frac{2 \times 10{,}000 \times 50}{2}}$$
$$EOQ = \sqrt{\frac{1{,}000{,}000}{2}}$$
$$EOQ = \sqrt{500{,}000}$$
$$EOQ = 707.11$$

Since the order quantity needs to be a whole number, we can round it to the nearest whole number:

$$EOQ \approx 707$$

Ordering Costs

$$Number\ of\ Orders\ per\ Year = \frac{D}{EOQ} = \frac{10{,}000}{707} \approx 14.14$$

$$Total\ Ordering\ Cost = Number\ of\ Orders\ per\ Year \times S = 14.14 \times 50 = 707$$

Holding Costs

$$Average\ Inventory\ Level = \frac{EOQ}{2} = \frac{707}{2} = 353.5$$

$$Total\ Holding\ Cost = Average\ Inventory\ Level \times H = 353.5 \times 2 = 707$$

Total Cost

$$Total\ Inventory\ Cost = Total\ Ordering\ Cost + Total\ Holding\ Cost$$

$$= 707 + 707 = 1414$$

Interpretation:

The EOQ and total cost for AJK Manufacturing Company is approximately 707 units and $1,414. This means the company should order 707 units of raw materials each time it places an order to minimize the total inventory costs of $1,414.

Example 2.7. *A company has an annual demand of 10,000 units for a product. The ordering cost is $50 per order, and the holding cost is $2 per unit per year. Calculate the EOQ.*

Solution 2.7. *Using the EOQ formula:*

$$EOQ = \sqrt{\frac{2 \times 10{,}000 \times 50}{2}} = \sqrt{\frac{1{,}000{,}000}{2}} = \sqrt{500{,}000} \approx 707$$

Therefore, the optimal order quantity is approximately 707 units.

Steps:

1. *Plug the values into the EOQ formula.*
2. *Calculate the EOQ by simplifying the expression under the square root.*

Example 2.8. *XYZ Electronics produces a specific type of microchip. The company needs to determine the optimal production quantity to minimize inventory costs. The details follow:*

- *The annual demand for microchips (D) is 50,000 units.*
- *The setup cost per production run (S) is $200.*

- *The holding cost per unit per year (H) is \$5.*
- *The production rate (p) is 10,000 units per month.*
- *The demand rate (d) is 4,167 units per month.*

Solution 2.8.

$$EPQ = \sqrt{\frac{2DS}{H}} \cdot \sqrt{\frac{p}{p-d}}$$

Here, $D = 50{,}000$, $S = 200$, $H = 5$, $p = 10{,}000 \times 12 = 120{,}000$ *units per year, and* $d = 4{,}167 \times 12 = 50{,}004$ *units per year.*

$$\begin{aligned}
EPQ &= \sqrt{\frac{2 \times 50{,}000 \times 200}{5}} \cdot \sqrt{\frac{120{,}000}{120{,}000 - 50{,}004}} \\
&= \sqrt{\frac{20{,}000{,}000}{5}} \cdot \sqrt{\frac{120{,}000}{69{,}996}} \\
&= \sqrt{4{,}000{,}000} \cdot \sqrt{1.7143} \\
&= 2{,}000 \cdot 1.3093 \\
&\approx 2{,}618.6 \ \mathit{units}
\end{aligned}$$

Since the production quantity needs to be a whole number, we can round it to the nearest whole number:

$$EPQ \approx 2{,}619 \ \mathit{units}$$

Setup Costs

$$\mathit{Number\ of\ Production\ Runs\ per\ Year} = \frac{D}{EPQ} = \frac{50{,}000}{2619} \approx 19.1$$

$$\begin{aligned}
&\mathit{Total\ Setup\ Cost} = \mathit{Number\ of\ Production\ Runs\ per\ Year} \times S \\
&= 19.1 \times 200 = \$3{,}820
\end{aligned}$$

Holding Costs

$$\textit{Average Inventory Level} = \frac{EPQ}{2} \times \left(1 - \frac{d}{p}\right)$$

$$= \frac{2619}{2} \times \left(1 - \frac{50,004}{120,000}\right)$$

$$\textit{Average Inventory Level} = 1{,}309.5 \times (1 - 0.4167)$$

$$= 1{,}309.5 \times 0.5833 \approx 763.9$$

$$\textbf{Total Holding Cost} = \textit{Average Inventory Level} \times H$$

$$= 763.9 \times 5 = \$3{,}819.5$$

Total Cost

$$\textit{Total Inventory Cost} = \textit{Total Setup Cost} + \textit{Total Holding Cost}$$

$$= 3820 + 3819.5 = \$7639.5$$

Interpretation:

The Economic Production Quantity for XYZ Electronics is approximately 2,619 units. This means the company should produce 2,619 units of microchips each time it runs a production batch to minimize the total inventory costs of $7,639.5.
The following are examples on how to use R Programming Language to plot EOQ and EPQ respectively. The results are illustrated in Figures 2.2, 2.3, and 2.4.

```
# Import SCperf package
library(SCperf)

# Set annual quantity demanded
D <- as.numeric(1000)

# Set ordering cost (fixed cost)
S <- as.numeric(5)

# Set holding cost (variable cost)
H <- as.numeric(1.25)

# Set shortage penalty cost per unit (variable cost)
b <- as.numeric(0)

# Set safety stock
ss <- as.numeric(0)

# Get EOQ Results
Q <- EOQ(D, S, H, b)
```

```

EOQ <- as.numeric(Q[1])
print(paste("Economic Order Quantity:", ceiling(EOQ)))

number_of_orders <- as.numeric(D / Q[1])
print(paste("Number of orders per year:",
    ceiling(number_of_orders)))

cycle_time <- as.numeric(Q[2] * 12)
print(paste("Time between cycles (months):",
    round(cycle_time, digits = 2)))

annual_ordering_cost <- as.numeric(D / Q[1] * S)
print(paste("Annual ordering cost:",
    round(annual_ordering_cost, digits = 2)))

annual_holding_cost <- as.numeric(Q[1] / 2 * H)
print(paste("Annual holding cost:",
    round(annual_holding_cost, digits = 2)))

total_cost <- as.numeric((D / Q[1] * S) + (Q[1] / 2 * H))
print(paste("Total cost:", round(total_cost, digits = 2)))

# Data frame construction
# Period column
period <- as.numeric(c(0, Q[2] * 12))

while (period[length(period)] < 12) {
  initial <- period[length(period)]
  period <- append(period, initial)
  final <- period[length(period)] + cycle_time
  period <- append(period, final)
}

# Inventory column
initial_inventory <- as.numeric(c(Q[1]))
inventory <- initial_inventory

while (length(inventory) != length(period)) {
  available_inventory <- 0
  inventory <- append(inventory, available_inventory)
  if (available_inventory == 0) {
    if (length(inventory) != length(period)) {
      available_inventory <- as.numeric(c(Q[1]))
      inventory <- append(inventory, available_inventory)
    }
  }
}

```

```
# Safety stock column
safety_stock <- c(rep(ss, length(period)))

# Add safety stock to available inventory
inventory <- inventory + safety_stock

# Construct inventory data frame
inventory_distribution <- data.frame(period, inventory, safety_stock)

# Import ggplot2 package
library(ggplot2)

# Build inventory graph without considering safety stock
p1 <- ggplot(data = inventory_distribution, aes(x = period)) +
  geom_line(aes(y = inventory)) +
  scale_x_continuous(breaks = c(0:12)) +
  ylim(c(0, max(inventory_distribution$inventory))) +
  labs(title = "Economic Order Quantity",
       subtitle = "Inventory Planning and Managing",
       x = "Month",
       y = "Inventory Level",
       caption = "No safety stock considered")

# Save the plot at 300 DPI
ggsave("inventory_without_safety_stock.png", plot = p1, dpi = 300)

# Build inventory graph considering safety stock
p2 <- ggplot(data = inventory_distribution,
             aes(x = period)) +
  geom_line(aes(y = inventory, col = "Available inventory")) +
  geom_line(aes(y = safety_stock, col = "Safety stock")) +
  scale_x_continuous(breaks = c(0:12)) +
  ylim(c(0, max(inventory_distribution$inventory))) +
  labs(title = "Economic Order Quantity",
       subtitle = "Inventory Planning and Managing",
       x = "Month",
       y = "Inventory Level",
       caption = "Safety stock considered",
       color = NULL) +
  scale_color_manual(labels = c("Available inventory", "Safety stock"),
                     values = c("Available inventory" = "black", "Safety stock" = "red"))

# Save the plot at 300 DPI
```

```
ggsave("inventory_with_safety_stock.png", plot = p2, dpi =
    300)
```

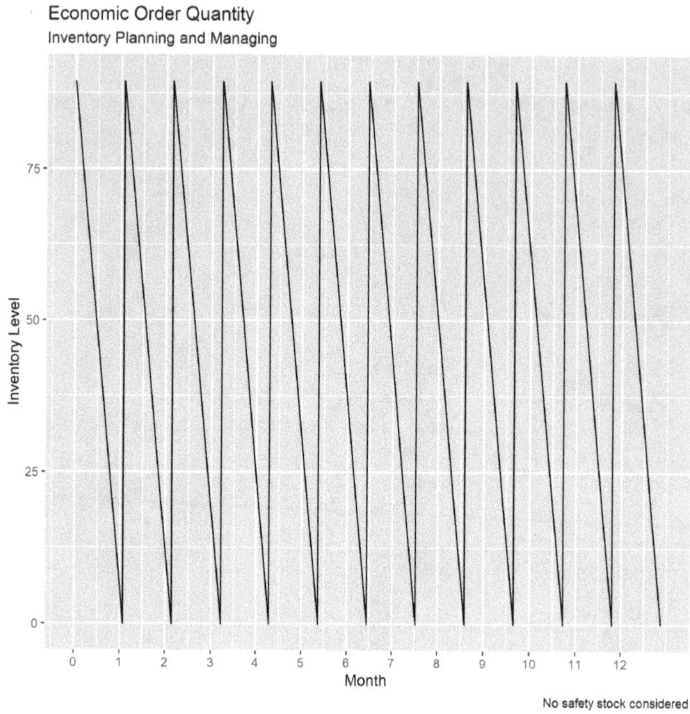

FIGURE 2.2
EOQ model (A).

```
# Install necessary packages if not already installed
if (!requireNamespace("SCperf", quietly = TRUE)) {
  install.packages("SCperf")
}
if (!requireNamespace("ggplot2", quietly = TRUE)) {
  install.packages("ggplot2")
}
if (!requireNamespace("xtable", quietly = TRUE)) {
  install.packages("xtable")
}

# Load the packages
library(SCperf)
library(ggplot2)
library(xtable)

```

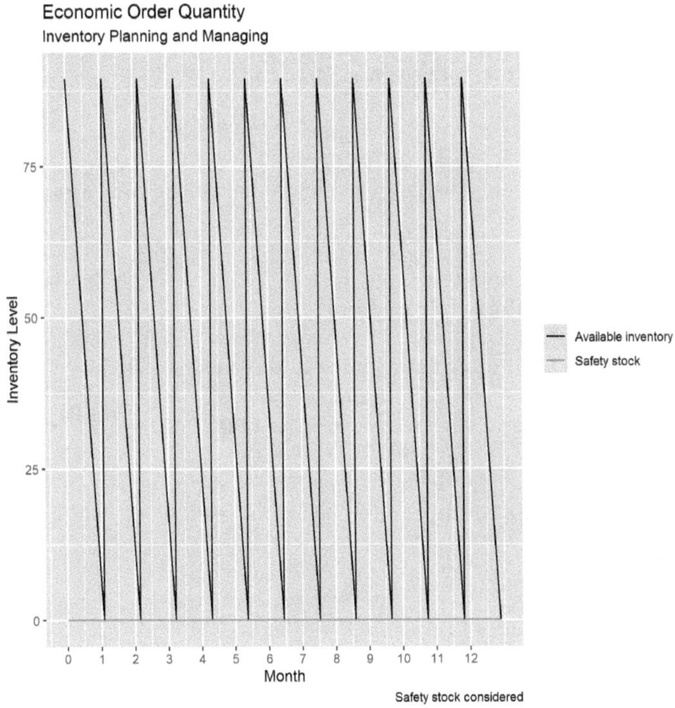

FIGURE 2.3
EOQ model (B).

```
# Set parameters
D <- as.numeric(200)  # Annual quantity demanded
P <- as.numeric(1000) # Annual production rate
K <- as.numeric(100)  # Cost per production run
H <- as.numeric(5)    # Unit holding cost per year

# Get EPQ Results
Q <- EPQ(D, P, K, H)
EPQ <- as.numeric(Q[1])
print(paste("Economic Production Quantity:", ceiling(EPQ)))

# Calculate various metrics
production_cycle_length <- as.numeric(Q[3] * 12)
print(paste("Production cycle length (months):",
    round(production_cycle_length, digits = 2)))

number_of_production_runs <- as.numeric(D / Q[1])
print(paste("Number of production runs per year:",
    ceiling(number_of_production_runs)))

```

```
production_run_length <- as.numeric(Q[2] * 12)
print(paste("Production run length (months):",
    round(production_run_length, digits = 2)))

demand_period_length <- as.numeric(Q[2] * 12)
print(paste("Demand period length (months):",
    round(production_cycle_length - production_run_length,
    digits = 2)))

annual_production_cost <-
    as.numeric(number_of_production_runs * K)
print(paste("Annual production run cost:",
    round(annual_production_cost, digits = 2)))

annual_holding_cost <- as.numeric((EPQ / 2) * (1 - (D /
    P)) * H)
print(paste("Annual holding cost:",
    round(annual_holding_cost, digits = 2)))

maximum_inventory <- as.numeric(Q[4])
print(paste("Maximum inventory per cycle:",
    maximum_inventory))

total_cost <- as.numeric(Q[5])
print(paste("Total cost:", round(total_cost, digits = 2)))

# Construct data frame for plotting
period <- as.numeric(c(0, Q[2] * 12, Q[3] * 12))
while (period[length(period)] < 12) {
  period <- append(period,
      as.numeric(period[length(period)] + Q[2] * 12))
  period <- append(period,
      as.numeric(period[length(period)] + ((Q[3] * 12) -
      (Q[2] * 12))))
}

inventory <- as.numeric(c(0, Q[4], 0))
while (length(inventory) < length(period)) {
  inventory <- append(inventory, as.numeric(Q[4]))
  inventory <- append(inventory, 0)
}

production <- c(0, as.numeric(Q[1]))
while (length(production) < length(period)) {
  production <- append(production, NA)
}

# Create data frame for inventory distribution
inventory_distribution <- data.frame(period, inventory,
```

```
    production)

# Plot the inventory distribution
p1 <- ggplot(data = inventory_distribution, aes(x =
    period)) +
  geom_line(aes(y = inventory, col = "Accumulated
      inventory")) +
  geom_line(aes(y = production, col = "Production rate"),
      linetype = "dashed") +
  scale_x_continuous(breaks = c(0:12)) +
  labs(title = "Economic Production Quantity",
       subtitle = "Inventory Planning and Managing",
       x = "Month",
       y = "Inventory Level",
       color = NULL) +
  scale_color_manual(labels = c("Accumulated inventory",
      "Production rate"),
                     values = c("Accumulated inventory" =
                         "black", "Production rate" =
                         "blue"))

# Save the plot at 300 DPI
ggsave("inventory_distribution.png", plot = p1, dpi = 300)

# Create a data frame with the results for LaTeX table
results <- data.frame(
  Metric = c("Economic Production Quantity",
             "Production cycle length (months)",
             "Number of production runs per year",
             "Production run length (months)",
             "Demand period length (months)",
             "Annual production run cost",
             "Annual holding cost",
             "Maximum inventory per cycle",
             "Total cost"),

  Value = c(ceiling(EPQ),
            round(production_cycle_length, digits = 2),
            ceiling(number_of_production_runs),
            round(production_run_length, digits = 2),
            round(production_cycle_length -
                production_run_length, digits = 2),
            round(annual_production_cost, digits = 2),
            round(annual_holding_cost, digits = 2),
            maximum_inventory,
            round(total_cost, digits = 2))
)

# Print the LaTeX tabular code
```

```
cat("\\begin{table}[ht]\n")
cat("\\centering\n")
cat("\\begin{tabular}{ll}\n")
cat("\\hline\n")
cat("Metric & Value \\\\\n")
cat("\\hline\n")
for (i in seq_len(nrow(results))) {
    cat(paste(results$Metric[i], "&", results$Value[i],
        "\\\\\n"))
}
cat("\\hline\n")
cat("\\end{tabular}\n")
cat("\\caption{EPQ Results Summary}\n")
cat("\\label{tab:epq_results}\n")
cat("\\end{table}\n")
```

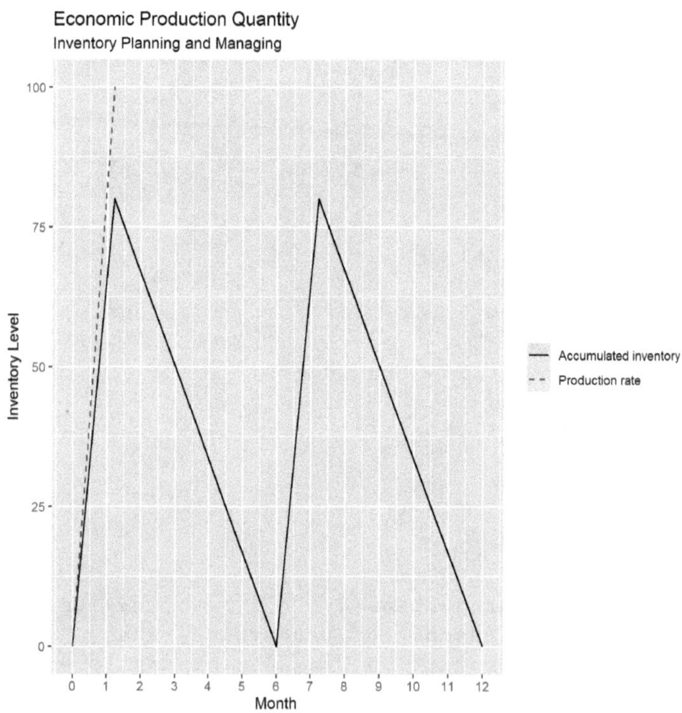

FIGURE 2.4
EPQ model (C).

TABLE 2.2
EPQ Results Summary

Metric	Value
Economic Production Quantity	100
Production cycle length (months)	6
Number of production runs per year	2
Production run length (months)	1.2
Demand period length (months)	4.8
Annual production run cost	200
Annual holding cost	200
Maximum inventory per cycle	80
Total cost	400

2.8.3 Advantages

- **Cost Minimization:** The EOQ/EPQ models helps in identifying the order/production quantity that minimizes the total cost of inventory, leading to significant cost savings.

- **Simplified Decision-Making:** The EOQ/EPQ formulae provide a clear and straightforward method for determining optimal order/production quantities, simplifying the decision-making process for inventory managers.

- **Efficiency Improvement:** Using the EOQ/EPQ models, organizations can streamline their inventory management processes, reducing the frequency of orders/productions and the amount of stock held, leading to improved operational efficiency.

- **Resource Optimization:** The EOQ/EPQ models aids in better allocation of resources, ensuring that capital is not unnecessarily tied up in inventory, allowing for more investment in other critical areas of the business.

2.8.4 Disadvantages

- **Simplistic Assumptions:** The EOQ/EPQ models are based on several simplifying assumptions (e.g., constant demand, fixed lead time) that may not hold true in real-world scenarios. Variations in demand and lead time can affect the accuracy of the EOQ/EPQ calculations.

- **Limited Flexibility:** The model does not account for situations where bulk discounts are available, or where the ordering/production cost and holding cost vary with the order/production quantity.

- **Static Nature:** The EOQ/EPQ models do not adapt to changes in market conditions, such as fluctuations in demand or changes in supplier pricing, requiring regular recalculations and adjustments.

- **Exclusion of Other Costs:** The models primarily focus on ordering/production and holding costs, potentially overlooking other relevant costs such as transportation, handling, and obsolescence costs.

The EOQ/EPQ models are valuable tools for optimizing inventory management, balancing ordering/production and holding costs to determine the most cost-effective order/production quantity. While they offers significant advantages in terms of cost minimization and operational efficiency, their applicability is limited by their underlying assumptions and the static nature of their calculations. To maximize the benefits of the EOQ/EPQ models, organizations should regularly review and adjust their calculations to reflect changing market conditions, and integrate the models with other inventory management techniques for a comprehensive inventory strategy.

2.9 Comparison of Inventory Control Techniques

Inventory control techniques are essential tools for managing inventory effectively and ensuring that the right products are available at the right time. In this section, we compare four commonly used inventory control techniques: ABC analysis, economic order/production quantity (EOQ/EPQ), safety stock, and reorder point (ROP).

2.9.1 ABC Analysis

ABC analysis is a method of categorizing inventory items based on their importance to the organization. Items are classified into three categories:

- **A Items:** High-value items with low frequency of sales.
- **B Items:** Moderate-value items with moderate frequency of sales.
- **C Items:** Low-value items with high frequency of sales.

Strengths:

- Prioritizes management efforts on the most critical items.
- Simplifies inventory management by focusing on a small percentage of items that account for the majority of the value.

Weaknesses:

- May overlook the importance of low-value but essential items.
- Requires regular updating and review to remain accurate.

2.9.2 Economic Order/Production Quantity (EOQ/EPQ)

The EOQ/EPQ models determines the optimal order/production quantity that minimizes total inventory costs, balancing ordering/production and holding costs.

Strengths:

- Provides a clear, quantitative basis for decision-making.
- Reduces total inventory costs by optimizing order/production size.

Weaknesses:

- Assumes constant demand and lead time, which may not be realistic.
- Not flexible enough to handle variability in demand or bulk discount opportunities.

2.9.3 Safety Stock

Safety stock is additional inventory kept to mitigate the risk of stockouts due to demand and supply variability.

Strengths:

- Provides a buffer against uncertainties in demand and lead time.
- Ensures high service levels and customer satisfaction.

Weaknesses:

- Increases holding costs due to higher inventory levels.
- Requires accurate demand forecasting and regular adjustments.

2.9.4 Reorder Point (ROP)

The reorder point is the inventory level at which a new order should be placed to replenish stock before it runs out.

Strengths:

- Ensures timely reordering, preventing stockouts.
- Can be automated, reducing manual oversight.

Weaknesses:

- Requires accurate data on lead time and demand.
- May not adapt quickly to sudden changes in demand patterns.

2.10 Criteria for Selecting Appropriate Techniques Based on Organizational Needs

Selecting the appropriate inventory control technique depends on several factors, including the nature of the products, demand variability, lead times, and organizational priorities. Here are some criteria to consider:

- **Demand Variability:** For products with high demand variability, techniques like safety stock and ROP are essential to buffer against uncertainties.
- **Order Frequency:** For items ordered frequently, EOQ can help optimize order quantities to minimize costs.
- **Value of Items:** ABC analysis is useful for categorizing items based on their value and prioritizing management efforts accordingly.
- **Service Level Requirements:** High service level requirements may necessitate higher safety stock and more frequent reorder points.
- **Resource Availability:** The complexity and resource requirements of each technique should align with the organization's capabilities and resources.

2.10.1 Integrated Inventory Management in a Retail Chain

A large retail chain implemented a coordinated approach using multiple inventory control techniques to manage its diverse product range effectively:

- **ABC Analysis:** The retail chain categorized its products into A, B, and C categories based on their sales value. This allowed the company to focus its management efforts and resources on high-value A items, while still maintaining adequate control over B and C items.
- **EOQ for High-Demand Products:** For high-demand A items, the retail chain used the EOQ model to determine the optimal order quantity. This helped in minimizing ordering and holding costs while ensuring sufficient stock availability.
- **Safety Stock for Critical Items:** The retail chain maintained safety stock for critical A and B items to prevent stockouts due to demand fluctuations and supply chain disruptions. This ensured high service levels and customer satisfaction.
- **ROP for Automated Reordering:** The retail chain set reorder points for all categories of items based on their lead time and average demand. This automated the reordering process, reducing manual oversight and ensuring timely replenishment.

Results:

- Improved inventory turnover and reduced holding costs for high-value items.
- Enhanced customer satisfaction due to better product availability.
- Streamlined inventory management processes through automation and prioritization.

2.10.2 Manufacturing Company Using Multiple Techniques

A manufacturing company producing electronic components can utilizes combination of inventory control techniques to manage its raw materials and finished goods:

- **ABC Analysis:** The company classified its raw materials and finished goods into A, B, and C categories. This helped prioritize the management of critical components that had a significant impact on production and sales.
- **EOQ for Raw Materials:** For high-value raw materials, the company used the EOQ model to determine the optimal order quantity, reducing ordering and holding costs.
- **Safety Stock for Key Components:** The company maintained safety stock for key components with long lead times or high demand variability. This ensured that production schedules were not disrupted due to shortages.
- **ROP for Finished Goods:** Reorder points were set for finished goods to ensure timely replenishment and fulfillment of customer orders. This automated the reordering process and improved customer satisfaction.

Results:

- Reduced production delays and stockouts, leading to smoother operations.
- Optimized inventory levels, balancing holding costs and service levels.
- Improved responsiveness to changes in demand and supply chain conditions.

The effective use of inventory control techniques requires a thorough understanding of their strengths and weaknesses, as well as the specific needs and circumstances of the organization. Comparing different techniques and selecting the most appropriate ones based on criteria such as demand variability, order frequency, and resource availability, companies can achieve better inventory management outcomes. Coordinating multiple techniques, as illustrated in the case studies, can lead to significant improvements in inventory control, cost management, and overall operational efficiency.

2.11 Importance of Inventory Control from an Organizational Perspective

Effective inventory control is a cornerstone of customer satisfaction, as it directly impacts the availability of products and services. When inventory is managed efficiently, organizations can ensure that the right products are available at the right time, leading to several key benefits:

- **Reduced Stockouts:** Effective inventory control helps to maintain adequate stock levels, ensuring that products are available when customers need them. This minimizes the occurrence of stockouts, which can lead to lost sales, decreased customer satisfaction, and potential loss of market share.
- **Improved Order Fulfillment:** Accurately tracking inventory levels and managing stock effectively, organizations can improve their order fulfillment processes. This leads to quicker response times and ensures that customer orders are filled accurately and promptly.
- **Consistent Product Quality:** Proper inventory control ensures that products are stored under appropriate conditions and rotated correctly. This helps maintain consistent product quality, reducing the risk of damage, spoilage, or obsolescence.
- **Enhanced Service Levels:** With efficient inventory control, organizations can maintain high service levels by meeting customer demand consistently. This leads to higher customer satisfaction and loyalty, as customers can rely on the company for timely and accurate delivery of products.
- **Lower Holding Costs:** Inventory control helps to minimize holding costs, which include storage, insurance, and taxes. Maintaining optimal inventory levels, organizations can reduce these costs and allocate resources more efficiently.
- **Reduced Ordering Costs:** Effective inventory control allows for better planning and scheduling of orders. This reduces the frequency of emergency orders and the associated costs, such as expedited shipping and administrative expenses.
- **Minimized Stockout Costs:** Stockouts can lead to significant costs, including lost sales and customer dissatisfaction. Preventing stockouts, inventory control minimizes these costs and ensures continuous sales and operations.
- **Improved Cash Flow:** Efficient inventory management ensures that capital is not unnecessarily tied up in excess inventory. This improves cash flow, allowing the organization to invest in other areas such as marketing, research, and development.

- **Improved Inventory Turnover:** Effective inventory control increases inventory turnover, which is a measure of how quickly inventory is sold and replaced. Higher turnover rates indicate efficient use of inventory and better alignment with customer demand.
- **Enhanced Customer Satisfaction:** When inventory is well-managed, customers are more likely to find the products they need, leading to higher levels of satisfaction and repeat business. Satisfied customers are also more likely to recommend the company to others.
- **Cost Savings:** Proper inventory control leads to cost savings by reducing excess inventory, minimizing holding and ordering costs, and preventing losses due to obsolescence or spoilage.
- **Increased Availability:** Efficient inventory management ensures that products are available when needed, increasing overall product availability and reducing the likelihood of lost sales.
- **Regulatory Compliance:** Many industries have regulations regarding inventory management, particularly for perishable goods or pharmaceuticals. Effective inventory control helps organizations comply with these regulations, avoiding fines and legal issues.
- **Operational Efficiency:** Streamlined inventory processes contribute to overall operational efficiency. This includes better coordination between purchasing, production, and sales, leading to smoother and more effective operations.
- **Reduced Lead Times:** Efficient inventory control can shorten lead times by ensuring that materials and products are available when needed. This improves the responsiveness of the supply chain and enhances the ability to meet customer demands quickly.
- **Lower Inventory Costs:** Optimizing inventory levels, organizations can reduce the costs associated with carrying inventory, including storage, handling, and insurance costs.
- **Enhanced Competitiveness:** Effective inventory control gives organizations a competitive edge by enabling them to respond quickly to market changes, meet customer demands efficiently, and maintain cost-effective operations.

Effective inventory control is vital for both customer satisfaction and organizational profitability. Adopting appropriate inventory control techniques and optimizing inventory management processes, companies can achieve significant improvements in operational efficiency, cost management, and overall business performance.

2.12 Functioning of Inventory Management Systems Within Organizations

Inventory management systems (IMS) are essential tools for managing the flow of goods within an organization, from procurement to sales. These systems help in tracking inventory levels, orders, sales, and deliveries, providing a comprehensive view of inventory status. Key functionalities of IMS include:

- **Inventory Tracking:** Real-time tracking of inventory levels, locations, and movements. This includes monitoring quantities on hand, quantities on order, and quantities allocated to customers.
- **Order Management:** Managing purchase orders, sales orders, and work orders. IMS helps streamline the ordering process, ensuring timely replenishment and fulfillment.
- **Demand Forecasting:** Predicting future inventory requirements based on historical data, market trends, and sales projections. Accurate demand forecasting helps in maintaining optimal inventory levels.
- **Supplier Management:** Maintaining information on suppliers, including lead times, order histories, and performance metrics. This ensures reliable sourcing of materials and products.
- **Reporting and Analytics:** Generating reports and analytics on inventory turnover, stock levels, and order statuses. These insights help in making informed decisions and identifying areas for improvement.
- **Automation:** Automating routine tasks such as reordering, inventory counts, and data entry. Automation reduces manual errors and increases operational efficiency.

Modern IMS are often integrated with other enterprise systems, such as enterprise resource planning (ERP) and customer relationship management (CRM), to provide a holistic view of the organization's operations.

2.12.1 Role of Coordination Among Different Departments in Inventory Management

Effective inventory management requires seamless coordination among various departments within an organization. Key departments involved include:

- **Procurement:** Responsible for sourcing and purchasing materials. Procurement needs accurate forecasts and inventory data to make informed purchasing decisions.

- **Sales and Marketing:** Provides insights into customer demand and market trends. Collaboration with sales and marketing ensures that inventory levels align with sales forecasts and promotional activities.
- **Operations/Production:** Manages the manufacturing process and needs a steady supply of materials to maintain production schedules. Coordination with inventory management ensures that raw materials and components are available when needed.
- **Warehouse Management:** Oversees storage and handling of inventory. Efficient warehouse operations are critical for timely order fulfillment and inventory accuracy.
- **Finance:** Tracks inventory costs and impacts on cash flow. Finance ensures that inventory investments are optimized and aligned with the organization's financial goals.

Effective communication and data sharing among these departments are vital for synchronizing activities and optimizing inventory levels. Integrated systems and collaborative processes help in aligning departmental goals and ensuring that inventory management supports the overall business strategy.

2.12.2 Strategies for Optimizing Expenditure on Materials and Inventory

Optimizing expenditure on materials and inventory involves several strategies aimed at reducing costs while maintaining adequate inventory levels to meet demand. Key strategies include:

- **Just-In-Time (JIT):** A strategy that reduces inventory levels by receiving goods only as they are needed in the production process. JIT minimizes holding costs and reduces waste but requires reliable suppliers and accurate demand forecasting.
- **Economic Order Quantity (EOQ):** A formula used to determine the optimal order quantity that minimizes total inventory costs, including ordering and holding costs. EOQ helps in making informed purchasing decisions and optimizing order sizes.
- **Vendor-Managed Inventory (VMI):** A collaborative strategy where suppliers manage inventory levels based on agreed-upon thresholds. VMI reduces the burden on the organization and leverages supplier expertise in inventory management.
- **ABC Analysis:** Categorizing inventory items based on their importance and value. Focusing on high-value (A) items for tighter control and management helps in prioritizing resources and reducing overall inventory costs.

- **Demand Forecasting and Planning:** Using advanced forecasting techniques and data analytics to predict future demand accurately. Improved forecasting reduces the risk of overstocking and stockouts, optimizing inventory levels.
- **Inventory Turnover Ratio:** Monitoring and improving the inventory turnover ratio, which measures how often inventory is sold and replaced over a period. Higher turnover indicates efficient inventory management and lower holding costs.
- **Automation and Technology:** Implementing automated systems and technologies, such as barcode scanning, RFID, and IoT, to enhance inventory accuracy, reduce manual errors, and improve tracking and visibility.

Inventory management systems play a crucial role in the smooth functioning of organizations, providing real-time data and insights necessary for informed decision-making. Effective coordination among different departments ensures that inventory management aligns with organizational goals and strategies. Implementing various strategies such as JIT, EOQ, and VMI, organizations can optimize their expenditure on materials and inventory, resulting in improved operational efficiency, cost savings, and enhanced profitability. The examples and case studies provided highlight the practical applications and benefits of these strategies in real-world scenarios.

2.13 Coordination and Optimal Expenditure on Materials

Effective inventory control is not just about maintaining optimal inventory levels but also about ensuring seamless coordination across various functions within an organization. Coordination in inventory control involves synchronizing activities and information flow among different departments such as procurement, production, sales, and finance. The importance of coordination in inventory control can be summarized as follows:

- **Aligning Goals and Objectives:** Ensures that all departments work towards common organizational goals, such as minimizing costs, maximizing customer satisfaction, and improving operational efficiency.
- **Improving Communication:** Enhances the communication between departments, reducing misunderstandings and delays. This is crucial for timely decision-making and efficient inventory management.
- **Optimizing Inventory Levels:** Helps in maintaining the right balance of inventory, preventing both overstocking and stockouts. Effective coordina-

tion ensures that inventory levels are adjusted based on real-time demand and supply information.

- **Reducing Costs:** Coordinating purchasing, production, and sales activities, organizations can reduce carrying costs, ordering costs, and stockout costs.
- **Enhancing Supplier Relationships:** Improved coordination with suppliers leads to better negotiation, timely deliveries, and reliable sourcing, which are essential for maintaining optimal inventory levels.

2.13.1 Techniques for Optimizing Material Expenditure

Optimizing material expenditure involves various techniques aimed at reducing costs while ensuring that inventory levels meet demand requirements. Key techniques include:

- **Just-In-Time (JIT):** This inventory strategy aims to reduce in-process inventory and associated carrying costs by receiving goods only as they are needed in the production process. JIT requires precise demand forecasting and reliable suppliers but can significantly reduce storage costs and waste.
- **Economic Order Quantity (EOQ):** The EOQ model helps determine the optimal order quantity that minimizes the total cost of inventory, including ordering and holding costs. Calculating the ideal order size, organizations can optimize their material expenditure.
- **Vendor Managed Inventory (VMI):** In VMI, suppliers are responsible for managing the inventory levels of their products at the customer's location. This reduces the burden on the organization and leverages the supplier's expertise in inventory management, leading to cost savings.
- **ABC Analysis:** This technique categorizes inventory items based on their importance and value. High-value items (A) are managed more closely than lower-value items (B and C). This prioritization helps in focusing resources on critical items and optimizing material expenditure.
- **Demand Forecasting and Planning:** Accurate demand forecasting using historical data, market trends, and advanced analytics helps in planning inventory levels and reducing excess inventory. Better forecasting leads to more efficient use of materials and lower costs.
- **Bulk Purchasing:** Leveraging economies of scale by purchasing materials in bulk can reduce per-unit costs. However, this needs to be balanced against the potential increase in holding costs.
- **Automation and Technology:** Implementing automated inventory management systems and technologies like RFID, barcode scanning, and IoT

enhances inventory accuracy and reduces manual errors, leading to cost savings.

Coordination and optimization of material expenditure are critical for effective inventory management. Ensuring seamless coordination among various departments and implementing techniques like JIT, EOQ, and VMI, organizations can achieve significant cost savings and operational efficiencies. The case studies and examples provided demonstrate the practical benefits of these strategies in real-world scenarios, highlighting the importance of integrated and coordinated inventory management practices.

2.14 Multi-Echelon Inventory Optimization (MEIO)

Multi-echelon inventory optimization (MEIO) is a sophisticated approach to inventory management that aims to optimize inventory levels across multiple stages or echelons within a supply chain. Unlike traditional single-echelon inventory models, which focus on optimizing inventory at individual locations, MEIO considers the entire supply chain network, including manufacturers, distribution centers, and retail outlets. This holistic view allows businesses to achieve greater efficiency, reduce costs, and improve service levels by aligning inventory policies and decisions across the entire supply chain [14].

2.14.1 Overview of MEIO

MEIO involves the strategic management of inventory across multiple interconnected levels within a supply chain. The key components of MEIO include:

- **Echelons:** Different stages or levels in the supply chain, such as suppliers, manufacturers, distribution centers, and retailers.
- **Inventory Policies:** Rules and guidelines for ordering, holding, and replenishing inventory at each echelon.
- **Coordination:** The synchronization of inventory decisions across echelons to achieve overall supply chain objectives.
- **Optimization Models:** Mathematical and algorithmic models used to determine optimal inventory levels, reorder points, and safety stock across the supply chain.

2.14.2 Key Benefits of MEIO

- **Cost Reduction:** Optimizing inventory levels across the entire supply chain, businesses can reduce holding costs, minimize excess inventory, and decrease stockouts.

- **Improved Service Levels:** Coordinated inventory policies ensure that products are available where and when they are needed, enhancing customer satisfaction.
- **Enhanced Efficiency:** MEIO enables better use of resources and more efficient operations by reducing redundancies and aligning inventory practices.
- **Risk Mitigation:** Considering interdependencies and variability across echelons, MEIO helps in managing risks related to demand fluctuations, lead times, and supply disruptions.
- **Increased Visibility:** MEIO provides a comprehensive view of inventory across the supply chain, enabling more informed decision-making and strategic planning.

2.14.3 Components of MEIO Models

MEIO models typically include several components to achieve optimal inventory levels and coordination across echelons:

- **Demand Forecasting:** Accurate prediction of demand at each echelon to inform inventory decisions.
- **Lead Time Analysis:** Consideration of lead times for replenishment at each stage to ensure timely availability of inventory.
- **Safety Stock Calculation:** Determination of appropriate safety stock levels to buffer against demand variability and supply disruptions.
- **Reorder Point Determination:** Setting reorder points for each echelon to trigger timely replenishment.
- **Inventory Holding Costs:** Analysis of holding costs at each stage to minimize overall inventory costs.
- **Service Level Targets:** Setting and achieving service level targets to meet customer expectations.

2.14.4 MEIO Techniques and Approaches

Several techniques and approaches are used in MEIO to achieve optimal inventory levels and coordination:

- **Mathematical Optimization:** Use of mathematical models and algorithms to determine optimal inventory policies and levels across echelons.
- **Simulation:** Simulation models to assess the performance of different inventory policies under various scenarios and uncertainties.

- **Heuristics:** Heuristic methods to find near-optimal solutions in complex and large-scale supply chain networks.
- **Statistical Methods:** Statistical techniques for demand forecasting, lead time analysis, and variability assessment.
- **Machine Learning:** Application of machine learning algorithms to improve demand forecasting accuracy and optimize inventory decisions.

2.14.5 Challenges in MEIO Implementation

Implementing MEIO can be challenging due to several factors:

- **Data Integration:** Integrating data from multiple sources and echelons to create a unified and accurate view of the supply chain.
- **Complexity:** The complexity of coordinating inventory policies across multiple levels and dealing with interdependencies.
- **Demand Variability:** Managing demand variability and uncertainty across different stages of the supply chain.
- **Change Management:** Ensuring that all stakeholders are aligned and committed to the new inventory policies and practices.
- **Technology and Tools:** Investing in the right technology and tools to support MEIO, such as advanced software and analytical platforms.

Multi-echelon inventory optimization (MEIO) is a powerful approach to inventory management that considers the entire supply chain network. Optimizing inventory levels across multiple echelons, businesses can achieve cost reduction, improved service levels, enhanced efficiency, risk mitigation, and increased visibility. Despite the challenges in implementation, the benefits of MEIO make it a valuable strategy for businesses aiming to optimize their supply chain operations and achieve a competitive advantage. With advancements in technology and analytical tools, the adoption of MEIO is likely to continue growing across various industries.

2.15 Conclusion

The techniques and concepts discussed in this chapter provide a comprehensive understanding of inventory control. Implementing these techniques, organizations can achieve significant improvements in inventory management, leading to enhanced operational efficiency, reduced costs, and improved customer satisfaction. The integration of these techniques into daily operations helps in maintaining optimal inventory levels, ensuring the availability of products, and minimizing the risks of overstocking and stockouts, contributing significantly to the overall efficiency and profitability of organizations.

Exercises

Exercise 2.1. *A factory produces a product at a rate of 10,000 units per month. The monthly demand for the product is 2,000 units. The setup cost for each production run is \$500, and the holding cost per unit per month is \$2. Determine the optimal production lot size using the EPQ model.*

Exercise 2.2. *A company manufactures widgets with an annual demand of 12,000 units. The production rate is 24,000 units per year, and the setup cost per production run is \$800. The holding cost is \$1.50 per unit per year. Calculate the optimal production quantity using the EPQ model.*

Exercise 2.3. *A firm produces a special component at a production rate of 8,000 units per quarter. The quarterly demand for the component is 2,000 units. The setup cost per production run is \$600, and the holding cost per unit per quarter is \$3. Determine the optimal production lot size.*

Exercise 2.4. *An assembly line produces parts at a rate of 15,000 units per year. The annual demand for these parts is 5,000 units. The setup cost per production run is \$1,000, and the holding cost per unit per year is \$2.50. Calculate the optimal production quantity using the EPQ model.*

Exercise 2.5. *A beverage company produces bottles at a rate of 20,000 units per month. The monthly demand for the bottles is 5,000 units. The setup cost per production run is \$700, and the holding cost per unit per month is \$1. Determine the optimal production lot size using the EPQ model.*

Exercise 2.6. *A company has the following inventory data for five items:*

Item	*Annual Demand (units)*
A	5,000
B	2,000
C	1,500
D	1,200
E	800

The unit price for each item is:

Item	*Unit Price ($)*
A	10
B	15
C	20
D	25
E	30

Calculate the annual inventory value for each item and perform an ABC analysis to classify the items into categories A, B, and C based on their contribution to the total inventory value.

Exercise 2.7. *A retailer has the following data for four products:*

Product	*Annual Demand (units)*	*Unit Cost ($)*
*P*1	4,000	12
*P*2	3,000	10
*P*3	2,000	8
*P*4	1,000	15

Calculate the total annual cost for each product and conduct an ABC analysis. Provide a summary of the results and classification of products into categories A, B, and C.

Exercise 2.8. *Consider a company with the following inventory data for six items:*

Item	*Annual Demand (units)*	*Unit Price ($)*
*I*1	6,000	5
*I*2	3,500	7
*I*3	2,500	10
*I*4	2,000	12
*I*5	1,500	8
*I*6	1,000	6

Determine the annual inventory value for each item, and perform an ABC analysis to categorize them. Show your calculations and final classification.

Exercise 2.9. *A warehouse holds the following inventory:*

SKU	*Annual Demand (units)*	*Cost per Unit ($)*
*X*1	5,000	20
*X*2	4,000	25
*X*3	3,000	15
*X*4	2,000	18
*X*5	1,000	22

Calculate the total inventory cost for each SKU and perform an ABC analysis. Provide a detailed explanation of your classification results.

Exercise 2.10. *A distribution center manages the following items:*

Item	*Annual Usage (units)*	*Unit Cost ($)*
*A*1	8,000	7
*A*2	6,000	6
*A*3	4,000	5
*A*4	3,000	8
*A*5	2,000	10

Determine the total annual cost for each item and classify them into A, B, and C categories based on the ABC analysis methodology.

Exercise 2.11. *A manufacturing company maintains the following items in their inventory:*

Item	*Usage Frequency*	*Criticality*
*M*1	5,000	V
*M*2	3,000	E
*M*3	2,000	D
*M*4	1,500	V
*M*5	1,000	E

Where the usage frequency is the total units used annually, and the criticality is classified as Vital (V), Essential (E), or Desirable (D), perform a VED analysis to classify the items into categories based on their usage frequency and criticality. Provide detailed calculations and reasoning for your classification.

Exercise 2.12. *A hospital has the following items:*

Item	*Annual Consumption*	*Importance*
*H*1	6,000	V
*H*2	4,000	E
*H*3	2,500	D
*H*4	1,800	V
*H*5	1,200	E

Classify the items into VED categories based on their annual consumption and importance levels. Show your calculations and provide a summary of the classification.

Exercise 2.13. *A warehouse stocks the following products:*

Product	*Annual Demand*	*Criticality*
*W*1	7,000	E
*W*2	3,500	D
*W*3	2,800	V
*W*4	2,000	E
*W*5	1,500	D

Perform a VED analysis to categorize the products. Provide a detailed explanation of the process and final classifications.

Exercise 2.14. *Consider a distribution center with the following items:*

Item	*Monthly Usage*	*Importance*
*D*1	8,000	*V*
*D*2	5,000	*E*
*D*3	3,000	*D*
*D*4	2,200	*V*
*D*5	1,500	*E*

Calculate the annual usage for each item and classify them into VED categories based on their monthly usage and importance. Provide a comprehensive analysis and justification for your classifications.

Exercise 2.15. *A retailer manages the following inventory items:*

Item	*Annual Usage (units)*	*Criticality*
*R*1	9,000	*D*
*R*2	6,000	*V*
*R*3	3,500	*E*
*R*4	2,500	*D*
*R*5	1,000	*V*

Classify these items into VED categories based on their annual usage and criticality. Provide detailed steps and reasoning behind your classifications.

Exercise 2.16. *A company aims to implement a just-in-time (JIT) inventory system for the following products:*

Product	*Annual Demand*	*Lead Time (days)*
*P*1	2,400	10
*P*2	1,800	15
*P*3	1,200	5
*P*4	1,500	8
*P*5	3,000	12

Calculate the reorder point for each product using the JIT approach. Assume a 365-day year and provide a detailed calculation for each product.

Exercise 2.17. *A manufacturing plant is adopting a JIT system. The following items are used in the production process:*

Item	*Monthly Usage*	*Reorder Point (units)*
*I*1	500	1,000
*I*2	300	500
*I*3	200	400
*I*4	700	1,400
*I*5	400	800

Determine the order quantity for each item based on the JIT inventory principles. Explain your calculations and approach.

Exercise 2.18. *An electronics retailer uses JIT inventory management. They have the following products:*

Product	*Daily Sales*	*Supplier Lead Time (days)*
*E*1	50	5
*E*2	30	7
*E*3	20	3
*E*4	40	4
*E*5	60	6

Calculate the safety stock level needed for each product to meet the JIT requirements. Provide a detailed explanation of your calculations.

Exercise 2.19. *A retailer is implementing a JIT system and needs to calculate the economic order quantity (EOQ) for their products. The following data is provided:*

Product	*Annual Demand*	*Ordering Cost*	*Holding Cost*
*R*1	10,000	200	50
*R*2	8,000	150	40
*R*3	6,000	100	30
*R*4	12,000	250	60
*R*5	9,000	180	45

Using the EOQ formula, calculate the optimal order quantity for each product under the JIT inventory model. Provide step-by-step calculations and explanations.

Exercise 2.20. *A company is transitioning to a JIT inventory system and has the following production data:*

Item	*Daily Production Rate*	*Daily Demand*
*C*1	100	80
*C*2	150	120
*C*3	200	150
*C*4	250	200
*C*5	300	250

Calculate the production cycle time for each item under the JIT system. Explain how the production cycle time affects inventory levels and overall efficiency.

Exercise 2.21. *A supply chain consists of three echelons: suppliers, distribution centers (DCs), and retail stores. The following data is provided:*

Echelon	*Demand (units/year)*	*Ordering Cost ($)*	*Holding Cost ($/unit/year)*
Suppliers	10,000	500	20
DCs	5,000	300	15
Retail Stores	2,000	200	10

Calculate the economic order quantity (EOQ) for each echelon using the MEIO principles. Provide a detailed step-by-step calculation.

Exercise 2.22. *Consider a three-echelon inventory system with the following lead times and demand distributions:*

Echelon	*Lead Time (days)*	*Annual Demand (units)*
Supplier	10	15,000
DC	7	8,000
Retailer	5	3,000

Determine the reorder point for each echelon, assuming a 365-day year. Provide the calculations for each echelon.

Exercise 2.23. *A company has a multi-echelon inventory system with two distribution centers and five retail stores. The following information is given:*

Location	*Annual Demand (units)*	*Safety Stock (% of demand)*
DC1	10,000	20%
DC2	8,000	15%
Store1	2,000	25%
Store2	2,500	30%
Store3	2,200	20%
Store4	2,700	25%
Store5	2,300	15%

Calculate the safety stock level for each location based on the provided percentages. Show the detailed calculations.

Exercise 2.24. *In a multi-echelon inventory system, the following parameters are provided:*

Echelon	*Annual Demand (units)*	*Lead Time (days)*	*Holding Cost ($/unit/year)*
Supplier	20,000	14	30
DC	12,000	10	25
Retail Store	6,000	7	20

Determine the total cost of inventory (TCI) for each echelon using the MEIO approach. Explain how the total cost is computed.

Exercise 2.25. *A company operates a multi-echelon inventory system with two levels: distribution centers (DCs) and retail stores. The following data is available:*

Level	*Annual Demand (units)*	*Ordering Cost (\$)*	*Holding Cost (\$/unit/year)*
DC	5,000	400	25
Retail Store	2,000	150	15

Calculate the Optimal Order Quantity (OOQ) for each level considering both ordering and holding costs. Provide a detailed explanation of the calculation.

3

Advanced Inventory Optimization Models

This chapter introduces advanced mathematical models and optimization techniques used in inventory management to determine optimal inventory levels and costs across various scenarios. It highlights the organizational significance of inventory optimization, focusing on its impact on customer satisfaction and profitability. The chapter reviews system-level and item-level inventory models, alongside sensitivity analyses, to assess their robustness. Additionally, it addresses the limitations of traditional models and the role of simulations in improving outcomes. Practical applications of these models are illustrated, providing valuable insights for practitioners, researchers, and students. Exploring effective inventory control methods through optimization, this chapter offers a thorough understanding of key inventory optimization strategies.

3.1 Introduction

Inventory optimization is a crucial aspect of inventory management that aims to determine the most efficient inventory levels for various products, taking into account the costs associated with holding, ordering, and shortage. Unlike traditional inventory control techniques that focus primarily on maintaining proper inventory levels, inventory optimization uses mathematical models and advanced techniques to achieve optimal solutions. This chapter delves into the different models and methodologies used for optimizing inventories.

3.1.1 Multi-Echelon Inventory Model

The multi-echelon inventory optimization (MEIO) model considers multiple stages in a supply chain, such as warehouses and retail outlets, and optimizes inventory levels across these stages to minimize total costs while meeting service level requirements. It accounts for lead time variability and demand variability at each stage.

DOI: 10.1201/9781003534860-3

- **Base-Stock Level Calculation:** For each stage i in the supply chain, the base-stock level S_i can be calculated as:

$$S_i = d_i \times L_i + Z \times \sigma_{i,LT}$$

where d_i is the average demand rate at stage i, L_i is the lead time at stage i, Z is the service level factor, and $\sigma_{i,LT}$ is the standard deviation of demand during lead time at stage i.

Example 3.1. *A supply chain consists of a central warehouse and several retail outlets. The average demand rate at a retail outlet (d_i) is 30 units per day, the lead time (L_i) is four days, the service level factor (Z) is 1.28, and the standard deviation of demand during lead time ($\sigma_{i,LT}$) is 15 units. What is the base-stock level for a retail outlet?*

Solution 3.1.

$$S_i = 30 \times 4 + 1.28 \times 15 = 120 + 19.2 = 139.2 \approx 140 \text{ units}$$

The retail outlet should maintain a base-stock level of 140 units.

- **Total Cost Minimization:** The total cost in a multi-echelon system typically includes holding costs, ordering costs, and stockout costs. The goal is to find the inventory levels that minimize these costs while maintaining the desired service level. The total cost function TC can be expressed as:

$$TC = \sum_{i=1}^{n} (H_i \times E[I_i] + O_i \times E[O_i] + S_i \times E[SL_i])$$

where H_i is the holding cost per unit at stage i, $E[I_i]$ is the expected inventory level at stage i, O_i is the ordering cost at stage i, $E[O_i]$ is the expected number of orders at stage i, S_i is the stockout cost at stage i, and $E[SL_i]$ is the expected stockout level at stage i.

Example 3.2. *A supply chain manager at ABC Manufacturing is tasked with minimizing the total cost for a system with a central warehouse and three retail outlets. The holding cost (H_i) at each retail outlet is \$2 per unit, the ordering cost (O_i) is \$50 per order, and the stockout cost (S_i) is \$10 per unit short. The expected inventory level ($E[I_i]$), the expected number of orders ($E[O_i]$), and the expected stockout level ($E[SL_i]$) are based on historical data. Specifically, the expected inventory level at the warehouse ($E[I_w]$) is 500 units, the expected number of orders at the warehouse ($E[O_w]$) is 20 orders, and the expected stockout level at the warehouse ($E[SL_w]$) is ten units. Additionally, the expected inventory level at each retail outlet ($E[I_i]$) is 100 units, the expected number of orders at each retail outlet ($E[O_i]$) is 15 orders, and the expected stockout level at each retail outlet ($E[SL_i]$) is five units. Calculate and minimize the total cost across the supply chain.*

Solution 3.2. *Given:*

- *Holding cost* (H_i) = *\$2 per unit*
- *Ordering cost* (O_i) = *\$50 per order*
- *Stockout cost* (S_i) = *\$10 per unit short*
- *Expected inventory level at the warehouse* ($E[I_w]$) = *500 units*
- *Expected number of orders at the warehouse* ($E[O_w]$) = *20 orders*
- *Expected stockout level at the warehouse* ($E[SL_w]$) = *10 units*
- *Expected inventory level at each retail outlet* ($E[I_i]$) = *100 units*
- *Expected number of orders at each retail outlet* ($E[O_i]$) = *15 orders*
- *Expected stockout level at each retail outlet* ($E[SL_i]$) = *5 units*

The total cost function (TC) *is:*

$$TC = H_w \times E[I_w] + O_w \times E[O_w] + S_w \times E[SL_w] + \sum_{i=1}^{3}$$
$$\times (H_i \times E[I_i] + O_i \times E[O_i] + S_i \times E[SL_i])$$

Simplifying:

$$TC = 2 \times 500 + 50 \times 20 + 10 \times 10 + 3 \times (2 \times 100 + 50 \times 15 + 10 \times 5)$$

Calculating:

$$\begin{aligned} TC &= 1000 + 1000 + 100 + 3 \times (200 + 750 + 50) \\ &= 1000 + 1000 + 100 + 3 \times 1000 \\ &= 1000 + 1000 + 100 + 3000 \\ &= 5100 \end{aligned}$$

Thus, the total cost for the supply chain is \$5,100 based on the provided expected values.

Example 3.3. *A supply chain manager at ABC Manufacturing is tasked with minimizing the total cost for a system with a central warehouse and three retail outlets. The holding cost* (H_i) *at each retail outlet is \$2 per unit, the ordering cost* (O_i) *is \$50 per order, and the stockout cost* (S_i) *is \$10 per unit short. The expected inventory level* ($E[I_i]$)*, the expected number of orders* ($E[O_i]$)*, and the expected stockout level* ($E[SL_i]$) *are based on historical data. Specifically, the expected inventory level at each retail outlet is* $E[I_1] = 100$, $E[I_2] = 150$, *and* $E[I_3] = 200$ *units; the expected number of orders at each retail outlet is* $E[O_1] = 10$, $E[O_2] = 12$, *and* $E[O_3] = 15$ *orders; and the expected stockout level at each retail outlet is* $E[SL_1] = 5$, $E[SL_2] = 8$, *and* $E[SL_3] = 10$ *units. Calculate and minimize the total cost across the supply chain.*

Solution 3.3. *Given:*

- *Holding cost* (H_i) = *\$2 per unit*
- *Ordering cost* (O_i) = *\$50 per order*
- *Stockout cost* (S_i) = *\$10 per unit short*
- *Expected inventory level at retail outlets* ($E[I_1]$ = *100 units*, $E[I_2]$ = *150 units*, $E[I_3]$ = *200 units*)
- *Expected number of orders at retail outlets* ($E[O_1]$ = *10 orders*, $E[O_2]$ = *12 orders*, $E[O_3]$ = *15 orders*)
- *Expected stockout level at retail outlets* ($E[SL_1]$ = *5 units*, $E[SL_2]$ = *8 units*, $E[SL_3]$ = *10 units*)

The total cost for each retail outlet is calculated as follows:

$$\begin{aligned}
Total\ Cost_1 &= 2 \times 100 + 50 \times 10 + 10 \times 5 = 200 + 500 + 50 = 750 \\
Total\ Cost_2 &= 2 \times 150 + 50 \times 12 + 10 \times 8 = 300 + 600 + 80 = 980 \\
Total\ Cost_3 &= 2 \times 200 + 50 \times 15 + 10 \times 10 = 400 + 750 + 100 = 1250
\end{aligned}$$

Therefore, the total cost across the supply chain is:

$$Total\ Cost = 750 + 980 + 1{,}250 = 2{,}980$$

Steps:

1. *Calculate the total cost for each retail outlet using the given data.*

2. *Sum the total costs to obtain the overall cost across the supply chain.*

3.2 Sensitivity Analysis

Sensitivity analysis involves testing the robustness of inventory optimization models by varying key parameters and observing the impact on the results. This helps in understanding how sensitive the model is to changes in demand, lead time, costs, and other factors.

3.2.1 Conducting Sensitivity Analysis

To perform sensitivity analysis, follow these steps:

1. Identify key parameters and their potential ranges of variation.

2. Modify one parameter at a time while keeping others constant.
3. Observe and record the impact on the inventory optimization model's outcomes.
4. Analyze the results to determine which parameters have the most significant impact.

Example 3.4. *Consider the EOQ model from Example 2.6. Conduct a sensitivity analysis by varying the holding cost (H) from \$1 to \$3 per unit per year and observing the impact on the EOQ.*

Solution 3.4. ***Steps:***

1. *Calculate the EOQ for* $H = 1$*:*

$$EOQ = \sqrt{\frac{2 \times 10{,}000 \times 50}{1}} = \sqrt{1{,}000{,}000} = 1{,}000$$

2. *Calculate the EOQ for* $H = 2$*:*

$$EOQ = \sqrt{\frac{2 \times 10{,}000 \times 50}{2}} = \sqrt{500{,}000} \approx 707$$

3. *Calculate the EOQ for* $H = 3$*:*

$$EOQ = \sqrt{\frac{2 \times 10{,}000 \times 50}{3}} = \sqrt{333{,}333.33} \approx 577$$

The EOQ decreases as the holding cost increases, indicating that the model is sensitive to changes in the holding cost parameter.

Analysis:

- *For* $H = 1$, EOQ = *1,000 units.*
- *For* $H = 2$, EOQ = *707 units.*
- *For* $H = 3$, EOQ = *577 units.*

3.3 Inventory Optimization for Seasonal Demand

Many industries experience seasonal demand patterns, which pose unique challenges for inventory management. Advanced techniques for optimizing inventory in the face of seasonal demand include:

- **Seasonal Forecasting Models:** Advanced forecasting models, such as seasonal decomposition and time-series analysis, help predict demand patterns and adjust inventory levels accordingly.
- **Dynamic Safety Stock Strategies:** Rather than maintaining static safety stock levels, dynamic safety stock strategies adjust inventory buffers based on seasonal demand fluctuations and lead time variability.
- **Flexible Capacity Planning:** Flexibility in production and distribution capacity allows businesses to ramp up operations during peak seasons and scale down during off-peak periods, minimizing excess inventory and costs.

3.3.1 Seasonal Forecasting Models

Advanced forecasting models, such as seasonal decomposition and time-series analysis, help predict demand patterns and adjust inventory levels accordingly.

3.3.1.1 Mathematical Formulation

A common approach for seasonal forecasting is to use a multiplicative seasonal model:

$$D_t = (T_t \times S_t \times C_t) + E_t$$

where:

D_t is the actual demand at time t.

T_t is the trend component.

S_t is the seasonal component.

C_t is the cyclical component.

E_t is the irregular component.

Example 3.5. *A retailer observes that demand for a product peaks every December due to holiday shopping. They use a multiplicative seasonal model to forecast demand. The trend component is an increasing linear trend, the seasonal component is high in December, and there are no significant cyclical components. Given the following values:*

$$T_t = 100 + 5t$$
$$S_{Dec} = 1.5$$
$$E_t = 0$$

Calculate the forecasted demand for December in year five.

Solution 3.5. *Using the multiplicative seasonal model:*

$$D_t = (T_t \times S_t \times C_t) + E_t$$

For December in year five:

$$\begin{aligned} t &= 5 \\ T_5 &= 100 + 5 \times 5 = 125 \\ S_{Dec} &= 1.5 \\ E_t &= 0 \end{aligned}$$

Therefore, the forecasted demand is:

$$D_{Dec,5} = 125 \times 1.5 + 0 = 187.5$$

Steps:

1. Calculate the trend component for year five.

2. Multiply the trend component by the seasonal factor for December.

3.3.2 Dynamic Safety Stock Strategies

Rather than maintaining static safety stock levels, dynamic safety stock strategies adjust inventory buffers based on seasonal demand fluctuations and lead time variability.

3.3.2.1 Mathematical Formulation

Dynamic safety stock can be calculated as:

$$SS = Z \times \sigma_{LT}$$

where:

SS is the safety stock.

Z is the z-score corresponding to the desired service level.

σ_{LT} is the standard deviation of demand during lead time, which can vary seasonally.

Example 3.6. *A company wants to maintain a 95% service level for a product with seasonal demand. The standard deviation of demand during lead time is 20 units in the off-season and 50 units in the peak season. Calculate the safety stock for both periods.*

Solution 3.6. *For a 95% service level, $Z \approx 1.645$.*

Off-Season:

$$SS = 1.645 \times 20 = 32.9$$

Peak Season:

$$SS = 1.645 \times 50 = 82.25$$

Steps:

1. *Determine the z-score for the desired service level.*
2. *Calculate the safety stock for the off-season.*
3. *Calculate the safety stock for the peak season.*

3.3.3 Flexible Capacity Planning

Flexibility in production and distribution capacity allows businesses to ramp up operations during peak seasons and scale down during off-peak periods, minimizing excess inventory and costs.

3.3.3.1 Mathematical Formulation

Flexible capacity planning can be modeled using linear programming to determine the optimal production and distribution schedule.

$$\min \sum_t (C_p P_t + C_h H_t + C_s S_t)$$

Subject to:

- $P_t + H_{t-1} - D_t = H_t + S_t$
- $P_t \leq$ max production capacity
- $H_t \geq 0$
- $S_t \geq 0$

where:

P_t is the production quantity at time t.

H_t is the inventory level at time t.

S_t is the shortage level at time t.

C_p is the cost of production.

C_h is the holding cost.

C_s is the shortage cost.

Example 3.7. *A toy manufacturer experiences peak demand in December and wants to optimize production and inventory levels for the last quarter of the year. The demand is 1,000 units in October, 2,000 units in November, and 5,000 units in December. The production cost is \$10 per unit, holding cost is \$2 per unit per month, and shortage cost is \$5 per unit. The maximum production capacity is 3,000 units per month.*

Solution 3.7. *Given:*

- *Demand in October (D_O) = 1,000 units*
- *Demand in November (D_N) = 2,000 units*
- *Demand in December (D_D) = 5,000 units*
- *Production cost (P_C) = \$10 per unit*
- *Holding cost (H_C) = \$2 per unit per month*
- *Shortage cost (S_C) = \$5 per unit*
- *Maximum production capacity = 3,000 units per month*

Let P_O, P_N, and P_D be the production quantities in October, November, and December respectively, and I_O, I_N, and I_D be the inventory levels at the end of October, November, and December respectively.

The objective is to minimize the total cost, which includes production cost, holding cost, and shortage cost.

The total cost function (TC) is:

$$TC = (P_C \times (P_O + P_N + P_D)) + (H_C \times (I_O + I_N + I_D)) + (S_C \times Shortages)$$

Constraints:

- *Production capacity:* $P_O \leq 3{,}000,\ P_N \leq 3{,}000,\ P_D \leq 3{,}000$
- *Inventory balance equations:*

$$I_O = P_O - D_O$$
$$I_N = P_N + I_O - D_N$$
$$I_D = P_D + I_N - D_D$$

- *Non-negativity:* $P_O, P_N, P_D, I_O, I_N, I_D \geq 0$

Assuming no initial inventory and no shortages:

$P_O = 1{,}000 + x$

(*to meet October demand and possibly create some inventory for November*)

$P_N = 2{,}000 + y$

(*to meet November demand and possibly create some inventory for December*)

$P_D = 5{,}000 + z$

(*to meet December demand*)

To satisfy demand and minimize costs:

$$P_O = 1{,}000 + x$$
$$I_O = x$$
$$P_N = 2{,}000 + (x - 2{,}000) + y \quad (\textit{as inventory from October will be used})$$
$$I_N = y$$
$$P_D = 5{,}000 + (y - 5{,}000) + z \quad (\textit{as inventory from November will be used})$$

From the production capacity constraints:

$$1{,}000 + x \leq 3{,}000 \quad \Rightarrow \quad x \leq 2{,}000$$
$$2{,}000 + y \leq 3{,}000 \quad \Rightarrow \quad y \leq 1{,}000$$
$$5{,}000 + z \leq 3{,}000 \quad \Rightarrow \quad z \leq -2{,}000 \quad (\textit{which is not possible})$$

So, we need to adjust production such that demand is met without exceeding production limits. Let's consider $x = 2{,}000$ *and* $y = 1{,}000$ *(maximum allowable within constraints):*

$$P_O = 1{,}000 + 2{,}000 = 3{,}000$$
$$I_O = 2{,}000$$
$$P_N = 2{,}000 + 2{,}000 - 2{,}000 = 2{,}000$$
$$I_N = 1{,}000$$
$$P_D = 5{,}000 + 1{,}000 - 5{,}000 = 1{,}000 \quad (\textit{Insufficient, must be corrected})$$

Given the constraints, we need to calculate for potential inventory carry-over:

Total production cost:

$$3{,}000 \times 10 + 2{,}000 \times 10 + 3{,}000 \times 10 = 30{,}000 + 20{,}000 + 30{,}000 = 80{,}000$$

Total holding cost:

$$2{,}000 \times 2 + 1{,}000 \times 2 = 4{,}000 + 2{,}000 = 6{,}000$$

Total shortage cost (correcting for December):

$$5{,}000 - 3{,}000 - 1{,}000 = 1{,}000 \quad (shortage for December) 1{,}000 \times 5 = 5{,}000$$

Total cost:

$$80{,}000 + 6{,}000 + 5{,}000 = 91{,}000$$

Thus, the minimum total cost for the given production and demand constraints is \$91,000.

Example 3.8. *A toy manufacturer faces seasonal demand with higher sales during the holiday season. The manufacturer wants to determine the optimal safety stock to maintain during the peak season to avoid stockouts. The average demand during the peak season is 1,000 units per week, with a standard deviation of 200 units. The lead time for replenishment is two weeks, and the desired service level is 95%.*

Solution 3.8. *To calculate the safety stock, we use the following formula:*

$$Safety\,Stock = Z \times \sigma_{LT}$$

where:

- *Z is the z-score corresponding to the desired service level (for 95% service level, $Z \approx 1.65$)*
- *σ_{LT} is the standard deviation of demand during lead time*

First, we calculate the standard deviation of demand during lead time (σ_{LT}):

$$\sigma_{LT} = \sigma \times \sqrt{L} = 200 \times \sqrt{2} \approx 283$$

Next, we calculate the safety stock:

$$Safety\,Stock = 1.65 \times 283 \approx 467$$

Therefore, the manufacturer should maintain approximately 467 units of safety stock during the peak season to achieve a 95% service level.

Steps:

1. *Calculate the standard deviation of demand during lead time: $\sigma_{LT} = 200 \times \sqrt{2} \approx 283$.*
2. *Determine the z-score for the desired service level: $Z \approx 1.65$.*
3. *Compute the safety stock: $Safety\,Stock = 1.65 \times 283 \approx 467$.*

3.4 Applications of Advanced Inventory Optimization Techniques

Advanced inventory optimization techniques are applicable across various industries, each with its own unique challenges and requirements. This section explores how these techniques can be tailored to specific sectors, enhancing efficiency and customer satisfaction.

3.4.1 Retail Industry

In the retail industry, optimizing inventory levels is crucial to meeting customer demand while minimizing costs. Advanced techniques employed include:

- **Demand-Driven Replenishment:** This technique ensures that inventory is replenished based on actual customer demand rather than forecasted demand, reducing excess inventory and stockouts.
- **RFID Technology:** Radio frequency identification (RFID) technology provides real-time inventory visibility, allowing retailers to track stock levels accurately and reduce shrinkage.
- **Dynamic Pricing Algorithms:** These algorithms adjust prices based on demand, competition, and other factors, helping to optimize inventory turnover and maximize profits.

Example 3.9. *A clothing retailer uses demand-driven replenishment to manage its inventory. The average weekly demand for a popular item is 500 units, with a standard deviation of 100 units. The retailer wants to maintain a 90% service level. The lead time for replenishment is one week.*

Solution 3.9. *To calculate the reorder point, including safety stock, we use the following formula:*

$$ROP = d \times L + Z \times \sigma_{LT}$$

where:

d *is the average demand per week (500 units).*

L *is the lead time in weeks (1 week).*

Z *is the z-score corresponding to the desired service level (for 90% service level,* $Z \approx 1.28$*).*

σ_{LT} *is the standard deviation of demand during lead time.*

Calculate the standard deviation of demand during lead time:

$$\sigma_{LT} = \sigma \times \sqrt{L} = 100 \times \sqrt{1} = 100$$

Calculate the reorder point:

$$ROP = 500 \times 1 + 1.28 \times 100 = 500 + 128 = 628$$

Therefore, the retailer should reorder the item when the inventory level drops to 628 units to maintain a 90% service level.

Steps:

1. *Calculate the standard deviation of demand during lead time:* $\sigma_{LT} = 100$.
2. *Determine the z-score for the desired service level:* $Z \approx 1.28$.
3. *Compute the reorder point:* $ROP = 500 + 128 = 628$.

3.4.2 Manufacturing Industry

Advanced manufacturing inventory optimization techniques focus on minimizing work-in-progress inventory and improving production efficiency. Key techniques include:

- **Lean Manufacturing Principles:** These principles aim to eliminate waste, streamline processes, and improve production flow.
- **Six Sigma Methodologies:** Six Sigma focuses on reducing variability and defects in manufacturing processes to enhance quality and efficiency.
- **Advanced Scheduling Algorithms:** These algorithms optimize production schedules, reducing lead times and work-in-progress inventory.

Example 3.10. *A manufacturer implements lean principles to reduce waste. The production process has an average cycle time of ten days, with a standard deviation of two days. The company wants to maintain a 95% service level. The lead time for raw materials is five days.*

Solution 3.10. *To calculate the safety stock for raw materials, we use the following formula:*

$$Safety\,Stock = Z \times \sigma_{LT}$$

where:

Z is the z-score corresponding to the desired service level (for 95% service level, $Z \approx 1.65$).

σ_{LT} is the standard deviation of demand during lead time.

Calculate the standard deviation of demand during lead time:

$$\sigma_{LT} = \sigma \times \sqrt{L} = 2 \times \sqrt{5} \approx 4.47$$

Calculate the safety stock:

$$Safety\,Stock = 1.65 \times 4.47 \approx 7.38$$

Therefore, the manufacturer should maintain approximately eight days' worth of raw materials as safety stock to achieve a 95% service level.

Steps:

1. Calculate the standard deviation of demand during lead time: $\sigma_{LT} \approx 4.47$.

2. Determine the z-score for the desired service level: $Z \approx 1.65$.

3. Compute the safety stock: $Safety\,Stock \approx 7.38$.

3.4.3 Healthcare Industry

In the healthcare industry, optimizing inventory levels is critical for ensuring patient safety and operational efficiency. Advanced techniques include:

- **Demand Forecasting:** Accurate demand forecasting helps manage medical supplies and pharmaceuticals to prevent shortages and overstock situations.
- **RFID Tracking:** RFID technology provides real-time visibility of inventory, improving tracking and management of medical equipment and supplies.
- **Just-in-Time Delivery Systems:** These systems ensure that medical supplies and pharmaceuticals are delivered as needed, reducing inventory holding costs.

Example 3.11. *A hospital uses demand forecasting to manage its inventory of surgical gloves. The average monthly demand is 2,000 units, with a standard deviation of 300 units. The lead time for replenishment is two weeks. The hospital wants to maintain a 99% service level.*

Solution 3.11. *To calculate the safety stock, we use the following formula:*

$$Safety\,Stock = Z \times \sigma_{LT}$$

where:

Z *is the z-score corresponding to the desired service level (for 99% service level,* $Z \approx 2.33$*).*

σ_{LT} *is the standard deviation of demand during lead time.*

Convert the lead time to months:

$$L = \frac{2}{4} = 0.5\,months$$

Calculate the standard deviation of demand during lead time:

$$\sigma_{LT} = \sigma \times \sqrt{L} = 300 \times \sqrt{0.5} \approx 212$$

Calculate the safety stock:

$$Safety\,Stock = 2.33 \times 212 \approx 494$$

Therefore, the hospital should maintain approximately 494 units of surgical gloves as safety stock to achieve a 99% service level.

Steps:

1. *Convert the lead time to months:* $L = 0.5$.
2. *Calculate the standard deviation of demand during lead time:* $\sigma_{LT} \approx 212$.
3. *Determine the z-score for the desired service level:* $Z \approx 2.33$.
4. *Compute the safety stock:* $Safety\,Stock \approx 494$.

3.4.4 E-commerce Industry

In the e-commerce industry, efficient inventory management is crucial for fulfilling customer orders rapidly and minimizing costs. Advanced techniques include:

- **Demand Forecasting Algorithms:** These algorithms predict future demand based on historical data, trends, and other factors, helping to optimize inventory levels [15].
- **Dynamic Pricing Strategies:** Prices are adjusted based on demand, competition, and other market conditions, helping to manage inventory turnover and profitability [16].
- **Warehouse Automation Technologies:** Automation in warehouses improves inventory accuracy, reduces labor costs, and speeds up order fulfillment processes [17].

Example 3.12. *An e-commerce company uses demand forecasting algorithms to manage its inventory of electronic gadgets. The average weekly demand for a popular gadget is 1,000 units, with a standard deviation of 150 units. The company wants to maintain a 97% service level. The lead time for replenishment is one week.*

Solution 3.12. *To calculate the reorder point, including safety stock, we use the following formula:*

$$ROP = d \times L + Z \times \sigma_{LT}$$

where:

d *is the average demand per week (1,000 units)*

L *is the lead time in weeks (1 week)*

Z is the z-score corresponding to the desired service level (for 97% service level, $Z \approx 1.88$*)*

σ_{LT} *is the standard deviation of demand during lead time*

Calculate the standard deviation of demand during lead time:

$$\sigma_{LT} = \sigma \times \sqrt{L} = 150 \times \sqrt{1} = 150$$

Calculate the reorder point:

$$ROP = 1{,}000 \times 1 + 1.88 \times 150 = 1{,}000 + 282 = 1{,}282$$

Therefore, the company should reorder the gadget when the inventory level drops to 1,282 units to maintain a 97% service level.

Steps:

1. *Calculate the standard deviation of demand during lead time:* $\sigma_{LT} = 150$.
2. *Determine the z-score for the desired service level:* $Z \approx 1.88$.
3. *Compute the reorder point:* $ROP = 1{,}000 + 282 = 1282$.

Inventory optimization techniques and models offer sophisticated approaches to managing inventory across various industries and organizational contexts. Leveraging multi-echelon optimization, seasonal demand forecasting, demand-driven inventory management, and industry-specific strategies, businesses can achieve optimal inventory levels, reduce costs, and enhance customer satisfaction. As technology advances and market dynamics evolve, continuous innovation and adoption of inventory optimization techniques are essential for staying competitive in today's dynamic business environment.

3.5 Conclusion

This chapter has provided an in-depth exploration of advanced inventory optimization techniques and models, including system-level and item-level analysis, sensitivity analysis, and strategies for seasonal demand. Understanding and applying these concepts, businesses can optimize their inventory management practices, reduce costs, and improve overall efficiency. These advanced techniques are essential for organizations aiming to remain competitive in today's dynamic market environment.

Exercises

Exercise 3.1. *Consider a multi-echelon inventory system with two echelons: a central warehouse and several retail stores. The following data is provided:*

- *The central warehouse has a fixed ordering cost of* $K_c = \$300$ *and a holding cost of* $h_c = \$1$ *per unit per period.*
- *Each retail store has a fixed ordering cost of* $K_r = \$50$ *and a holding cost of* $h_r = \$2$ *per unit per period.*
- *The demand at each retail store follows a Poisson distribution with a mean rate of* $\lambda = 80$ *units per period.*
- *The lead time from the warehouse to each retail store is* $L = 3$ *periods.*
- *There are* $n = 10$ *retail stores.*

Derive the optimal order quantities for both the central warehouse and the retail stores that minimize the total cost. Use the economic order quantity (EOQ) model for both echelons and calculate the optimal order quantities.

Exercise 3.2. *In a three-echelon inventory system with suppliers, distribution centers, and retail stores, the following data is provided:*

- *Supplier's ordering cost:* $K_s = \$400$*; holding cost:* $h_s = \$2$ *per unit per period.*
- *Distribution center's ordering cost:* $K_d = \$250$*; holding cost:* $h_d = \$3$ *per unit per period.*
- *Retail store's ordering cost:* $K_r = \$100$*; holding cost:* $h_r = \$4$ *per unit per period.*
- *Demand at each echelon follows a normal distribution with the following parameters:*
  - *Suppliers: Mean demand* $\mu_s = 2{,}000$ *units; Variance* $\sigma_s^2 = 40{,}000$ *units*^{2}.
  - *Distribution centers: Mean demand* $\mu_d = 5{,}000$ *units; Variance* $\sigma_d^2 = 90{,}000$ *units*^{2}.
  - *Retail stores: Mean demand* $\mu_r = 10{,}000$ *units; Variance* $\sigma_r^2 = 160{,}000$ *units*^{2}.

Develop a linear programming model to determine the optimal order quantities at each echelon that minimize the total expected inventory cost. Include the objective function and constraints in your formulation.

Exercise 3.3. *Analyze the bullwhip effect in a multi-echelon inventory system with the following parameters:*

- *Retail stores use an (s, S) policy with* $s = 60$ *and* $S = 120$.
- *Distribution centers use an (s, S) policy with* $s = 180$ *and* $S = 360$.
- *Demand at retail stores follows a Poisson distribution with a mean rate of* $\lambda = 100$ *units per period.*
- *Lead time is* $L = 2$ *periods.*

Using a simulation approach, compare the variance in orders and inventory levels across echelons. Calculate the variance of inventory levels at both the retail stores and the distribution centers. Discuss how the choice of inventory policy affects the bullwhip effect.

Exercise 3.4. *Develop a Monte Carlo simulation for a multi-echelon inventory system with the following data:*

- *The system includes a central warehouse and* $n = 15$ *retail stores.*
- *Demand at each retail store follows a normal distribution with a mean* $\mu = 150$ *units and standard deviation* $\sigma = 30$ *units.*
- *Holding cost is* $h = \$1$ *per unit per period.*
- *Ordering cost at the central warehouse is* $K_c = \$400$ *and at each retail store is* $K_r = \$80$.
- *Lead time from the central warehouse to the retail stores is* $L = 2$ *periods.*

Describe the steps in setting up the simulation, including demand generation, inventory policy implementation, and performance evaluation. Perform the simulation to estimate the total cost and service levels. Provide numerical results from your simulation.

Exercise 3.5. *Investigate the impact of lead time variability in a multi-echelon inventory system with the following parameters:*

- *The system includes suppliers, distribution centers, and retail stores.*
- *Demand follows a normal distribution with a mean* $\mu = 200$ *units and standard deviation* $\sigma = 40$ *units at each echelon.*
- *Lead time at suppliers varies between two and five periods, at distribution centers varies between three and six periods, and at retail stores varies between one and four periods.*

- *Holding cost is* $h = \$2$ *per unit per period at each echelon.*
- *Ordering cost is* $K = \$300$ *per order at each echelon.*

Develop a sensitivity analysis model to quantify the effect of lead time variability on total inventory costs and service levels. Use numerical examples to illustrate how changes in lead time variability impact the system's performance.

4

Inventory Management in the Manufacturing Sector

This chapter explores the key concepts and principles of inventory management within the manufacturing sector, a critical component of supply chain management. It covers the management, planning, and budgeting of stock to balance demand and supply for various products. The chapter discusses the objectives, costs, and challenges specific to inventory management in manufacturing, while also addressing types of inventories and methods for their control and optimization. Additionally, it reviews essential practices like inventory analytics and optimization models, offering a foundational understanding of inventory management's role in manufacturing. The chapter is a valuable resource for practitioners, researchers, and students seeking insights into how inventory management impacts operational efficiency and prepares readers for more advanced optimization techniques.

4.1 Introduction to Inventory Management in Manufacturing

Inventory management in manufacturing involves the systematic oversight and control of raw materials, components, work-in-progress items, and finished products. It is essential for maintaining operational efficiency, reducing production costs, and ensuring timely delivery of products to customers. Effective inventory management helps manufacturers balance the need for sufficient inventory to meet demand while minimizing the costs associated with excess stock. It encompasses a range of activities, including demand forecasting, inventory tracking, procurement, storage, and distribution, all aimed at optimizing the flow of goods through the production process. Implementing robust inventory management practices, manufacturers can enhance production planning, reduce waste, improve cash flow, and achieve a competitive advantage in the market.

DOI: 10.1201/9781003534860-4

4.1.1 Definition and Scope of Inventory Management Specific to the Manufacturing Sector

Inventory management in the manufacturing sector involves the systematic oversight and control of goods, materials, and products used in the production process. This includes raw materials, work-in-progress (WIP) inventory, and finished goods. The primary goal is to ensure that the right quantity and quality of materials are available at the right time and place to meet production schedules and customer demand while minimizing costs associated with holding and ordering inventory. The scope of inventory management in manufacturing is extensive, encompassing activities such as:

- Demand Forecasting: Predicting future product demand to determine the necessary inventory levels.
- Procurement and Purchasing: Sourcing and acquiring raw materials and components needed for production.
- Inventory Control: Monitoring and regulating inventory levels to prevent overstocking or stockouts.
- Warehouse Management: Organizing and managing the storage of inventory to ensure efficient retrieval and distribution.
- Quality Control: Ensuring that inventory meets quality standards and specifications required for production.
- Inventory Accounting: Tracking inventory costs and valuing inventory for financial reporting purposes.

4.1.2 Importance of Inventory Management in Manufacturing Operations

Effective inventory management is critical to the success of manufacturing operations. It plays a vital role in ensuring smooth and efficient production processes, maintaining product quality, and meeting customer demands. The key benefits of effective inventory management in manufacturing include:

- Cost Reduction: Optimizing inventory levels, manufacturers can reduce carrying costs, minimize waste, and avoid the costs associated with stockouts and expedited orders.
- Improved Cash Flow: Effective inventory management ensures that capital is not tied up in excess inventory, freeing up cash for other operational needs.
- Enhanced Production Efficiency: With the right materials available at the right time, production processes can proceed without interruptions, reducing downtime and increasing output.

- Better Customer Satisfaction: Maintaining appropriate inventory levels helps ensure timely delivery of products to customers, enhancing satisfaction and loyalty.
- Risk Mitigation: Proper inventory management can buffer against supply chain disruptions, demand fluctuations, and production delays, reducing the risk of operational inefficiencies.

4.1.2.1 Warehouse Management

Warehouse management encompasses the organization, storage, and retrieval of inventory within a manufacturing facility. Efficient warehouse management practices, such as proper layout design, use of technology (e.g., barcode scanning, warehouse management systems), and effective inventory tracking, enhance the speed and accuracy of inventory handling. This directly impacts production efficiency by reducing the time needed to locate and transport materials.

4.1.2.2 Quality Control and Assurance

Ensuring the quality of inventory is critical in manufacturing, where defective materials can lead to production delays and compromised product quality. Quality control processes involve inspecting and testing incoming materials and in-process inventory to ensure they meet required specifications. Maintaining high-quality standards minimizes rework and scrap, contributing to overall production efficiency.

Quality Control Cost:

The cost associated with quality control can be calculated using the following formula:

$$\text{Quality Control Cost} = \sum_{i=1}^{n} (Q_i \times C_i) \tag{4.1}$$

where:

Q_i is the quantity of material i inspected.

C_i is the cost per inspection for material i.

Example 4.1. *Calculate the quality control cost for inspecting 150 units of Material X at \$4 per unit, and 250 units of Material Y at \$2.50 per unit.*

Solution 4.1. *The quality control cost can be computed as follows:*

$$\begin{aligned} \textit{Quality Control Cost} &= (150 \times 4) + (250 \times 2.50) && (4.2) \\ &= 600 + 625 && (4.3) \\ &= 1{,}225 && (4.4) \end{aligned}$$

Thus, the total quality control cost is \$1,225.

Example 4.2. *Determine the quality control cost for inspecting 80 units of Material A at \$6 per unit and 120 units of Material B at \$3.75 per unit.*

Solution 4.2. *The quality control cost calculation is:*

$$\begin{aligned} \textit{Quality Control Cost} &= (80 \times 6) + (120 \times 3.75) && (4.5) \\ &= 480 + 450 && (4.6) \\ &= 930 && (4.7) \end{aligned}$$

Therefore, the total quality control cost is \$930.

4.1.2.3 Inventory Accounting and Valuation

Inventory accounting involves tracking the costs associated with inventory, including procurement, holding, and production costs. Accurate inventory valuation is essential for financial reporting and for assessing the financial health of the manufacturing operation. Methods such as first-in, first-out (FIFO), last-in, first-out (LIFO), and weighted average cost are commonly used for inventory valuation.

Inventory Value (FIFO):

The inventory value using the FIFO method can be calculated using the following formula:

$$\text{Inventory Value (FIFO)} = \sum_{i=1}^{n} (\text{Quantity purchased}_i \times \text{Purchase price}_i) \quad (4.8)$$

Example 4.3. *Calculate the inventory value using FIFO for the following purchases:*

- *100 units purchased at \$5 each*
- *150 units purchased at \$6 each*
- *200 units purchased at \$7 each*

Solution 4.3. *The inventory value using FIFO is calculated as follows:*

$$\begin{aligned} \textit{Inventory Value} &= (100 \times 5) + (150 \times 6) + (200 \times 7) && (4.9) \\ &= 500 + 900 + 1{,}400 && (4.10) \\ &= 2{,}800 && (4.11) \end{aligned}$$

Thus, the total inventory value using FIFO is \$2,800.

Example 4.4. *Determine the inventory value using FIFO for the following purchases:*

- *50 units purchased at \$8 each*
- *75 units purchased at \$9 each*
- *125 units purchased at \$10 each*

Solution 4.4. *The inventory value using FIFO is calculated as:*

$$\begin{aligned} \textit{Inventory Value} &= (50 \times 8) + (75 \times 9) + (125 \times 10) && (4.12) \\ &= 400 + 675 + 1{,}250 && (4.13) \\ &= 2{,}325 && (4.14) \end{aligned}$$

Therefore, the total inventory value using FIFO is \$2,325.

4.1.3 Impact on Production

Effective inventory management has a direct impact on production by:

- Reducing Production Downtime: Ensuring that the necessary materials are available when needed prevents production delays and downtime.
- Enhancing Production Scheduling: Accurate inventory data allows for better production planning and scheduling, ensuring efficient use of resources and labor.
- Improving Production Flexibility: With optimal inventory levels, manufacturers can quickly respond to changes in demand or production schedules, enhancing overall agility and responsiveness.
- Maintaining Product Quality: Proper inventory management ensures that high-quality materials are used in production, maintaining consistent product quality and reducing defects.

Inventory management in manufacturing is a comprehensive and dynamic process that significantly influences production efficiency, cost control, and customer satisfaction. Understanding and implementing effective inventory management practices, manufacturers can achieve a competitive edge and drive operational success.

4.2 Significance of Inventory Management in Manufacturing

Inventory management is a critical component of the manufacturing sector, directly influencing the efficiency, cost-effectiveness, and overall success of production operations. Effective inventory management ensures that the right quantity of raw materials, work-in-progress goods, and finished products are available at the right time and place, minimizing costs associated with overstocking and stockouts. It plays a pivotal role in maintaining smooth production processes, reducing waste, improving customer satisfaction, and ultimately enhancing the profitability and competitiveness of manufacturing firms. Optimizing inventory levels and streamlining supply chain operations, manufacturers can respond swiftly to market demands, reduce lead times, and achieve better resource utilization.

4.3 Role of Inventory Management

Inventory management in manufacturing plays a critical role in overseeing, strategizing, and allocating resources to ensure smooth and efficient operations. It involves the following key activities:

4.3.1 Strategizing Inventory Policies

Effective inventory management requires the development of strategic policies that dictate how inventory is ordered, stored, and used. This includes setting reorder points, determining safety stock levels, and deciding on the use of inventory control techniques such as just-in-time (JIT) and economic order quantity (EOQ).

$$ROP = d \times L + Z \times \sigma_{LT}$$

where ROP is the reorder point, d is the average daily demand, L is the lead time, Z is the service level factor, and σ_{LT} is the standard deviation of demand during lead time.

Example 4.5. *Calculate the reorder point (ROP) given the following data:*

- *Average daily demand (d): 50 units*
- *Lead time (L): 5 days*
- *Service level factor (Z): 1.65 (for a 95% service level)*
- *Standard deviation of demand during lead time (σ_{LT}): 20 units*

Solution 4.5. *The reorder point (ROP) is calculated as follows:*

$$\begin{aligned} ROP &= d \times L + Z \times \sigma_{LT} \\ &= 50 \times 5 + 1.65 \times 20 \\ &= 250 + 33 \\ &= 283 \; units \end{aligned} \tag{4.15}$$

Therefore, the reorder point (ROP) is 283 units.

Example 4.6. *Determine the reorder point (ROP) for the following parameters:*

- *Average daily demand (d): 80 units*
- *Lead time (L): 7 days*
- *Service level factor (Z): 1.96 (for a 97.5% service level)*
- *Standard deviation of demand during lead time (σ_{LT}): 15 units*

Solution 4.6. *The reorder point (ROP) is calculated as:*

$$\begin{aligned} ROP &= d \times L + Z \times \sigma_{LT} \\ &= 80 \times 7 + 1.96 \times 15 \\ &= 560 + 29.4 \\ &= 589.4 \; units \end{aligned} \tag{4.16}$$

Thus, the reorder point (ROP) is approximately 589 units.

4.3.2 Allocating Resources Efficiently

Inventory management involves the efficient allocation of resources, including space, labor, and capital. This ensures that inventory is stored properly, handled efficiently, and capital is not tied up in excess inventory.

$$\text{Total Cost} = \text{Ordering Cost} + \text{Holding Cost} + \text{Stockout Cost}$$

where the total cost is minimized through efficient resource allocation.

Example 4.7. *Calculate the total cost given the following data:*

- *Ordering cost: $200 per order*
- *Number of orders per year: 10*
- *Holding cost: $5 per unit per year*

- *Average inventory: 100 units*
- *Stockout cost: $50 per unit*
- *Number of stockouts: 2*

Solution 4.7. *The total cost is calculated as follows:*

$$\begin{aligned}
\textit{Ordering Cost} &= \textit{Ordering cost per order} \times \textit{Number of orders} \qquad (4.17)\\
&= \$200 \times 10\\
&= \$2{,}000\\
\textit{Holding Cost} &= \textit{Holding cost per unit per year} \times \textit{Average inventory}\\
&= \$5 \times 100\\
&= \$500\\
\textit{Stockout Cost} &= \textit{Stockout cost per unit} \times \textit{Number of stockouts}\\
&= \$50 \times 2\\
&= \$100\\
\textit{Total Cost} &= \textit{Ordering Cost} + \textit{Holding Cost} + \textit{Stockout Cost}\\
&= \$2{,}000 + \$500 + \$100\\
&= \$2{,}600
\end{aligned}$$

Therefore, the total cost is $2,600.

Example 4.8. *Determine the total cost given the following parameters:*

- *Ordering cost: $150 per order*
- *Number of orders per year: 15*
- *Holding cost: $3 per unit per year*
- *Average inventory: 200 units*
- *Stockout cost: $40 per unit*
- *Number of stockouts: 3*

Solution 4.8. *The total cost is calculated as:*

$$\begin{aligned}
\textit{Ordering Cost} &= \textit{Ordering cost per order} \times \textit{Number of orders} \qquad (4.18)\\
&= \$150 \times 15\\
&= \$2{,}250\\
\textit{Holding Cost} &= \textit{Holding cost per unit per year} \times \textit{Average inventory}\\
&= \$3 \times 200\\
&= \$600\\
\textit{Stockout Cost} &= \textit{Stockout cost per unit} \times \textit{Number of stockouts}\\
&= \$40 \times 3\\
&= \$120\\
\textit{Total Cost} &= \textit{Ordering Cost} + \textit{Holding Cost} + \textit{Stockout Cost}\\
&= \$2{,}250 + \$600 + \$120\\
&= \$2{,}970
\end{aligned}$$

Thus, the total cost is \$2,970.

4.3.3 Ensuring the Availability of Materials for Continuous Production

A primary function of inventory management in manufacturing is to ensure the continuous availability of materials needed for production. This involves:

4.3.3.1 Demand Forecasting and Planning

Accurate demand forecasting helps predict future material requirements, ensuring that inventory levels are sufficient to meet production needs without interruption. Techniques such as time-series analysis, moving averages, and exponential smoothing are used for forecasting.

$$D_t = \alpha S_t + (1 - \alpha) D_{t-1}$$

where D_t is the forecasted demand for period t, S_t is the actual sales for period t, and α is a smoothing constant.

Example 4.9. *Forecast the demand for the next period using exponential smoothing with* $\alpha = 0.3$. *The actual sales for the current period is 150 units, and the forecasted demand for the current period is 140 units.*

Solution 4.9. *Using the exponential smoothing formula:*

$$\begin{aligned} D_t &= \alpha S_t + (1-\alpha)D_{t-1} \\ &= 0.3 \times 150 + (1-0.3) \times 140 \\ &= 0.3 \times 150 + 0.7 \times 140 \\ &= 45 + 98 \\ &= 143 \end{aligned} \tag{4.19}$$

Therefore, the forecasted demand for the next period is 143 units.

Example 4.10. *Using exponential smoothing with $\alpha = 0.5$, calculate the forecasted demand for the next period if the actual sales for the current period is 200 units, and the forecasted demand for the current period is 180 units.*

Solution 4.10. *Applying the exponential smoothing formula:*

$$\begin{aligned} D_t &= \alpha S_t + (1-\alpha)D_{t-1} \\ &= 0.5 \times 200 + (1-0.5) \times 180 \\ &= 0.5 \times 200 + 0.5 \times 180 \\ &= 100 + 90 \\ &= 190 \end{aligned} \tag{4.20}$$

Thus, the forecasted demand for the next period is 190 units.

4.3.3.2 Cycle Counting

Cycle counting is a periodic inventory auditing process where a subset of inventory is counted on a regular basis, improving accuracy and control over inventory levels.

$$\text{Cycle Count Frequency} = \frac{\text{Total Inventory Items}}{\text{Counted Items per Period}}$$

Example 4.11. *A warehouse has a total of 10,000 inventory items and plans to count 200 items per day. Calculate the cycle count frequency.*

Solution 4.11. *Using the cycle count frequency formula:*

$$\begin{aligned} \textit{Cycle Count Frequency} &= \frac{\textit{Total Inventory Items}}{\textit{Counted Items per Period}} \\ &= \frac{10,000}{200} \\ &= 50 \end{aligned} \tag{4.21}$$

Therefore, the cycle count frequency is 50 days, meaning the entire inventory will be counted every 50 days.

Example 4.12. *A company has 5,000 items in its inventory and wants to complete a full count of the inventory in 25 days. Determine how many items need to be counted each day.*

Solution 4.12. *Rearranging the cycle count frequency formula to solve for the number of items counted per period:*

$$\begin{aligned} \textit{Counted Items per Period} &= \frac{\textit{Total Inventory Items}}{\textit{Cycle Count Frequency}} \\ &= \frac{5{,}000}{25} \\ &= 200 \end{aligned} \tag{4.22}$$

Thus, the company needs to count 200 items each day to complete the cycle count in 25 days.

The significance of inventory management in manufacturing lies in its ability to oversee and strategize inventory policies, ensure material availability for continuous production, and balance inventory levels to minimize costs and maximize efficiency. Implementing these techniques and using mathematical models, manufacturers can achieve optimal inventory levels, reduce costs, and enhance overall operational performance.

4.3.4 Practical Applications

Implementing these inventory management principles and practices in manufacturing settings can lead to significant improvements in efficiency and cost-effectiveness.

Example 4.13. *A manufacturing company uses historical sales data to forecast future demand for its products. The historical sales data for the past six months is as follows: January: 120 units, February: 135 units, March: 150 units, April: 160 units, May: 170 units, June: 180 units. The company applies the exponential smoothing method with a smoothing constant* $\alpha = 0.3$ *to forecast demand for July.*

Solution 4.13. *Using exponential smoothing, the forecast is updated using the formula:*

$$F_{t+1} = \alpha \cdot D_t + (1 - \alpha) \cdot F_t$$

where F_{t+1} *is the forecast for the next period,* D_t *is the actual demand in the current period, and* F_t *is the forecast for the current period.*

Assuming the initial forecast for January (F_{Jan}) *is the same as the actual demand (120 units):*

$$F_{Feb} = 0.3 \cdot 135 + 0.7 \cdot 120 = 124.5 \textit{ units}$$

$$F_{Mar} = 0.3 \cdot 150 + 0.7 \cdot 124.5 = 132.15 \textit{ units}$$

$$F_{Apr} = 0.3 \cdot 160 + 0.7 \cdot 132.15 = 140.505 \text{ units}$$

$$F_{May} = 0.3 \cdot 170 + 0.7 \cdot 140.505 = 149.3535 \text{ units}$$

$$F_{Jun} = 0.3 \cdot 180 + 0.7 \cdot 149.3535 = 158.54745 \text{ units}$$

$$F_{Jul} = 0.3 \cdot 180 + 0.7 \cdot 158.54745 = 167.98322 \text{ units}$$

Thus, the forecasted demand for July is approximately 167.98 units.

Example 4.14. *A company that produces electronic components determines its EOQ to minimize the total cost of ordering and holding inventory. Given an annual demand of 10,000 units, an ordering cost of \$100 per order, and a holding cost of \$2 per unit per year, calculate the EOQ.*

Solution 4.14. *The EOQ is calculated using the formula:*

$$EOQ = \sqrt{\frac{2DS}{H}}$$

where D is the annual demand, S is the ordering cost per order, and H is the holding cost per unit per year.

Plugging in the given values:

$$EOQ = \sqrt{\frac{2 \times 10{,}000 \times 100}{2}} = \sqrt{1{,}000{,}000} = 1{,}000 \text{ units}$$

This means the company should order 1,000 units each time to minimize the total ordering and holding costs.

Example 4.15. *A manufacturing company categorizes its inventory into A, B, and C classes based on the annual consumption value. The annual consumption values for five items are: Item 1: \$30,000, Item 2: \$20,000, Item 3: \$10,000, Item 4: \$5,000, Item 5: \$1,000. Classify these items into A, B, and C categories.*

Solution 4.15. *ABC analysis categorizes items as follows:*

- *Class A: Top 70–80% of annual consumption value*
- *Class B: Next 15–25% of annual consumption value*
- *Class C: Last 5% of annual consumption value*

Total annual consumption value: \$30,000 + \$20,000 + \$10,000 + \$5,000 + \$1,000 = \$66,000

Calculating the cumulative percentages:

- *Item 1: (\$30,000 / \$66,000) ≈ 45.45%*
- *Item 2: (\$20,000 / \$66,000) ≈ 30.30%*

- *Item 3: (\$10,000 / \$66,000) ≈ 15.15%*
- *Item 4: (\$5,000 / \$66,000) ≈ 7.58%*
- *Item 5: (\$1,000 / \$66,000) ≈ 1.52%*

Cumulative percentages:

- *Item 1: 45.45%*
- *Item 2: 75.75%*
- *Item 3: 90.90%*
- *Item 4: 98.48%*
- *Item 5: 100%*

Classification:

- *Class A: Item 1 (45.45%)*
- *Class B: Items 2 (30.30%) and 3 (15.15%)*
- *Class C: Items 4 (7.58%) and 5 (1.52%)*

This classification helps prioritize resource allocation and management efforts.

Example 4.16. *A car manufacturer adopts JIT inventory management to reduce holding costs and improve production efficiency. Explain how JIT can help the company and provide an example scenario.*

Solution 4.16. *JIT inventory management minimizes inventory levels by receiving goods only as they are needed in the production process. This reduces holding costs and increases efficiency.*

For example, the car manufacturer coordinates with suppliers to deliver components like engines, transmissions, and tires just in time for their assembly. Instead of storing large quantities of these components, the manufacturer receives daily shipments based on the production schedule. This approach ensures that parts are available when needed, reduces storage costs, and minimizes waste due to overproduction or obsolete inventory.

Example 4.17. *A food processing company uses the continuous review model to manage its inventory of perishable goods. With an average daily demand of 200 units, a lead time of five days, a service level factor corresponding to a 95% service level, and a standard deviation of demand during lead time of 50 units, calculate the reorder point.*

Solution 4.17. *The reorder point (R) is calculated using the formula (1.65 for 95):*

$$R = d \cdot L + z \cdot \sigma_L$$

where d is the average daily demand, L is the lead time in days, z is the service level factor

Given:

- *Average daily demand (d): 200 units*
- *Lead time (L): 5 days*
- *Service level factor (z): 1.65*
- *Standard deviation of demand during lead time (σ_L): 50 units*

Calculating the reorder point:

$$R = 200 \times 5 + 1.65 \times 50 = 1{,}000 + 82.5 = 1,082.5 \text{ units}$$

The company places a new order when the inventory level reaches 1,082.5 units to maintain service levels and avoid stockouts.

The core principles and practices of inventory management, including demand forecasting, EOQ, ABC analysis, JIT, and continuous review models, are essential for optimizing inventory levels and minimizing costs in manufacturing settings. Applying these techniques, manufacturers can achieve greater efficiency, reduce waste, and improve overall operational performance.

4.4 Inventory Analytics

Data analytics plays a crucial role in manufacturing inventory management by providing insights that help in making informed decisions. It involves collecting, processing, and analyzing data to optimize inventory levels, reduce costs, and improve efficiency.

4.4.1 Descriptive Analytics

Descriptive analytics helps in understanding the current state of inventory by summarizing historical data. It involves techniques such as data visualization and reporting.

$$\text{Inventory Accuracy} = \frac{\text{Accurate Inventory Records}}{\text{Total Inventory Records}} \times 100$$

Example 4.18. *A warehouse has 9,500 accurate inventory records out of a total of 10,000 inventory records. Calculate the inventory accuracy percentage.*

Solution 4.18. *Using the inventory accuracy formula:*

$$\begin{aligned} \textit{Inventory Accuracy} &= \frac{\textit{Accurate Inventory Records}}{\textit{Total Inventory Records}} \times 100 \\ &= \frac{9,500}{10,000} \times 100 \\ &= 95\% \end{aligned} \tag{4.23}$$

Therefore, the inventory accuracy percentage is 95%.

Example 4.19. *A company conducted an inventory audit and found that 7,800 records were accurate out of a total of 8,000 records. Determine the inventory accuracy percentage.*

Solution 4.19. *Applying the inventory accuracy formula:*

$$\begin{aligned} \textit{Inventory Accuracy} &= \frac{\textit{Accurate Inventory Records}}{\textit{Total Inventory Records}} \times 100 \\ &= \frac{7,800}{8,000} \times 100 \\ &= 97.5\% \end{aligned} \tag{4.24}$$

Thus, the inventory accuracy percentage is 97.5%.

Descriptive analytics provides valuable insights into inventory management by summarizing historical data and highlighting areas that require attention. Maintaining high inventory accuracy, manufacturers can reduce errors, enhance efficiency, and make better-informed decisions.

4.4.2 Predictive Analytics

Predictive analytics forecasts future inventory needs based on historical data, trends, and patterns. It uses statistical models and machine learning algorithms to predict demand and lead times.

$$F_t = \alpha D_{t-1} + (1 - \alpha) F_{t-1}$$

where F_t is the forecasted demand for period t, D_{t-1} is the actual demand for the previous period, and α is a smoothing constant.

Example 4.20. *Given the actual demand D_{t-1} of 120 units for the previous period and a forecasted demand F_{t-1} of 115 units, with a smoothing constant $\alpha = 0.5$, calculate the forecasted demand F_t for the current period.*

Solution 4.20. *Using the predictive analytics formula:*

$$\begin{aligned} F_t &= \alpha D_{t-1} + (1-\alpha)F_{t-1} \\ &= 0.5 \times 120 + (1-0.5) \times 115 \\ &= 60 + 57.5 \\ &= 117.5 \end{aligned} \tag{4.25}$$

Therefore, the forecasted demand F_t for the current period is 117.5 units.

Example 4.21. *For the previous period, the actual demand D_{t-1} was 150 units, and the forecasted demand F_{t-1} was 140 units. If the smoothing constant α is 0.3, determine the forecasted demand F_t for the current period.*

Solution 4.21. *Applying the predictive analytics formula:*

$$\begin{aligned} F_t &= \alpha D_{t-1} + (1-\alpha)F_{t-1} \\ &= 0.3 \times 150 + (1-0.3) \times 140 \\ &= 45 + 98 \\ &= 143 \end{aligned} \tag{4.26}$$

Thus, the forecasted demand F_t for the current period is 143 units.

Predictive analytics provides valuable insights by forecasting future inventory requirements, helping manufacturers to optimize inventory levels, reduce costs, and improve efficiency through data-driven decision-making.

4.4.3 Key Performance Indicators (KPIs) and Metrics for Inventory Performance in Manufacturing

Key performance indicators (KPIs) are essential for measuring the effectiveness of inventory management. Common KPIs include:

4.4.3.1 Inventory Turnover Ratio

The inventory turnover ratio measures how often inventory is sold and replaced over a specific period.

$$\text{Inventory Turnover Ratio} = \frac{\text{Cost of Goods Sold (COGS)}}{\text{Average Inventory}}$$

Example 4.22. *Given the cost of goods sold (COGS) of \$500,000 and an average inventory of \$100,000, calculate the inventory turnover ratio.*

Solution 4.22. *Using the inventory turnover ratio formula:*

$$\begin{aligned} \textit{Inventory Turnover Ratio} &= \frac{\textit{COGS}}{\textit{Average Inventory}} \\ &= \frac{500{,}000}{100{,}000} \\ &= 5 \end{aligned} \tag{4.27}$$

Therefore, the inventory turnover ratio is 5, indicating that the inventory is sold and replaced five times during the period.

Example 4.23. *For a company with a cost of goods sold (COGS) of $750,000 and an average inventory of $150,000, determine the inventory turnover ratio.*

Solution 4.23. *Applying the inventory turnover ratio formula:*

$$\begin{aligned} Inventory\ Turnover\ Ratio &= \frac{COGS}{Average\ Inventory} \\ &= \frac{750{,}000}{150{,}000} \\ &= 5 \end{aligned} \tag{4.28}$$

Thus, the inventory turnover ratio is 5, meaning the inventory is sold and replaced five times during the period.

4.4.3.2 Days of Inventory on Hand (DOH)

DOH measures the number of days inventory remains in stock before being sold.

$$\text{DOH} = \frac{\text{Average Inventory}}{\text{Cost of Goods Sold (COGS)}} \times 365$$

Example 4.24. *Given an average inventory of $150,000 and a cost of goods sold (COGS) of $600,000, calculate the days of inventory on hand (DOH).*

Solution 4.24. *Using the DOH formula:*

$$\begin{aligned} DOH &= \frac{Average\ Inventory}{COGS} \times 365 \\ &= \frac{150{,}000}{600{,}000} \times 365 \\ &= 0.25 \times 365 \\ &= 91.25 \end{aligned} \tag{4.29}$$

Therefore, the days of inventory on hand (DOH) is approximately 91.25 days.

Example 4.25. *For a company with an average inventory of $200,000 and a cost of goods sold (COGS) of $800,000, determine the days of inventory on hand (DOH).*

Solution 4.25. *Applying the DOH formula:*

$$\begin{aligned} DOH &= \frac{Average\ Inventory}{COGS} \times 365 \\ &= \frac{200{,}000}{800{,}000} \times 365 \\ &= 0.25 \times 365 \\ &= 91.25 \end{aligned} \tag{4.30}$$

Thus, the days of inventory on hand (DOH) is approximately 91.25 days.

4.4.3.3 Stockout Rate

The stockout rate measures the frequency of stockouts during a specific period.

$$\text{Stockout Rate} = \frac{\text{Number of Stockouts}}{\text{Total Orders}} \times 100$$

Example 4.26. *Suppose a company experienced 15 stockouts out of 500 total orders during the year. Calculate the stockout rate.*

Solution 4.26. *Using the stockout rate formula:*

$$\begin{aligned} Stockout\ Rate &= \frac{Number\ of\ Stockouts}{Total\ Orders} \times 100 \\ &= \frac{15}{500} \times 100 \\ &= 0.03 \times 100 \\ &= 3 \end{aligned} \tag{4.31}$$

Therefore, the stockout rate is 3%.

Example 4.27. *For a different scenario, a retailer had 25 stockouts out of 1,000 total orders in a quarter. Determine the stockout rate.*

Solution 4.27. *Applying the stockout rate formula:*

$$\begin{aligned} Stockout\ Rate &= \frac{Number\ of\ Stockouts}{Total\ Orders} \times 100 \\ &= \frac{25}{1{,}000} \times 100 \\ &= 0.025 \times 100 \\ &= 2.5 \end{aligned} \tag{4.32}$$

Thus, the stockout rate is 2.5%.

4.4.3.4 Order Fulfillment Rate

The order fulfillment rate measures the percentage of customer orders that are fulfilled from available inventory.

$$\text{Order Fulfillment Rate} = \frac{\text{Orders Fulfilled from Inventory}}{\text{Total Orders}} \times 100$$

Example 4.28. *Consider a company that had 450 orders fulfilled from inventory out of a total of 500 orders. Calculate the order fulfillment rate.*

Solution 4.28. *Using the order fulfillment rate formula:*

$$\begin{aligned} \textit{Order Fulfillment Rate} &= \frac{\textit{Orders Fulfilled from Inventory}}{\textit{Total Orders}} \times 100 \\ &= \frac{450}{500} \times 100 \\ &= 0.9 \times 100 \\ &= 90 \end{aligned} \tag{4.33}$$

Therefore, the order fulfillment rate is 90%.

Example 4.29. *A retailer managed to fulfill 300 orders from inventory out of 400 total orders in a month. Determine the order fulfillment rate.*

Solution 4.29. *Applying the order fulfillment rate formula:*

$$\begin{aligned} \textit{Order Fulfillment Rate} &= \frac{\textit{Orders Fulfilled from Inventory}}{\textit{Total Orders}} \times 100 \\ &= \frac{300}{400} \times 100 \\ &= 0.75 \times 100 \\ &= 75 \end{aligned} \tag{4.34}$$

Thus, the order fulfillment rate is 75%.

4.4.4 Techniques for Forecasting Demand and Lead Times Specific to Manufacturing

Accurate forecasting of demand and lead times is critical for maintaining optimal inventory levels in manufacturing.

4.4.4.1 Time Series Analysis

Time series analysis is used to forecast future demand based on historical data. It includes methods such as moving averages and exponential smoothing.

$$F_t = \alpha D_{t-1} + (1 - \alpha) F_{t-1}$$

where F_t is the forecasted demand for period t, D_{t-1} is the actual demand for the previous period, and α is a smoothing constant.

Example 4.30. *Suppose the actual demand D_{t-1} for the previous period is 200 units, and the forecasted demand F_{t-1} is 180 units. If the smoothing constant α is 0.4, calculate the forecasted demand F_t for the current period.*

Solution 4.30. *Using the exponential smoothing formula:*

$$\begin{aligned} F_t &= \alpha D_{t-1} + (1-\alpha)F_{t-1} \\ &= 0.4 \times 200 + (1-0.4) \times 180 \\ &= 80 + 108 \\ &= 188 \end{aligned} \tag{4.35}$$

Thus, the forecasted demand F_t for the current period is 188 units.

Example 4.31. *Consider a situation where the previous period's actual demand D_{t-1} was 150 units, and the forecasted demand F_{t-1} was 160 units. With a smoothing constant α of 0.6, determine the forecasted demand F_t for the current period.*

Solution 4.31. *Applying the exponential smoothing formula:*

$$\begin{aligned} F_t &= \alpha D_{t-1} + (1-\alpha)F_{t-1} \\ &= 0.6 \times 150 + (1-0.6) \times 160 \\ &= 90 + 64 \\ &= 154 \end{aligned} \tag{4.36}$$

Therefore, the forecasted demand F_t for the current period is 154 units.

4.4.4.2 Seasonal Decomposition

Seasonal decomposition involves breaking down time series data into trend, seasonal, and residual components to forecast demand more accurately.

$$D_t = T_t + S_t + R_t$$

where D_t is the actual demand, T_t is the trend component, S_t is the seasonal component, and R_t is the residual component.

Example 4.32. *Suppose the actual demand D_t for a given period is 500 units. If the trend component T_t is 450 units, the seasonal component S_t is 50 units, and the residual component R_t is 0 units, calculate the actual demand D_t.*

Solution 4.32. *Using the seasonal decomposition formula:*

$$\begin{aligned} D_t &= T_t + S_t + R_t \\ &= 450 + 50 + 0 \\ &= 500 \end{aligned} \tag{4.37}$$

Thus, the actual demand D_t is 500 units.

Example 4.33. *For another period, the actual demand D_t is 620 units. If the trend component T_t is 580 units, the seasonal component S_t is 40 units, and the residual component R_t is 0 units, determine the actual demand D_t.*

Solution 4.33. *Applying the seasonal decomposition formula:*

$$\begin{aligned} D_t &= T_t + S_t + R_t \\ &= 580 + 40 + 0 \\ &= 620 \end{aligned} \tag{4.38}$$

Thus, the actual demand D_t is 620 units.

4.4.4.3 Lead Time Demand Forecasting

Forecasting lead time demand involves predicting the demand during the lead time period.

$$\text{Lead Time Demand} = d \times L$$

where d is the average daily demand and L is the lead time in days.

Example 4.34. *Assume the average daily demand d is 100 units and the lead time L is ten days. Calculate the lead time demand.*

Solution 4.34. *Using the lead time demand forecasting formula:*

$$\begin{aligned} \textit{Lead Time Demand} &= d \times L \\ &= 100 \times 10 \\ &= 1{,}000 \end{aligned} \tag{4.39}$$

Thus, the lead time demand is 1,000 units.

Example 4.35. *For a different scenario, suppose the average daily demand d is 150 units and the lead time L is seven days. Determine the lead time demand.*

Solution 4.35. *Applying the lead time demand forecasting formula:*

$$\begin{aligned} \textit{Lead Time Demand} &= d \times L \\ &= 150 \times 7 \\ &= 1{,}050 \end{aligned} \tag{4.40}$$

Therefore, the lead time demand is 1,050 units.

Example 4.36. *A manufacturing company uses exponential smoothing to forecast demand for its products. Historical sales data shows the following demand for the past five periods: 100, 120, 130, 150, and 140 units. Using a smoothing constant (α) of 0.5, the forecast for the next period is calculated as follows:*

Solution 4.36.

$$F_6 = \alpha D_5 + (1 - \alpha) F_5$$

$$F_6 = 0.5 \times 140 + 0.5 \times 150 = 145 \textit{ units}$$

Example 4.37. *A company with an average daily demand of 200 units and a lead time of five days calculates the lead time demand as follows:*

Solution 4.37.

$$Lead\ Time\ Demand = d \times L = 200 \times 5 = 1{,}000\ units$$

This means the company should have 1,000 units available to meet demand during the lead time.

Inventory analytics involves leveraging data to optimize inventory management in manufacturing. Utilizing descriptive, predictive, and prescriptive analytics, along with key performance indicators and accurate forecasting techniques, manufacturers can enhance efficiency, reduce costs, and improve overall inventory performance.

4.4.4.4 Food and Beverage Industry

In the food and beverage industry, inventory management focuses on minimizing waste and ensuring product freshness. Companies often use FIFO inventory methods to rotate stock effectively.

$$\text{FIFO Inventory Method} = \frac{\text{Cost of Oldest Inventory}}{\text{Total Inventory Cost}}$$

A dairy company reduced waste by 20% by implementing FIFO, ensuring older products were sold first.

4.4.4.5 Pharmaceutical Industry

Pharmaceutical companies prioritize inventory accuracy and regulatory compliance. Advanced inventory management systems with RFID tracking ensure real-time inventory visibility and reduce the risk of stockouts and overstocking.

$$\text{Stockout Rate} = \frac{\text{Number of Stockouts}}{\text{Total Orders}} \times 100$$

A pharmaceutical company decreased its stockout rate by 15% after implementing RFID tracking and automated inventory systems.

4.4.4.6 Textile Industry

In the textile industry, managing raw materials and finished goods inventory is critical. Using EOQ and safety stock calculations, a textile manufacturer optimized its inventory levels, balancing ordering and holding costs.

$$EOQ = \sqrt{\frac{2DS}{H}}$$

$$Safety\,Stock = Z \times \sigma_{LT}$$

With an annual demand (D) of 10,000 units, an ordering cost (S) of \$50 per order, and a holding cost (H) of \$2 per unit per year, the optimal order quantity (EOQ) was calculated to be approximately 500 units. The safety stock level, considering a service level factor (Z) of 1.65 and a standard deviation of demand during lead time (σ_{LT}) of 50 units, was 82.5 units.

Example: Textile Industry Calculation

Given:

- Annual demand (D): 10,000 units
- Ordering cost (S): \$50 per order
- Holding cost (H): \$2 per unit per year
- Service level factor (Z): 1.65
- Standard deviation of demand during lead time (σ_{LT}): 50 units

Calculations:

$$EOQ = \sqrt{\frac{2 \times 10{,}000 \times 50}{2}} = \sqrt{500{,}000} \approx 707 \text{ units}$$

$$Safety\,Stock = 1.65 \times 50 = 82.5 \text{ units}$$

The optimal order quantity is approximately 707 units, and the safety stock level is 82.5 units.

4.5 Conclusion

This chapter provided a comprehensive overview of inventory management in the manufacturing sector, highlighting its definition, scope, and significance. Key principles and practices were discussed, including the importance of data analytics, forecasting techniques, and mathematical models in optimizing inventory levels. Real-world case studies and practical applications illustrated the successful implementation of inventory management strategies across various industries. It explores various effective methods for inventory control and optimization, making it an essential resource for practitioners, researchers, and students seeking to deepen their knowledge and understanding of inventory management in manufacturing.

Exercises

Exercise 4.1. *Given a workforce scheduling problem with the following data:*

- *Worker costs:* $c = [10, 20, 15, 25]$
- *Task requirements:* $b = [1, 1, 1, 1]$
- *Assignment matrix:* $a = \begin{bmatrix} 1 & 0 & 0 & 0 \\ 0 & 1 & 0 & 0 \\ 0 & 0 & 1 & 0 \\ 0 & 0 & 0 & 1 \end{bmatrix}$

Solve the problem using the Branch and Bound method.

Exercise 4.2. *Given an average daily demand* $d = 30$ *units and a lead time* $L = 5$ *days, calculate the lead time demand.*

Exercise 4.3. *A manufacturing company has a cost of goods sold (COGS) of $1,200,000 and an average inventory of $300,000. Calculate the inventory turnover ratio.*

Exercise 4.4. *For a given time series data, the trend component* $T_t = 100$*, seasonal component* $S_t = 20$*, and residual component* $R_t = 5$*. Calculate the actual demand* D_t*.*

Exercise 4.5. *For the previous period, the actual demand* D_{t-1} *was 200 units, and the forecasted demand* F_{t-1} *was 190 units. If the smoothing constant* α *is 0.4, determine the forecasted demand* F_t *for the current period.*

Exercise 4.6. *A company has 950 accurate inventory records out of 1,000 total inventory records. Calculate the inventory accuracy.*

Exercise 4.7. *In a month, a company experiences three stockouts out of 150 total orders. Calculate the stockout rate.*

Exercise 4.8. *A manufacturer inspects 1,000 units of a material with an inspection cost of $5 per unit. Calculate the quality control cost.*

Exercise 4.9. *A company fulfills 450 orders from inventory out of 500 total orders. Calculate the order fulfillment rate.*

5

Emerging Trends in Inventory Management

This chapter explores the emerging trends reshaping the field of inventory management, highlighting the integration of advanced technologies such as artificial intelligence, machine learning, and robotics. It examines how these innovations are transforming traditional practices, with a focus on cloud-based systems that enhance remote accessibility, flexibility, and collaboration. The chapter also discusses the role of robotics in automating warehouse operations, improving efficiency, and reducing errors. Additionally, it covers distributed inventory management and the use of blockchain technology in optimizing stock control and logistics. Offering insights into these trends, the chapter equips businesses with the knowledge to streamline operations, improve efficiency, and reduce costs, guiding strategic decisions in the evolving inventory management landscape.

5.1 Introduction

Inventory management is a critical component of business operations, directly impacting efficiency, cost-effectiveness, and customer satisfaction. With the rapid advancement of technology, traditional inventory management practices are undergoing a significant transformation. This chapter explores these emerging trends, providing a comprehensive overview of the developments shaping the future of inventory management.

5.2 Integration of Cutting-Edge Technologies

The integration of artificial intelligence (AI), machine learning (ML), and robotics into inventory management systems marks a significant departure from conventional practices. These technologies enable enhanced data analysis

DOI: 10.1201/9781003534860-5

capabilities and the automation of repetitive tasks, significantly improving efficiency and reducing human error.

5.2.1 Artificial Intelligence and Machine Learning in Inventory Management

AI and ML are transforming inventory management by enabling more sophisticated analysis and decision-making processes. These technologies can process vast amounts of data to identify patterns, predict outcomes, and optimize operations.

5.2.2 Enhanced Data Analysis Capabilities

AI and ML algorithms can analyze complex datasets to uncover hidden patterns and correlations that are not immediately apparent through traditional methods. This advanced data analysis allows businesses to:

- Predict future inventory needs based on historical trends and external factors.
- Optimize stock levels to minimize costs while ensuring product availability.
- Identify slow-moving or obsolete inventory to improve turnover rates.

Demand Forecasting

One of the key applications of AI and ML in inventory management is demand forecasting. Accurate demand forecasting helps businesses maintain optimal inventory levels, reducing the risk of stockouts and overstocking. The exponential smoothing method is a widely used technique for demand forecasting. It was introduced in the previous chapter:

$$F_{t+1} = \alpha D_t + (1 - \alpha) F_t$$

where:

F_{t+1} is the forecast for the next period.

D_t is the actual demand in the current period.

F_t is the forecast for the current period.

α is the smoothing constant $0 < \alpha < 1$.

The smoothing constant α determines the weight given to the most recent demand data. A higher α value gives more weight to recent observations, making the forecast more responsive to changes in demand patterns.

5.2.3 Inventory Optimization Using Machine Learning Algorithms

ML algorithms play a crucial role in inventory optimization. These algorithms can analyze vast amounts of historical data to determine optimal inventory levels, minimizing the total cost associated with holding, ordering, and stock-outs. We will discuss the following ML techniques in detail: linear regression, decision trees, and neural networks. ML involves the use of algorithms that can learn from and make predictions or decisions based on data. In inventory optimization, ML algorithms analyze historical sales data, demand patterns, and other relevant factors to forecast future demand and optimize stock levels. This results in reduced costs and improved efficiency in inventory management.

5.2.4 Linear Regression

Linear regression is a fundamental machine learning algorithm used for predictive analysis. In the context of inventory optimization, linear regression can model the relationship between demand and various influencing factors, such as time, price, and promotions.

The linear regression model can be expressed as:

$$Y = \beta_0 + \beta_1 X_1 + \beta_2 X_2 + \cdots + \beta_n X_n + \epsilon$$

where:

Y is the dependent variable (e.g., demand).

β_0 is the intercept.

$\beta_1, \beta_2, \ldots, \beta_n$ are the coefficients of the independent variables.

$X_1, X_2, \ldots, X_n$ are the independent variables (e.g., time, price, promotions).

ϵ is the error term.

Example 5.1. *A retail company wants to predict the demand for a specific product based on historical sales data. They collect data on weekly sales (dependent variable) and advertising spend (independent variable). Using linear regression, they build the following model:*

$$Sales = \beta_0 + \beta_1 \times Advertising\ Spend + \epsilon$$

The coefficients estimated from the data are $\beta_0 = 100$ and $\beta_1 = 0.5$. If the company plans to spend \$2,000 on advertising next week, what is the predicted sales volume?

Solution 5.1. *Substitute the values into the linear regression model:*

$$Sales = 100 + 0.5 \times 2{,}000 = 100 + 1{,}000 = 1{,}100$$

Therefore, the predicted sales volume is 1,100 units.

Example 5.2. *An e-commerce platform wants to optimize its inventory of a seasonal product. They use historical sales data (dependent variable) and pricing strategies (independent variable) to build a linear regression model:*

$$Sales = \beta_0 + \beta_1 \times Price + \epsilon$$

From their data analysis, they find $\beta_0 = 500$ *and* $\beta_1 = -2$*. If they set the price at \$300 for the upcoming season, what is the predicted sales volume?*

Solution 5.2. *Apply the price to the regression model:*

$$Sales = 500 + (-2) \times 300 = 500 - 600 = -100$$

The negative predicted sales volume suggests that they may need to reconsider their pricing strategy or other influencing factors before finalizing their inventory plans.

In inventory optimization, linear regression can be used to forecast demand based on historical data and external factors. Understanding the relationship between these factors and demand, businesses can make informed decisions about how much inventory to keep in stock.

5.2.5 Decision Trees

Decision trees are another popular machine learning technique used for both classification and regression tasks.

5.2.5.1 What Is a Decision Tree?

Imagine you're trying to make a decision about something, like what to have for dinner. You might start with a broad question and then narrow down your options based on your answers. For example:

1. **Do I want to eat at home or go out?**
   - **If at home:**
     - Do I want something quick or something fancy?
     - If quick, you might decide:
       * Should I make a sandwich or heat up leftovers?
   - **If going out:**
     - Do I want something cheap or am I willing to spend more?
     - If cheap, you might consider:
       * Fast food or a casual diner?

This process of asking questions to narrow down your choices is similar to how a decision tree works in machine learning.

5.2.5.2 How Does a Decision Tree Work?

A decision tree is a tool that helps computers make decisions by following a series of questions (called nodes) that leads to an answer (called a leaf). It's a tree-like model, where each question splits into branches leading to other questions or answers.

1. **Start With a Question (Root Node):**
    - This is the main question that starts the decision-making process. For example, in a classification task to decide if a person likes sports, the first question might be, "Do they watch sports on TV?"
2. **Branch Out (Decision Nodes):**
    - Based on the answer, the tree branches out to more specific questions. For instance, if they do watch sports on TV, the next question might be, "Do they watch football?"
3. **Continue Branching (More Decision Nodes):**
    - This process continues, with each answer leading to more branches until the questions can no longer be divided further.
4. **Reach a Conclusion (Leaf Nodes):**
    - The end of a branch is where a decision is made. For example, if a person watches sports on TV and enjoys football, the leaf node might say, "This person likes sports."

5.2.6 Classification and Regression

1. **Classification:**
    - Used to categorize things into different groups. For example, deciding if an email is spam or not.
2. **Regression:**
    - Used to predict a continuous value. For example, predicting the price of a house based on features like size, location, and age.

5.2.7 Visualization

A decision tree can be visualized as a flowchart:

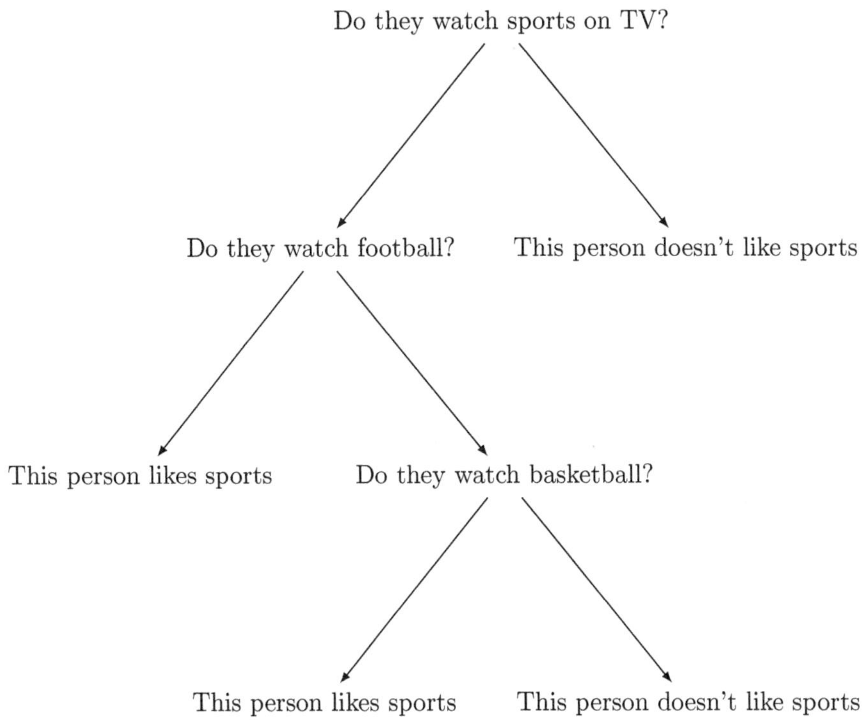

Example 5.3. *The company has a threshold for minimum stock levels and needs to decide if it's time to reorder a product. The decision depends on the current inventory level and whether the sales forecast suggests a potential stockout.*

1. ***Question 1:*** *Is the current inventory level below the minimum threshold?*
   - ***Yes:*** *Proceed to the next question.*
   - ***No:*** *No need to reorder.*
2. ***Question 2:*** *Is the projected demand (based on sales forecast) greater than the current inventory?*
   - ***Yes:*** *Reorder the product.*
   - ***No:*** *No need to reorder.*

Solution 5.3. *Let's visualize this decision-making process using a decision tree:*

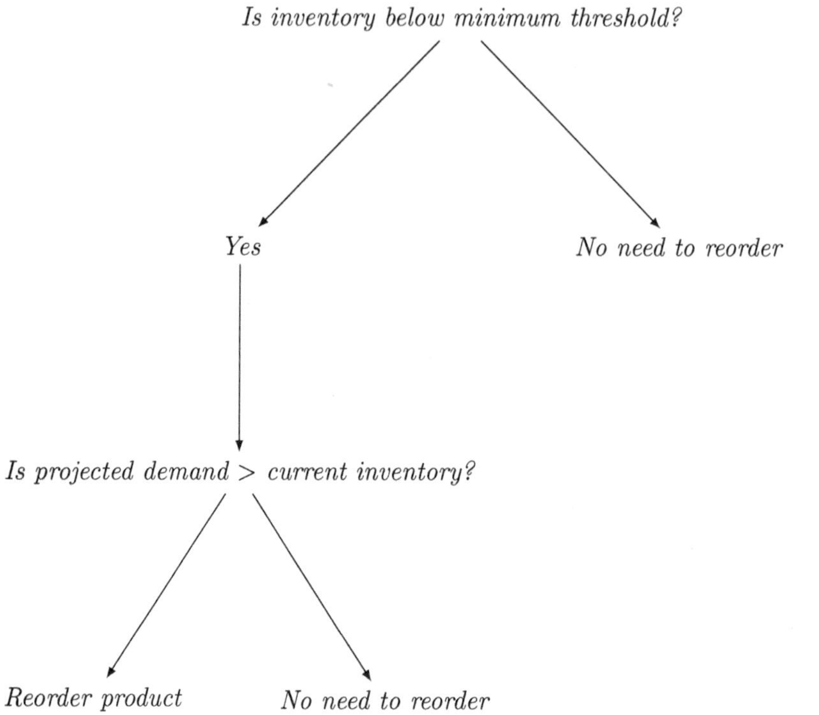

Explanation:

- *The root node of the tree asks if the inventory level is below the minimum threshold.*
- *If the answer is "Yes," the decision tree moves to the next question: whether the projected demand exceeds the current inventory level.*
- *Based on this second question, the tree decides whether to reorder the product or not.*
- *If the inventory level is not below the minimum threshold, the decision is to not reorder the product.*

Example 5.4. *The company needs to decide whether to reorder a product. They use the following data:*

- ***Current Inventory (I):*** *150 units*
- ***Average Monthly Demand (D):*** *30 units*
- ***Lead Time (L):*** *2 months*
- ***Reorder Point (R):*** *This is calculated as* $R = D \times L$

The decision rule is:

- *If $I \leq R$, reorder the product.*
- *If $I > R$, no need to reorder.*

Solution 5.4. *Let's perform the calculation to determine the reorder point:*

$$\begin{aligned} R &= D \times L \\ &= 30\,\textit{units/month} \times 2\,\textit{months} \\ &= 60\,\textit{units} \end{aligned}$$

The reorder point R is 60 units.

Now, we compare the current inventory (150 units) with the reorder point (60 units):

- *Since $I = 150\,\textit{units} > R = 60\,\textit{units}$, there is no need to reorder.*

Let's visualize this decision-making process using a decision tree:

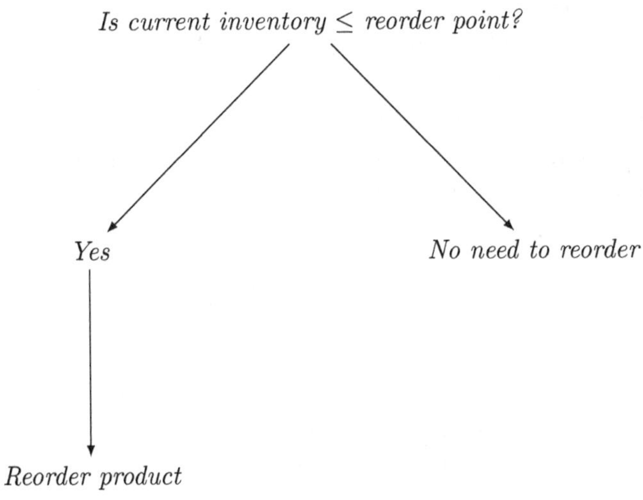

Explanation:

- *The root node of the tree asks if the current inventory is less than or equal to the reorder point.*
- *If the answer is "Yes," the decision is to reorder the product.*
- *If the answer is "No," the decision is not to reorder the product.*

5.2.8 Neural Networks

Neural networks are a set of algorithms modeled after the human brain, designed to recognize patterns. They are particularly effective for complex, nonlinear relationships in data. A neural network consists of layers of nodes (neurons), including an input layer, hidden layers, and an output layer. Each node in a layer is connected to nodes in the subsequent layer with associated weights. Here, we discuss the mathematical concepts behind neural networks.

1. **Neuron and Activation Function**: Each neuron computes a weighted sum of its inputs, adds a bias, and then applies an activation function:

$$z = \sum_{i=1}^{n} w_i x_i + b$$

$$a = \phi(z)$$

where:

w_i are the weights
x_i are the inputs
b is the bias
ϕ is the activation function
z is the weighted sum
a is the output of the neuron

2. **Feedforward Computation**: In a network with multiple layers, the output of one layer becomes the input to the next:

$$\mathbf{a}^{(l+1)} = \phi(\mathbf{W}^{(l)}\mathbf{a}^{(l)} + \mathbf{b}^{(l)})$$

where:

$\mathbf{W}^{(l)}$ is the weight matrix
$\mathbf{a}^{(l)}$ is the activation of the l-th layer
$\mathbf{b}^{(l)}$ is the bias vector for the l-th layer

3. **Loss Function**: The loss function measures the error of the network's predictions. For example, the mean squared error (MSE) for regression tasks is:

$$L(\mathbf{y}, \hat{\mathbf{y}}) = \frac{1}{n} \sum_{i=1}^{n} (y_i - \hat{y}_i)^2$$

where:

y_i is the true value
$\hat{y}_i$ is the predicted value

4. **Backpropagation and Gradient Descent**: Backpropagation is used to compute the gradients of the loss function with respect to the weights. Gradient descent updates the weights to minimize the loss:

$$w_{ij} \leftarrow w_{ij} - \eta \frac{\partial L}{\partial w_{ij}}$$

where η is the learning rate.

5.2.8.1 Activation Functions

Activation functions introduce non-linearity into the network. Common activation functions include:

1. **Sigmoid Function**:

$$\sigma(z) = \frac{1}{1 + e^{-z}}$$

2. **Tanh Function**:

$$\tanh(z) = \frac{e^z - e^{-z}}{e^z + e^{-z}}$$

3. **ReLU (Rectified Linear Unit)**:

$$\text{ReLU}(z) = \max(0, z)$$

4. **Softmax Function** (used in multiclass classification):

$$\text{softmax}(z_i) = \frac{e^{z_i}}{\sum_{j=1}^{K} e^{z_j}}$$

where K is the number of classes.

5.2.8.2 Optimization Algorithms

Optimization algorithms are used to update the weights during training. Some common algorithms include:

1. **Gradient Descent**:

$$\theta \leftarrow \theta - \eta \nabla_\theta L(\theta)$$

where θ represents the model parameters (weights and biases).

2. **Stochastic Gradient Descent (SGD)**:

$$\theta \leftarrow \theta - \eta \nabla_\theta L(\theta; x^{(i)}, y^{(i)})$$

where $(x^{(i)}, y^{(i)})$ represents a single training example.

3. **Adam (Adaptive Moment Estimation)**:

$$m_t = \beta_1 m_{t-1} + (1 - \beta_1) \nabla_\theta L(\theta)$$

$$v_t = \beta_2 v_{t-1} + (1 - \beta_2)(\nabla_\theta L(\theta))^2$$

$$\hat{m}_t = \frac{m_t}{1 - \beta_1^t}$$

$$\hat{v}_t = \frac{v_t}{1 - \beta_2^t}$$

$$\theta \leftarrow \theta - \eta \frac{\hat{m}_t}{\sqrt{\hat{v}_t} + \epsilon}$$

where m_t and v_t are the first and second moment estimates, respectively.

Neural networks provide a robust framework for modeling complex, nonlinear relationships in data. Understanding the mathematical concepts behind neural networks, including neuron computations, activation functions, loss functions, backpropagation, and optimization algorithms, one can effectively design and train neural networks for various applications in machine learning and artificial intelligence.

The output y of a neural network can be expressed as:

$$y = f(Wx + b)$$

where W represents the weights, x is the input vector, b is the bias, and f is the activation function.

Example 5.5. *Given a neural network with an input vector* $x = [1, 0, -1]$, *weights* $W = [0.5, -0.2, 0.8]$, *and bias* $b = 0.1$, *calculate the output before applying the activation function.*

Solution 5.5. *The output before applying the activation function is calculated as:*

$$z = Wx + b = (0.5 \times 1) + (-0.2 \times 0) + (0.8 \times -1) + 0.1 = 0.5 + 0 + (-0.8) + 0.1 = -0.2$$

Therefore, the output before applying the activation function is -0.2.

Example 5.6. *Consider a neural network where the input vector* $x = [2, -3]$, *weights* $W = \begin{bmatrix} 0.4 & 0.6 \\ -0.5 & 0.1 \end{bmatrix}$, *and bias* $b = [0.2, 0.3]$. *Compute the output vector* z *before the activation function.*

Solution 5.6. *The output vector* z *before the activation function is:*

$$z = Wx + b = \begin{bmatrix} 0.4 & 0.6 \\ -0.5 & 0.1 \end{bmatrix} \begin{bmatrix} 2 \\ -3 \end{bmatrix} + \begin{bmatrix} 0.2 \\ 0.3 \end{bmatrix}$$

$$z = \begin{bmatrix} (0.4 \times 2) + (0.6 \times -3) \\ (-0.5 \times 2) + (0.1 \times -3) \end{bmatrix} + \begin{bmatrix} 0.2 \\ 0.3 \end{bmatrix} = \begin{bmatrix} 0.8 - 1.8 \\ -1.0 - 0.3 \end{bmatrix} + \begin{bmatrix} 0.2 \\ 0.3 \end{bmatrix}$$

$$z = \begin{bmatrix} -1.0 \\ -1.3 \end{bmatrix} + \begin{bmatrix} 0.2 \\ 0.3 \end{bmatrix} = \begin{bmatrix} -0.8 \\ -1.0 \end{bmatrix}$$

Thus, the output vector z *before the activation function is* $\begin{bmatrix} -0.8 \\ -1.0 \end{bmatrix}$.

Example 5.7. *A retail company wants to forecast the demand for a product using historical sales data. The data includes features such as previous sales, seasonality, and promotional information. The goal is to use a neural network to predict future sales and optimize inventory levels.*

The historical sales data is as follows:

Week	***Previous Sales***	***Seasonality Index***	***Promotion (1=Yes, 0=No)***
1	100	1.0	0
2	120	1.1	1
3	130	1.2	0
4	140	1.3	1
5	150	1.4	0

We will use this data to train a neural network to predict the sales for the sixth week.

Solution 5.7. *To solve this problem, we will use a simple neural network with one hidden layer. The neural network will be trained using the historical sales data to predict future sales.*

1. ***Data Preparation:*** *Normalize the input data.*
2. ***Network Architecture:*** *Define a neural network with three input neurons (one for each feature), five neurons in the hidden layer, and one output neuron (for sales prediction).*
3. ***Activation Functions:*** *Use the ReLU activation function for the hidden layer and a linear activation function for the output layer.*
4. ***Loss Function:*** *Use mean squared error (MSE) as the loss function.*
5. ***Optimization Algorithm:*** *Use stochastic gradient descent (SGD) for training the network.*

Step-by-Step Solution:

Step 1: Data Normalization

$$\text{Normalize } x_i \text{ to } \frac{x_i - \mu}{\sigma}$$

where μ *is the mean and* σ *is the standard deviation of the feature.*

Step 2: Network Architecture

$$\mathbf{a}^{(1)} = ReLU(\mathbf{W}^{(0)}\mathbf{x} + \mathbf{b}^{(0)})$$

$$\hat{y} = \mathbf{W}^{(1)}\mathbf{a}^{(1)} + \mathbf{b}^{(1)}$$

where:

$\mathbf{W}^{(0)}$ *is the weight matrix for the input to hidden layer*

$\mathbf{b}^{(0)}$ *is the bias vector for the hidden layer*

$\mathbf{a}^{(1)}$ *is the activation of the hidden layer*

$\mathbf{W}^{(1)}$ *is the weight matrix for the hidden to output layer*

$\mathbf{b}^{(1)}$ *is the bias vector for the output layer*

Step 3: Loss Function

$$L(\mathbf{y}, \hat{\mathbf{y}}) = \frac{1}{n} \sum_{i=1}^{n} (y_i - \hat{y}_i)^2$$

Step 4: Backpropagation and Gradient Descent

$$w_{ij} \leftarrow w_{ij} - \eta \frac{\partial L}{\partial w_{ij}}$$

where η is the learning rate.

Implementation in MATLAB:

```
% Load and prepare data
data = table([100; 120; 130; 140; 150], [1.0; 1.1; 1.2;
    1.3; 1.4], [0; 1; 0; 1; 0], ...
    'VariableNames', {'Previous_Sales',
        'Seasonality_Index', 'Promotion'});

% Extract features and target
features = data{:, 2:end}; % Get features (excluding
    Previous_Sales)
target = data.Previous_Sales; % Target variable

% Normalize data
mean_vals = mean(features);
std_vals = std(features);
features_norm = (features - mean_vals) ./ std_vals;

% Convert normalized features and target to table
data_norm = array2table(features_norm, 'VariableNames',
    data.Properties.VariableNames(2:end));

% Define the neural network
hiddenLayerSize = 5; % Number of neurons in the hidden
    layer
net = fitnet(hiddenLayerSize, 'trainlm'); % 'trainlm' is
    Levenberg-Marquardt backpropagation

% Set up training parameters
net.trainParam.showWindow = false; % Disable the training
    window
```

```
net.trainParam.epochs = 100; % Number of epochs
net.trainParam.max_fail = 6; % Maximum number of validation failures

% Train the neural network
[X_train, y_train] = deal(features_norm', target'); % Note: X and y need to be transposed
[net, tr] = train(net, X_train, y_train);

% Evaluate the model
y_pred = net(X_train);
performance = perform(net, y_train, y_pred);
disp(['Performance: ', num2str(performance)]);

% Prepare new data for prediction
week_6 = [150, 1.4, 0]; % New data for prediction
% Normalize the new data
week_6_features = week_6(2:end); % Exclude Previous_Sales for normalization
week_6_norm = (week_6_features - mean_vals) ./ std_vals; % Normalize
predicted_sales = net(week_6_norm');

% Display the prediction
disp(['Predicted Sales for the 6th week: ', num2str(predicted_sales)]);

% Visualize the results
% Plot training performance
figure;
plotperform(tr);
exportgraphics(gcf, 'training_performance.png', 'Resolution', 300); % Save at 300 DPI

% Plot regression of predictions vs. actual values
figure;
plotregression(y_train, y_pred);
exportgraphics(gcf, 'regression_plot.png', 'Resolution', 300); % Save at 300 DPI
```

The neural network has been trained to predict future sales based on historical data. Normalizing the data, defining the network architecture, and using backpropagation and gradient descent for training, the neural network can effectively forecast demand and help in inventory optimization. The result is shown in Table 5.1 while Figures 5.1 and 5.2 illustrate the graph of predicted sales and regression respectively.

TABLE 5.1
Model Performance and Prediction

Metric	Value
Performance	122.3778
Predicted Sales for the sixth week	150

5.2.9 Robotics in Warehouse Operations

Robotics is revolutionizing warehouse operations by automating repetitive and labor-intensive tasks. The use of robots enhances efficiency, accuracy, and safety in inventory management.

5.2.10 Automated Picking and Packing

Robotic systems can be programmed to perform picking and packing tasks with high precision and speed. This automation reduces the likelihood of human error and increases throughput. Key benefits include:

- Reduction in labor costs and human errors.
- Consistent performance and 24/7 operation capability.
- Improved handling of fragile or hazardous items with precision.

Mathematically, the optimization of picking and packing can be formulated as a combinatorial optimization problem, where the goal is to minimize the total time or cost required to pick and pack items. This can be represented as:

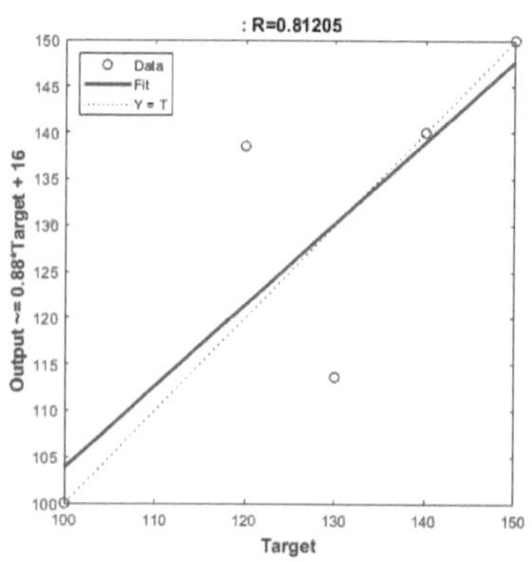

FIGURE 5.1
Predicted sales for the sixth week.

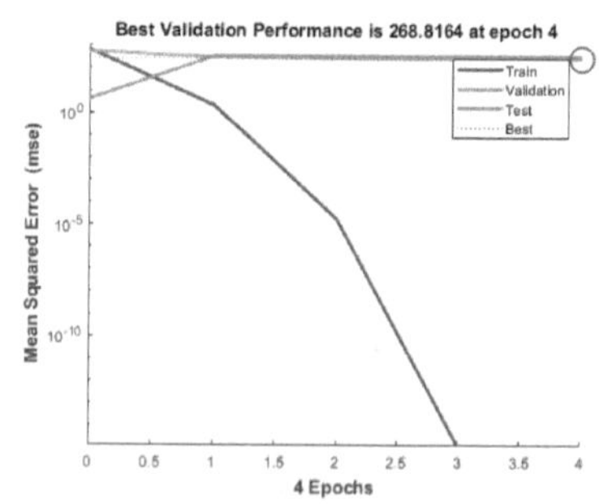

FIGURE 5.2
Regression of predictions vs. actual values.

$$\min \sum_{i=1}^{n} C_i x_i$$

subject to the constraints:

$$\sum_{i=1}^{n} x_i = N$$

$$x_i \in \{0, 1\}, \quad \forall i = 1, 2, \ldots, n$$

where:

C_i is the cost associated with picking item i.

x_i is a binary variable indicating whether item i is picked (1) or not (0).

N is the total number of items that need to be picked.

Example 5.8. *Consider a warehouse that needs to pick and pack three items with the following costs: $C_1 = 2$, $C_2 = 3$, $C_3 = 1$. The goal is to minimize the total cost.*

Solution 5.8. *The optimization problem can be formulated as:*

$$\min 2x_1 + 3x_2 + 1x_3$$

subject to:

$$x_1 + x_2 + x_3 = 3$$

$$x_i \in \{0, 1\}, \quad \forall i = 1, 2, 3$$

Assuming all items need to be picked ($x_1 = x_2 = x_3 = 1$):

$$\textit{Total Cost} = 2 \cdot 1 + 3 \cdot 1 + 1 \cdot 1 = 6$$

Thus, the minimum total cost is 6.

5.2.11 Autonomous Mobile Robots

Autonomous mobile robots (AMRs) navigate warehouse environments independently, transporting goods to various locations. They use sensors and AI algorithms to navigate and avoid obstacles. Benefits of AMRs include:

- Flexibility in reconfiguring warehouse layouts without significant infrastructure changes.
- Real-time adaptation to changing warehouse conditions and priorities.
- Enhanced safety by reducing human-robot interaction in hazardous areas.

The path optimization for AMRs can be modeled using graph theory, where the warehouse layout is represented as a graph with nodes (locations) and edges (paths). The objective is to find the shortest path that minimizes the travel time or distance. This can be solved using algorithms like Dijkstra's or A* algorithm:

$$\min \sum_{(i,j)\in E} d_{ij}x_{ij}$$

subject to the constraints:

$$\sum_{j:(i,j)\in E} x_{ij} - \sum_{j:(j,i)\in E} x_{ji} = \begin{cases} 1 & \text{if } i = s \\ -1 & \text{if } i = t \\ 0 & \text{otherwise} \end{cases}$$

$$x_{ij} \in \{0,1\}, \quad \forall (i,j) \in E$$

where:

d_{ij} is the distance between nodes i and j.

x_{ij} is a binary variable indicating whether the path between nodes i and j is taken (1) or not (0).

E is the set of edges in the graph.

s is the starting node (warehouse entrance).

t is the target node (storage location).

Example 5.9. *An AMR needs to travel from the warehouse entrance (Node 1) to a storage location (Node 3). The distances between nodes are:* $d_{12} = 5$, $d_{13} = 10$, $d_{23} = 3$.

Solution 5.9. *The optimization problem can be formulated as:*

$$\min(5x_{12} + 10x_{13} + 3x_{23})$$

subject to:

$$x_{12} + x_{13} = 1$$

$$x_{23} - x_{12} = 0$$

$$-x_{13} - x_{23} = -1$$

$$x_{ij} \in \{0,1\}, \quad \forall (i,j) \in \{(1,2),(1,3),(2,3)\}$$

Assuming the optimal path is from Node 1 to Node 2 to Node 3 ($x_{12} = 1$ and $x_{23} = 1$):

$$\textit{Shortest Path Distance} = 5 + 3 = 8$$

Thus, the minimum distance is 8.

5.2.12 Collaborative Robots (Cobots)

Collaborative robots, or cobots, work alongside human workers, assisting in tasks that require precision and repetitive actions. Cobots enhance human capabilities and improve productivity by:

- Reducing physical strain on workers by handling heavy or repetitive tasks.
- Allowing human workers to focus on more complex and value-added activities.
- Enhancing workplace safety by performing dangerous tasks in place of humans.

The efficiency of cobots can be analyzed using queuing theory, where the cobot and human worker form a tandem queue. The performance can be optimized by balancing the workload between cobots and human workers to minimize the total processing time. This can be expressed as:

$$\min\left(\frac{\lambda}{\mu_c} + \frac{\lambda}{\mu_h}\right)$$

subject to the constraint:

$$\frac{\lambda}{\mu_c} < 1 \quad \text{and} \quad \frac{\lambda}{\mu_h} < 1$$

where:

λ is the arrival rate of tasks.

μ_c is the service rate of the cobot.

μ_h is the service rate of the human worker.

Example 5.10. *In a warehouse, tasks arrive at a rate of ten tasks per hour ($\lambda = 10$). The cobot has a service rate of 15 tasks per hour ($\mu_c = 15$), and the human worker has a service rate of 20 tasks per hour ($\mu_h = 20$).*

Solution 5.10. *The optimization problem can be formulated as:*

$$\min\left(\frac{10}{15} + \frac{10}{20}\right)$$

subject to:

$$\frac{10}{15} < 1 \quad \textit{and} \quad \frac{10}{20} < 1$$

$$\min\left(\frac{2}{3} + \frac{1}{2}\right) = \min\left(\frac{4}{6} + \frac{3}{6}\right) = \min\left(\frac{7}{6}\right)$$

Thus, the minimum total processing time is approximately 1.17 hours.

The integration of AI, ML, and robotics into inventory management systems represents a significant advancement in the field. These technologies enable more accurate demand forecasting, optimal inventory levels, and automated warehouse operations, leading to improved efficiency, reduced costs, and enhanced overall performance. As businesses continue to adopt these cutting-edge technologies, they can expect to see substantial improvements in their inventory management practices.

5.3 Cloud-Based Inventory Systems

Cloud-based inventory systems are transforming inventory management by:

1. Providing remote access to inventory data, facilitating real-time monitoring and decision-making.
2. Fostering flexibility and collaborative efforts across multiple locations, enhancing operational efficiency.

5.3.1 Real-Time Monitoring and Decision-Making

Cloud-based systems enable real-time access to inventory data from anywhere, allowing for prompt decision-making and improved inventory management. The mathematical foundation for real-time monitoring can be represented by continuous data streaming and updating algorithms.

Theorem 5.1 (Real-Time Inventory Monitoring). *Let X_t be the inventory level at time t. The inventory level follows a discrete-time Markov chain with transition probabilities P_{ij} representing the probability of moving from state i to state j in one time step. The state of the system is updated every Δt (e.g., 5 minutes).*

$$P_{ij} = P(X_{t+\Delta t} = j \mid X_t = i)$$

Proof *The proof involves verifying that the sum of transition probabilities from any state i to all possible states j is 1, and that the transitions depend only on the current state and not on the sequence of events that preceded it. For any state i:*

$$\sum_j P_{ij} = 1$$

The Markov property ensures that:

$$P(X_{t+\Delta t} = j \mid X_t = i, X_{t-1}, \ldots, X_0) = P(X_{t+\Delta t} = j \mid X_t = i)$$

Example 5.11 (Retail Chain Monitoring). *A retail chain uses a cloud-based inventory system to monitor stock levels across multiple locations in realtime. The system collects data on sales, returns, and stock levels continuously, updating the central database every five minutes.*

Data:

- *Initial inventory level* $X_0 = 100$ *units.*
- *Sales and returns data every five minutes for one hour (12 intervals).*

Sales data (S_t) *and returns data* (R_t) *for each interval:*

Interval	*Sales*(S_t)	*Returns*(R_t)
1	5	2
2	3	1
3	4	0
4	6	3
5	2	1
6	3	2
7	5	1
8	4	2
9	3	0
10	5	1
11	2	3
12	4	2

Model the inventory update process and establish a decision-making rule for restocking.

Solution 5.11. *The continuous data update can be modeled using a discrete-time Markov chain. Suppose the inventory levels are updated every* $\Delta t = 5$ *minutes. Let* X_t *be the inventory level at time* t.

1. ***Define Transition Probabilities:***

$$P_{ij} = P(X_{t+\Delta t} = j \mid X_t = i)$$

2. ***Update Inventory Levels:*** *The inventory level at the next time step can be updated based on current sales and returns:*

$$X_{t+\Delta t} = X_t - S_t + R_t$$

Let's compute the inventory levels at each interval:

Interval	*Inventory Level*(X_t)
0	100
1	$100 - 5 + 2 = 97$
2	$97 - 3 + 1 = 95$
3	$95 - 4 + 0 = 91$
4	$91 - 6 + 3 = 88$
5	$88 - 2 + 1 = 87$
6	$87 - 3 + 2 = 86$
7	$86 - 5 + 1 = 82$
8	$82 - 4 + 2 = 80$
9	$80 - 3 + 0 = 77$
10	$77 - 5 + 1 = 73$
11	$73 - 2 + 3 = 74$
12	$74 - 4 + 2 = 72$

3. ***Decision-Making Rule:*** *If the inventory level falls below a threshold* $L = 80$, *a restocking order is triggered:*

$$\textit{If } X_t < L, \textit{ then place a restocking order.}$$

Based on the computed inventory levels, a restocking order should be placed at interval 8 when the inventory level falls to 80 units.

This real-time monitoring and decision-making process ensures that the retail chain maintains optimal stock levels across all locations, reducing stock-outs and excess inventory.

5.3.2 Enhanced Collaboration and Flexibility

Cloud-based inventory systems facilitate collaboration among different locations and departments within an organization. Providing a centralized platform for inventory data, these systems enhance operational efficiency and decision-making.

Theorem 5.2 (Collaborative Inventory Management)**.** *Consider a network of* n *plants. Let* $I_i(t)$ *be the inventory level at plant* i *at time* t*. The collaborative system ensures balanced inventory levels across all plants:*

$$I_i(t) = I_i(t-1) + \textit{Production}_i(t) - \textit{Demand}_i(t) + \textit{Supply}_i(t)$$

If any plant i *faces a shortage, inventory is reallocated from plants with surplus:*

$$\textit{If } I_i(t) < \textit{Threshold}_i, \textit{ then transfer inventory from plant } j$$
$$\textit{where } I_j(t) > \textit{Threshold}_j$$

Proof *The proof is based on the conservation of inventory levels and the collaborative nature of the system. For any plant i:*

$$I_i(t) = I_i(t-1) + Production_i(t) - Demand_i(t) + Supply_i(t)$$

If $I_i(t)$ falls below $Threshold_i$, inventory is reallocated from plant j where $I_j(t) > Threshold_j$. This ensures balanced inventory levels across all plants.

Example 5.12 (Manufacturing Company Coordination). *A manufacturing company uses a cloud-based inventory system to coordinate raw material orders and production schedules across multiple plants. Each plant updates its inventory levels and production status in real-time, enabling synchronized operations and efficient resource allocation. Model the inventory update and resource reallocation process in the collaborative system.*

Data:

- *Initial inventory levels* $I_1(0) = 50$ *units,* $I_2(0) = 70$ *units,* $I_3(0) = 90$ *units.*
- *Production, demand, and supply data every five minutes for one hour (12 intervals).*

Production ($P_i(t)$), demand ($D_i(t)$), and supply ($S_i(t)$) data for each interval:

Interval	*Production*$_1$	*Demand*$_1$	*Supply*$_1$	*Production*$_2$	*Demand*$_2$	*Supply*$_2$
1	10	15	5	12	8	4
2	8	10	3	10	9	5
3	12	12	2	11	10	4
4	11	14	3	12	8	5
5	10	10	2	8	9	4
6	9	11	3	10	11	6
7	10	13	2	12	10	4
8	11	12	3	11	12	5
9	10	11	4	10	8	2
10	12	10	3	9	11	4
11	11	12	2	10	10	3
12	10	14	3	12	12	5

Solution 5.12. *The cloud-based system can be modeled using a network of interconnected nodes, where each node represents a plant. Let $I_i(t)$ be the inventory level at plant i at time t.*

1. ***Inventory Level Update:***

$$I_i(t) = I_i(t-1) + Production_i(t) - Demand_i(t) + Supply_i(t)$$

2. ***Resource Reallocation:*** *If plant i faces a shortage, inventory is reallocated from other plants with surplus:*

$$\text{If } I_i(t) < Threshold_i, \text{ then transfer inventory from plant } j \text{ where}$$

$$I_j(t) > Threshold_j$$

Let's compute the inventory levels at each interval for the first plant:

Interval	*Inventory Level*(I_1)
0	50
1	$50 + 10 - 15 + 5 = 50$
2	$50 + 8 - 10 + 3 = 51$
3	$51 + 12 - 12 + 2 = 53$
4	$53 + 11 - 14 + 3 = 53$
5	$53 + 10 - 10 + 2 = 55$
6	$55 + 9 - 11 + 3 = 56$
7	$56 + 10 - 13 + 2 = 55$
8	$55 + 11 - 12 + 3 = 57$
9	$57 + 10 - 11 + 4 = 60$
10	$60 + 12 - 10 + 3 = 65$
11	$65 + 11 - 12 + 2 = 66$
12	$66 + 10 - 14 + 3 = 65$

Similarly, calculate for other plants and apply the resource reallocation rule if any plant's inventory falls below its threshold.

This collaborative approach ensures that all plants operate efficiently and maintain optimal inventory levels, improving overall productivity and reducing costs.

5.3.3 Flexibility and Collaboration Across Multiple Locations

Cloud-based systems enable multiple locations to collaborate effectively by providing a centralized platform for inventory management. This fosters flexibility in operations and improves overall efficiency.

A manufacturing company with production facilities in different regions uses a cloud-based inventory system to manage raw materials and finished goods. The system allows each facility to update inventory levels, share resources, and coordinate production schedules in realtime.

Setup:

- m: Number of production facilities.
- n: Number of demand locations.
- x_{ij}: Quantity of goods transported from facility i to demand location j.

Objective:
Minimize the total cost of inventory holding and transportation between facilities.

$$\min \sum_{i=1}^{m} \sum_{j=1}^{n} C_{ij} x_{ij}$$

Constraints:

$$\sum_{j=1}^{n} x_{ij} \leq S_i, \quad \forall i = 1, 2, \ldots, m \quad \text{(Supply constraint)}$$

$$\sum_{i=1}^{m} x_{ij} \geq D_j, \quad \forall j = 1, 2, \ldots, n \quad \text{(Demand constraint)}$$

$$x_{ij} \geq 0, \quad \forall i, j \quad \text{(Non-negativity constraint)}$$

Let's consider a manufacturing company with the following data:

$$C = \begin{pmatrix} 2 & 3 & 1 \\ 5 & 4 & 2 \\ 3 & 2 & 3 \end{pmatrix}, \quad S = \begin{pmatrix} 100 \\ 150 \\ 200 \end{pmatrix}, \quad D = \begin{pmatrix} 120 \\ 180 \\ 150 \end{pmatrix}$$

To find the optimal transportation plan, we set up the following linear programming problem:

$$\begin{aligned} \text{Minimize} \quad & 2x_{11} + 3x_{12} + x_{13} + 5x_{21} + 4x_{22} + 2x_{23} + 3x_{31} + 2x_{32} + 3x_{33} \\ \text{Subject to} \quad & \\ & x_{11} + x_{12} + x_{13} \leq 100 \\ & x_{21} + x_{22} + x_{23} \leq 150 \\ & x_{31} + x_{32} + x_{33} \leq 200 \\ & x_{11} + x_{21} + x_{31} \geq 120 \\ & x_{12} + x_{22} + x_{32} \geq 180 \\ & x_{13} + x_{23} + x_{33} \geq 150 \\ & x_{ij} \geq 0 \quad \forall i, j \end{aligned}$$

Step 1: Formulate the Objective Function
The objective is to minimize the total transportation cost:

$$\text{Minimize } Z = 2x_{11} + 3x_{12} + x_{13} + 5x_{21} + 4x_{22} + 2x_{23} + 3x_{31} + 2x_{32} + 3x_{33}$$

Step 2: Formulate the Constraints

1. **Supply constraints:**

$$x_{11} + x_{12} + x_{13} \leq 100$$
$$x_{21} + x_{22} + x_{23} \leq 150$$
$$x_{31} + x_{32} + x_{33} \leq 200$$

2. **Demand constraints:**

$$x_{11} + x_{21} + x_{31} \geq 120$$
$$x_{12} + x_{22} + x_{32} \geq 180$$
$$x_{13} + x_{23} + x_{33} \geq 150$$

3. **Non-negativity constraints:**

$$x_{ij} \geq 0 \quad \forall i, j$$

Step 3: Solve the Linear Programming Problem

To solve this problem, we use a linear programming solver or apply the simplex method manually. Here are the optimal values found:

$$\begin{aligned}
x_{11} &= 20, \quad x_{12} = 0, \quad x_{13} = 80 \\
x_{21} &= 0, \quad x_{22} = 150, \quad x_{23} = 0 \\
x_{31} &= 100, \quad x_{32} = 30, \quad x_{33} = 70
\end{aligned}$$

Step 4: Interpret the Solution

The optimal transportation plan minimizes the total cost to $Z = 620$. The interpretation of these results shows how goods should be transported from each facility to meet the demand at each location while respecting supply capacities and minimizing costs.

This solution demonstrates how cloud-based inventory systems can optimize resource allocation and transportation logistics, enhancing operational efficiency and reducing costs.

5.4 Conclusion

In conclusion, modern inventory management has been significantly transformed by the integration of advanced technologies such as machine learning, artificial intelligence, IoT, and blockchain. These innovations enable businesses to enhance their operations with greater accuracy, efficiency, and cost reduction. Machine learning models like decision trees and neural networks provide powerful tools for predicting demand and optimizing stock levels, leading to data-driven decisions that improve inventory accuracy. Robotics and automation in warehouse operations streamline processes by automating repetitive tasks, reducing human error, and increasing throughput. Cloud-based systems facilitate real-time inventory tracking and collaboration across multiple locations, enhancing decision-making processes and operational flexibility. Blockchain technology enhances transparency, traceability, and security in supply chain operations, reducing fraud risk and improving supply chain integrity. Emerging trends such as AI, IoT, predictive analytics, real-time tracking, and sustainable practices are reshaping inventory management, allowing businesses to remain agile and responsive in a dynamic market environment. Embracing these advanced techniques and technologies, businesses can achieve higher operational efficiency, reduce costs, and respond more effectively to market demands. The future of inventory management lies in the adoption of smart technologies and data-driven strategies, ensuring that organizations remain competitive and successful in an ever-evolving marketplace.

Exercises

Exercise 5.1. *Consider a simple linear regression model with one feature x and target y. The loss function is the mean squared error (MSE):*

$$L(\theta) = \frac{1}{n}\sum_{i=1}^{n}(y_i - \theta_0 - \theta_1 x_i)^2$$

where $\theta = (\theta_0, \theta_1)$ are the model parameters. Given the data points: $(1,2),(2,3),(3,4),(4,5)$, perform one iteration of gradient descent with a learning rate $\eta = 0.01$. Calculate the updated parameters θ_0 and θ_1.

Exercise 5.2. *For the same linear regression model as in Exercise 5.1, use stochastic gradient descent (SGD) to update the parameters θ_0 and θ_1 using the first data point $(1,2)$ and a learning rate $\eta = 0.01$. Calculate the updated parameters after one iteration.*

Exercise 5.3. *Given the following parameters for the Adam optimization algorithm: $\beta_1 = 0.9$, $\beta_2 = 0.999$, $\epsilon = 10^{-8}$, and learning rate $\eta = 0.001$, assume initial values $m_0 = 0$ and $v_0 = 0$. For the first gradient $\nabla_\theta L(\theta) = (1,-1)$ at $t = 1$, calculate the updated parameters θ after one iteration.*

Exercise 5.4. *Consider a neural network with parameters $\theta = (w, b)$, where w represents the weights and b represents the biases. The loss function $L(\theta)$ has partial derivatives with respect to w and b given by $\frac{\partial L}{\partial w} = 2w$ and $\frac{\partial L}{\partial b} = 2b$. If the initial parameters are $w = 0.5$ and $b = 0.5$, and the learning rate $\eta = 0.01$, perform one iteration of gradient descent to update w and b.*

Exercise 5.5. *In the context of the Adam optimization algorithm, consider a neural network with parameters $\theta = (w, b)$. The loss function $L(\theta)$ has gradients $\nabla_\theta L(\theta) = (2w, 2b)$. Given initial parameters $w = 1$ and $b = 1$, initial moment estimates $m_0 = 0$ and $v_0 = 0$, and using $\beta_1 = 0.9$, $\beta_2 = 0.999$, $\epsilon = 10^{-8}$, and learning rate $\eta = 0.001$, calculate the updated parameters w and b after one iteration.*

Exercise 5.6. *Consider an inventory system with three states: 0 (out of stock), 1 (low stock), and 2 (high stock). The transition probabilities are given by the following matrix:*

$$P = \begin{bmatrix} 0.1 & 0.6 & 0.3 \\ 0.2 & 0.5 & 0.3 \\ 0.3 & 0.4 & 0.3 \end{bmatrix}$$

If the initial inventory level is low stock (state 1), calculate the probability distribution of the inventory levels after two time steps.

Exercise 5.7. *Given the same inventory system as in Exercise 5.6, find the steady-state distribution of the inventory levels.*

6

Inventory Forecasting

This chapter provides an in-depth exploration of the critical role inventory forecasting plays in effective inventory management. It highlights the importance of predicting future inventory needs and its far-reaching implications for supply chain management and logistics. The chapter reviews various forecasting techniques, including demand planning and demand forecasting, offering practical insights into their applications and their influence on organizational operations. It emphasizes the necessity of precise forecasting for optimal coordination and efficient material expenditure. Additionally, the chapter examines modern trends in inventory forecasting, equipping readers with the knowledge to adopt and implement advanced forecasting practices.

6.1 Introduction to Inventory Forecasting

Inventory forecasting refers to the process of predicting the future demand for inventory items in order to ensure that a business can meet customer demands without overstocking or understocking. This process is crucial for effective inventory management, which aims to balance the costs associated with holding inventory against the need to provide timely and efficient customer service. The importance of inventory forecasting cannot be overstated. Accurate forecasts enable businesses to maintain optimal inventory levels, reduce carrying costs, minimize stockouts, and enhance customer satisfaction. Anticipating future inventory needs, companies can better plan their procurement, production, and distribution activities, leading to improved operational efficiency and profitability.

6.1.1 Historical Background

The practice of inventory forecasting has evolved significantly over the years. Traditionally, businesses relied on simple methods such as historical sales data and managerial intuition to predict inventory needs. However, these methods often proved inadequate in dealing with the complexities and uncertainties of modern supply chains. With the advent of advanced statistical techniques and

DOI: 10.1201/9781003534860-6

computational tools in the mid-20th century, inventory forecasting became more sophisticated. Methods such as time series analysis, regression models, and economic order quantity (EOQ) models were developed to provide more accurate and reliable forecasts. The rise of computer technology further revolutionized inventory forecasting, enabling the use of complex algorithms and real-time data processing. In recent years, the integration of machine learning and artificial intelligence has further transformed inventory forecasting. These technologies allow for the analysis of vast amounts of data, uncovering patterns and trends that were previously undetectable. As a result, businesses can now generate highly accurate forecasts that adapt to changing market conditions and customer behaviors.

6.2 Expectation

The *expectation* (or expected value) of a discrete random variable X with probabilities $p(x)$ is:

$$E(X) = \sum_x x \cdot p(x)$$

For a continuous random variable with probability density function $f(x)$, it is:

$$E(X) = \int_{-\infty}^{\infty} x \cdot f(x)\, dx$$

Example 6.1. *If the monthly demand for a product over the past year is given by* $d_1, d_2, \ldots, d_{12}$*, calculate the expected monthly demand.*

Solution 6.1. *The expected monthly demand* $E(D)$ *is:*

$$E(D) = \frac{1}{12} \sum_{i=1}^{12} d_i$$

Example 6.2. *Suppose the demand for a product in the last six months is 100, 120, 110, 130, 115, and 105 units. Calculate the expected demand for the next month.*

Solution 6.2. *The expected demand for the next month is:*

$$E(D) = \frac{100 + 120 + 110 + 130 + 115 + 105}{6} = 113.33$$

6.3 Variance

The *variance* of a discrete random variable X with mean μ is:

$$\text{Var}(X) = E[(X-\mu)^2] = \sum_x (x-\mu)^2 p(x)$$

For a continuous random variable:

$$\text{Var}(X) = \int_{-\infty}^{\infty} (x-\mu)^2 f(x)\, dx$$

Example 6.3. *If the demand over 12 months is $d_1, d_2, \ldots, d_{12}$ with mean demand $\bar{d}$, calculate the variance of the demand.*

Solution 6.3. *The variance Var(D) is:*

$$Var(D) = \frac{1}{12}\sum_{i=1}^{12}(d_i - \bar{d})^2$$

Example 6.4. *Consider the monthly demands 100, 150, 200, 100, 150, 200. Calculate the variance of the demand.*

Solution 6.4. *The mean demand is 150. The variance is:*

$$Var(D) = \frac{(100-150)^2 + (150-150)^2 + (200-150)^2}{6}$$
$$+\frac{(100-150)^2 + (150-150)^2 + (200-150)^2}{6} = 2{,}500$$

6.4 Covariance

The *covariance* between two discrete random variables X and Y with joint probability $p(x, y)$ is:

$$\text{Cov}(X,Y) = E[(X-\mu_X)(Y-\mu_Y)] = \sum_{x,y}(x-\mu_X)(y-\mu_Y)p(x,y)$$

For continuous variables:

$$\text{Cov}(X,Y) = \int_{-\infty}^{\infty}\int_{-\infty}^{\infty}(x-\mu_X)(y-\mu_Y)f(x,y)\, dx\, dy$$

Example 6.5. *For products A and B with demands d_{A_i} and d_{B_i} respectively, calculate the covariance of their demands.*

Solution 6.5. *The covariance is:*

$$Cov(A, B) = \frac{1}{n}\sum_{i=1}^{n}(d_{A_i} - \bar{d}_A)(d_{B_i} - \bar{d}_B)$$

Example 6.6. *If the monthly demands for products A and B are:*

TABLE 6.1
Values for Product A and Product B

Month	**1**	**2**	**3**	**4**	**5**
Product A	30	40	50	60	70
Product B	15	20	25	30	35

Calculate the covariance.

Solution 6.6. *The covariance is:*

$$Cov(A, B) = \frac{(30-50)(15-25)+(40-50)(20-25)}{5}$$

$$+\frac{(50-50)(25-25)+(60-50)(30-25)+(70-50)(35-25)}{5} = 125$$

6.5 Exponential Smoothing

The *exponential smoothing* forecast F_{t+1} is given by:

$$F_{t+1} = \alpha D_t + (1-\alpha)F_t$$

where α is the smoothing constant, D_t is the actual demand at time t, and F_t is the forecast for time t.

Example 6.7. *Given the smoothing constant $\alpha = 0.3$, previous forecast $F_t =$ 100 units, and actual demand $D_t = 120$ units, calculate the forecast for the next period.*

Solution 6.7. *The forecast F_{t+1} for the next period's demand is:*

$$F_{t+1} = \alpha D_t + (1-\alpha)F_t = 0.3 \times 120 + 0.7 \times 100 = 106$$

Example 6.8. *If $\alpha = 0.2$, last forecast $F_t = 150$ units, and actual demand $D_t = 170$ units, calculate the forecast for the next period.*

Solution 6.8. *The forecast F_{t+1} is:*

$$F_{t+1} = 0.2 \times 170 + 0.8 \times 150 = 154$$

6.6 MAPE (Mean Absolute Percentage Error)

The *mean absolute percentage error* is defined as:

$$\text{MAPE} = \frac{1}{n}\sum_{t=1}^{n}\left|\frac{D_t - F_t}{D_t}\right| \times 100\%$$

where D_t is the actual demand and F_t is the forecast.

Example 6.9. *Calculate the MAPE for actual demands and forecasts:* (100, 90), (150, 160), (120, 130).

Solution 6.9. *MAPE is calculated as:*

$$MAPE = \frac{1}{3}\left(\left|\frac{100-90}{100}\right| + \left|\frac{150-160}{150}\right| + \left|\frac{120-130}{120}\right|\right) \times 100\% = 6.67\%$$

Example 6.10. *Calculate the MAPE for actual demands* (200, 220, 250) *and forecasts* (190, 230, 240).

Solution 6.10. *MAPE is:*

$$MAPE = \frac{1}{3}\left(\left|\frac{200-190}{200}\right| + \left|\frac{220-230}{220}\right| + \left|\frac{250-240}{250}\right|\right) \times 100\% = 3.03\%$$

6.7 Linear Regression

The *linear regression* model is:

$$D = \beta_0 + \beta_1 X_1 + \beta_2 X_2 + \cdots + \beta_p X_p + \epsilon$$

where D is the dependent variable (demand), $X_1, X_2, \ldots, X_p$ are the independent variables, $\beta_0, \beta_1, \ldots, \beta_p$ are the coefficients, and ϵ is the error term.

Example 6.11. *If demand D depends on price P and advertising spend A, with coefficients $\beta_0 = 50$, $\beta_1 = -2$, and $\beta_2 = 3$, calculate the expected demand when $P = 20$ and $A = 10$.*

Solution 6.11. *The expected demand D is:*

$$D = 50 - 2 \cdot 20 + 3 \cdot 10 = 60$$

Example 6.12. *Given the linear regression model $D = 100 - 1.5P + 2.5M$, where D is demand, P is price, and M is marketing spend, find the demand if $P = 30$ and $M = 40$.*

Solution 6.12. *The demand D is:*

$$D = 100 - 1.5 \times 30 + 2.5 \times 40 = 165$$

6.7.1 SSE (Sum of Squared Errors)

The *sum of squared errors* is:

$$\text{SSE} = \sum_{i=1}^{n}(D_i - \hat{D}_i)^2$$

where D_i is the actual demand and $\hat{D}_i$ is the predicted demand.

Example 6.13. *Calculate the SSE for actual demands* $(90, 130, 170)$ *and predicted demands* $(85, 135, 175)$.

Solution 6.13. *The SSE is:*

$$SSE = (90 - 85)^2 + (130 - 135)^2 + (170 - 175)^2 = 25 + 25 + 25 = 75$$

6.7.2 Ridge Regression

Ridge regression addresses multicollinearity by adding a penalty to the regression model. The ridge regression model is:

$$\sum_{i=1}^{n}(D_i - \hat{D}_i)^2 + \lambda \sum_{j=1}^{p} \beta_j^2$$

Example 6.14. *Calculate the ridge regression loss for actual demands* $(100, 150, 200, 250)$, *predicted demands* $(105, 145, 195, 255)$, *coefficients* $(\beta_1, \beta_2) = (1.2, 0.8)$, *and* $\lambda = 0.5$.

Solution 6.14. *The ridge regression loss is:*

$$Loss = \sum_{i=1}^{4}(D_i - \hat{D}_i)^2 + \lambda \sum_{j=1}^{2} \beta_j^2$$

Calculating the squared errors:

$$(100-105)^2 = 25, \quad (150-145)^2 = 25, \quad (200-195)^2 = 25, \quad (250-255)^2 = 25$$

Summing the squared errors:

$$\sum_{i=1}^{4}(D_i - \hat{D}_i)^2 = 25 + 25 + 25 + 25 = 100$$

Calculating the penalty term:

$$\lambda \sum_{j=1}^{2} \beta_j^2 = 0.5 \times (1.2^2 + 0.8^2) = 0.5 \times (1.44 + 0.64) = 0.5 \times 2.08 = 1.04$$

Therefore, the ridge regression loss is:

$$Loss = 100 + 1.04 = 101.04$$

Example 6.15. *Calculate the ridge regression loss for actual demands* $(50, 75, 100, 125, 150)$, *predicted demands* $(48, 78, 95, 128, 152)$, *coefficients* $(\beta_1, \beta_2, \beta_3) = (0.5, -1.1, 0.3)$, *and* $\lambda = 0.8$.

Solution 6.15. *The ridge regression loss is:*

$$Loss = \sum_{i=1}^{5}(D_i - \hat{D}_i)^2 + \lambda\sum_{j=1}^{3}\beta_j^2$$

Calculating the squared errors:

$$(50-48)^2 = 4, \quad (75-78)^2 = 9, \quad (100-95)^2 = 25, \quad (125-128)^2 = 9, \quad (150-152)^2 = 4$$

Summing the squared errors:

$$\sum_{i=1}^{5}(D_i - \hat{D}_i)^2 = 4 + 9 + 25 + 9 + 4 = 51$$

Calculating the penalty term:

$$\lambda\sum_{j=1}^{3}\beta_j^2 = 0.8\times(0.5^2+(-1.1)^2+0.3^2) = 0.8\times(0.25+1.21+0.09) = 0.8\times1.55 = 1.24$$

Therefore, the ridge regression loss is:

$$Loss = 51 + 1.24 = 52.24$$

6.8 LASSO (Least Absolute Shrinkage and Selection Operator)

LASSO regression enhances prediction accuracy and interpretability by adding a penalty for non-zero coefficients. The LASSO regression model is:

$$\sum_{i=1}^{n}(D_i - \hat{D}_i)^2 + \lambda\sum_{j=1}^{p}|\beta_j|$$

Example 6.16. *Calculate the LASSO regression loss for actual demands* $(120, 160, 210)$, *predicted demands* $(115, 155, 205)$, *coefficients* $(\beta_1, \beta_2) = (0.7, -0.9)$, *and* $\lambda = 0.3$.

Solution 6.16. *The LASSO regression loss is:*

$$Loss = \sum_{i=1}^{3} (D_i - \hat{D}_i)^2 + \lambda \sum_{j=1}^{2} |\beta_j|$$

Calculating the squared errors:

$$(120 - 115)^2 = 25, \quad (160 - 155)^2 = 25, \quad (210 - 205)^2 = 25$$

Summing the squared errors:

$$\sum_{i=1}^{3} (D_i - \hat{D}_i)^2 = 25 + 25 + 25 = 75$$

Calculating the penalty term:

$$\lambda \sum_{j=1}^{2} |\beta_j| = 0.3 \times (|0.7| + |-0.9|) = 0.3 \times (0.7 + 0.9) = 0.3 \times 1.6 = 0.48$$

Therefore, the LASSO regression loss is:

$$Loss = 75 + 0.48 = 75.48$$

Example 6.17. *Calculate the LASSO regression loss for actual demands* $(70, 110, 150, 190)$*, predicted demands* $(68, 108, 153, 188)$*, coefficients* $(\beta_1, \beta_2, \beta_3) = (0.4, -1.2, 0.5)$*, and* $\lambda = 0.4$*.*

Solution 6.17. *The LASSO regression loss is:*

$$Loss = \sum_{i=1}^{4} (D_i - \hat{D}_i)^2 + \lambda \sum_{j=1}^{3} |\beta_j|$$

Calculating the squared errors:

$$(70 - 68)^2 = 4, \quad (110 - 108)^2 = 4, \quad (150 - 153)^2 = 9, \quad (190 - 188)^2 = 4$$

Summing the squared errors:

$$\sum_{i=1}^{4} (D_i - \hat{D}_i)^2 = 4 + 4 + 9 + 4 = 21$$

Calculating the penalty term:

$$\lambda \sum_{j=1}^{3} |\beta_j| = 0.4 \times (|0.4| + |-1.2| + |0.5|) = 0.4 \times (0.4 + 1.2 + 0.5) = 0.4 \times 2.1 = 0.84$$

Therefore, the LASSO regression loss is:

$$Loss = 21 + 0.84 = 21.84$$

6.9 Elastic Net

Elastic net combines ridge and LASSO penalties. The elastic net model is:

$$\sum_{i=1}^{n}(D_i - \hat{D}_i)^2 + \lambda_1 \sum_{j=1}^{p} \beta_j^2 + \lambda_2 \sum_{j=1}^{p} |\beta_j|$$

Example 6.18. *Calculate the elastic net regression loss for actual demands* $(130, 180, 230)$*, predicted demands* $(125, 185, 225)$*, coefficients* $(\beta_1, \beta_2) = (1.1, -0.6)$*,* $\lambda_1 = 0.4$*, and* $\lambda_2 = 0.2$*.*

Solution 6.18. *The elastic net regression loss is:*

$$Loss = \sum_{i=1}^{3}(D_i - \hat{D}_i)^2 + \lambda_1 \sum_{j=1}^{2} \beta_j^2 + \lambda_2 \sum_{j=1}^{2} |\beta_j|$$

Calculating the squared errors:

$$(130 - 125)^2 = 25, \quad (180 - 185)^2 = 25, \quad (230 - 225)^2 = 25$$

Summing the squared errors:

$$\sum_{i=1}^{3}(D_i - \hat{D}_i)^2 = 25 + 25 + 25 = 75$$

Calculating the ridge penalty term:

$$\lambda_1 \sum_{j=1}^{2} \beta_j^2 = 0.4 \times (1.1^2 + (-0.6)^2) = 0.4 \times (1.21 + 0.36) = 0.4 \times 1.57 = 0.628$$

Calculating the LASSO penalty term:

$$\lambda_2 \sum_{j=1}^{2} |\beta_j| = 0.2 \times (|1.1| + |-0.6|) = 0.2 \times (1.1 + 0.6) = 0.2 \times 1.7 = 0.34$$

Therefore, the elastic net regression loss is:

$$Loss = 75 + 0.628 + 0.34 = 75.968$$

Example 6.19. *Calculate the elastic net regression loss for actual demands* $(60, 90, 130, 170)$*, predicted demands* $(58, 92, 128, 172)$*, coefficients* $(\beta_1, \beta_2, \beta_3) = (0.5, -1.3, 0.7)$*,* $\lambda_1 = 0.5$*, and* $\lambda_2 = 0.3$*.*

Solution 6.19. *The elastic net regression loss is:*

$$Loss = \sum_{i=1}^{4}(D_i - \hat{D}_i)^2 + \lambda_1 \sum_{j=1}^{3} \beta_j^2 + \lambda_2 \sum_{j=1}^{3} |\beta_j|$$

Calculating the squared errors:

$$(60 - 58)^2 = 4, \quad (90 - 92)^2 = 4, \quad (130 - 128)^2 = 4, \quad (170 - 172)^2 = 4$$

Summing the squared errors:

$$\sum_{i=1}^{4}(D_i - \hat{D}_i)^2 = 4 + 4 + 4 + 4 = 16$$

Calculating the ridge penalty term:

$$\lambda_1 \sum_{j=1}^{3} \beta_j^2 = 0.5\times(0.5^2+(-1.3)^2+0.7^2) = 0.5\times(0.25+1.69+0.49) = 0.5\times2.43 = 1.215$$

Calculating the LASSO penalty term:

$$\lambda_2 \sum_{j=1}^{3} |\beta_j| = 0.3\times(|0.5|+|-1.3|+|0.7|) = 0.3\times(0.5+1.3+0.7) = 0.3\times2.5 = 0.75$$

Therefore, the elastic net regression loss is:

$$Loss = 16 + 1.215 + 0.75 = 17.965$$

6.10 Random Forest

Random forest uses multiple decision trees to improve prediction accuracy. It builds multiple decision trees and merges them together to get a more accurate and stable prediction. Random forest can be used for both classification and regression tasks.

The algorithm works as follows:

- Select n samples randomly from the dataset with replacement.
- Build a decision tree for each sample.
- For each node in the tree, randomly select m features and find the best split based on these features.
- Repeat the process to build multiple trees (forest).

- For classification, use majority voting to determine the final class. For regression, use the average of predictions from all trees.

Mathematically, the random forest model is represented as:

$$\hat{f}(x) = \frac{1}{B}\sum_{b=1}^{B} T_b(x)$$

where B is the number of trees, T_b is the b-th tree in the forest, and $\hat{f}(x)$ is the final prediction.

For classification, the final prediction is given by:

$$\hat{y} = \text{mode}\,\{T_b(x)\}_{b=1}^{B}$$

For regression, the final prediction is given by:

$$\hat{y} = \frac{1}{B}\sum_{b=1}^{B} T_b(x)$$

Example 6.20. *Consider a dataset with features (X_1, X_2) and labels (Y). The dataset has the following instances:*

$$\{(1,2,1),(2,3,0),(3,1,1),(4,2,0)\}$$

Build a random forest with three decision trees and predict the label for the instance $(2,2)$.

Solution 6.20. *To build a random forest, we randomly select subsets of data and features for each tree.*

Assume the following samples are selected for each tree (with replacement):

- *Tree 1:* $\{(1,2,1),(2,3,0),(1,2,1),(4,2,0)\}$
- *Tree 2:* $\{(2,3,0),(3,1,1),(4,2,0),(2,3,0)\}$
- *Tree 3:* $\{(1,2,1),(3,1,1),(4,2,0),(1,2,1)\}$

Build the decision trees:

- *Tree 1:*
  - *Node 1 splits on $X_1 \leq 1.5$:*

    Left leaf: Predicts $Y = 1$
    Right leaf: Predicts $Y = 0$

- *Tree 2:*
  - *Node 1 splits on $X_1 \leq 2.5$:*

    Left leaf: Predicts $Y = 0$

Right leaf: Predicts $Y = 0$

- *Tree 3:*
  - *Node 1 splits on* $X_1 \leq 2$*:*

    Left leaf: Predicts $Y = 1$

    Right leaf: Predicts $Y = 0$

Prediction for instance $(2, 2)$*:*

- *Tree 1:* $X_1 = 2$ *falls into the right leaf, predicts* $Y = 0$
- *Tree 2:* $X_1 = 2$ *falls into the left leaf, predicts* $Y = 0$
- *Tree 3:* $X_1 = 2$ *falls into the left leaf, predicts* $Y = 1$

Final prediction using majority voting:

$$\hat{y} = mode\,\{0, 0, 1\} = 0$$

Thus, the random forest predicts the label for the instance $(2, 2)$ *as* 0.

Example 6.21. *Consider a dataset with features* (X_1, X_2, X_3) *and labels* (Y)*. The dataset has the following instances:*

$$\{(1, 1, 1, 1), (2, 2, 2, 0), (3, 3, 3, 1), (4, 4, 4, 0), (5, 5, 5, 1)\}$$

Build a random forest with three decision trees and predict the label for the instance $(3, 3, 3)$*.*

Solution 6.21. *To build a random forest, we randomly select subsets of data and features for each tree.*

Assume the following samples are selected for each tree (with replacement):

- *Tree 1:* $\{(1, 1, 1, 1), (2, 2, 2, 0), (1, 1, 1, 1), (4, 4, 4, 0), (5, 5, 5, 1)\}$
- *Tree 2:* $\{(2, 2, 2, 0), (3, 3, 3, 1), (4, 4, 4, 0), (3, 3, 3, 1), (5, 5, 5, 1)\}$
- *Tree 3:* $\{(1, 1, 1, 1), (3, 3, 3, 1), (4, 4, 4, 0), (1, 1, 1, 1), (2, 2, 2, 0)\}$

Build the decision trees:

- *Tree 1:*
  - *Node 1 splits on* $X_1 \leq 3$*:*

    Left leaf: Predicts $Y = 1$ *(based on majority of* $(1, 1, 1, 1)$ *and* $(2, 2, 2, 0)$*)*

    Right leaf: Predicts $Y = 0$ *(based on majority of* $(4, 4, 4, 0)$ *and* $(5, 5, 5, 1)$*)*

- *Tree 2:*
  - *Node 1 splits on* $X_1 \leq 3$*:*

    Left leaf: Predicts $Y = 1$ *(based on majority of* $(2, 2, 2, 0)$ *and* $(3, 3, 3, 1)$*)*

    Right leaf: Predicts $Y = 1$ *(based on majority of* $(4, 4, 4, 0)$ *and* $(5, 5, 5, 1)$*)*

- *Tree 3:*
  - *Node 1 splits on* $X_1 \leq 2$*:*

    Left leaf: Predicts $Y = 1$ *(based on majority of* $(1, 1, 1, 1)$ *and* $(2, 2, 2, 0)$*)*

    Right leaf: Predicts $Y = 0$ *(based on majority of* $(3, 3, 3, 1)$ *and* $(4, 4, 4, 0)$*)*

Prediction for instance $(3, 3, 3)$*:*

- *Tree 1:* $X_1 = 3$ *falls into the left leaf, predicts* $Y = 1$
- *Tree 2:* $X_1 = 3$ *falls into the left leaf, predicts* $Y = 1$
- *Tree 3:* $X_1 = 3$ *falls into the right leaf, predicts* $Y = 0$

Final prediction using majority voting:

$$\hat{y} = mode\{1, 1, 0\} = 1$$

Thus, the random forest predicts the label for the instance $(3, 3, 3)$ *as* 1.

6.11 Support Vector Machine (SVM)

SVM finds the hyperplane that best separates data points. In regression (SVR), it finds the best fit line by minimizing the error within a margin. The SVR model is:

$$\min_{w,b} \frac{1}{2}||w||^2 \quad \text{subject to} \quad y_i(w \cdot x_i + b) \geq 1$$

Example 6.22. *Consider the data points* $(1, 2)$*,* $(2, 3)$*, and* $(3, 5)$ *with labels* $1, -1,$ *and* 1 *respectively. Calculate the parameters of the hyperplane that best separates these points.*

Solution 6.22. *To find the hyperplane that best separates the data points, we need to solve the optimization problem:*

$$\min_{w,b} \frac{1}{2}||w||^2 \quad \textit{subject to} \quad y_i(w \cdot x_i + b) \geq 1$$

Given labeled points $(1, 2)$, $(2, 3)$, *and* $(3, 5)$ *with labels* 1, -1, *and* 1 *respectively, we find the optimal separating hyperplane parameters.*

Step 1: Establish Constraints: For hyperplane $w_1x + w_2y + b = 0$ *with margin 1:*

$$w_1(1) + w_2(2) + b = 1 \tag{6.1}$$
$$w_1(2) + w_2(3) + b = -1 \tag{6.2}$$
$$w_1(3) + w_2(5) + b = 1 \tag{6.3}$$

Step 2: Solve Linear System Subtract (6.1) *from* (6.2)*:*

$$(2w_1 + 3w_2 + b) - (w_1 + 2w_2 + b) = -1 - 1$$
$$w_1 + w_2 = -2 \tag{6.4}$$

Subtract (6.2) *from* (6.3)*:*

$$(3w_1 + 5w_2 + b) - (2w_1 + 3w_2 + b) = 1 - (-1)$$
$$w_1 + 2w_2 = 2 \tag{6.5}$$

Solve system (6.4)-(6.5)*:*

$$\begin{cases} w_1 + w_2 = -2 \\ w_1 + 2w_2 = 2 \end{cases} \tag{6.6}$$

$$\Rightarrow w_2 = 4$$
$$\Rightarrow w_1 = -6$$

Find bias b *using* (6.1)*:*

$$-6(1) + 4(2) + b = 1$$
$$-6 + 8 + b = 1$$
$$b = -1$$

Step 3: Final Parameters

$$\textit{Weight vector: } \mathbf{w} = \begin{bmatrix} -6 \\ 4 \end{bmatrix}$$

$$\textit{Bias: } b = -1$$

$$\textit{Hyperplane: } -6x + 4y - 1 = 0$$

$$\textit{Margin: } \frac{1}{||\mathbf{w}||} = \frac{1}{\sqrt{(-6)^2 + 4^2}} = \frac{1}{2\sqrt{13}}$$

Verification:

$$\begin{aligned} (1,2): & \quad -6(1)+4(2)-1=1 \quad (\checkmark) \\ (2,3): & \quad -6(2)+4(3)-1=-1 \quad (\checkmark) \\ (3,5): & \quad -6(3)+4(5)-1=1 \quad (\checkmark) \end{aligned}$$

Final Answer:

$$\boxed{\mathbf{w}=\begin{bmatrix}-6\\4\end{bmatrix}, \quad b=-1}$$

Example 6.23. *Calculate the SVR parameters for the data points* $(2,3)$, $(4,5)$, $(6,7)$ *with labels* $1,-1,1$ *using the linear kernel. Simplify assumptions as needed.*

Solution 6.23. *Given points* $(2,3)$, $(4,5)$, $(6,7)$ *with labels* 1, -1, *and* 1, *we find the linear SVR parameters with* $\epsilon=0$ *assumption.*

Assumptions:

- *Linear kernel:* $f(x,y)=w_1x+w_2y+b$
- $\epsilon=0$ *tolerance*
- *Soft margin with least squares approximation*

System of Equations:

$$2w_1+3w_2+b=1 \tag{6.7}$$

$$4w_1+5w_2+b=-1 \tag{6.8}$$

$$6w_1+7w_2+b=1 \tag{6.9}$$

Inconsistency Resolution: Subtract (6.7) *from* (6.8)*:*

$$\begin{aligned} (4w_1+5w_2+b)-(2w_1+3w_2+b) &= -1-1 \\ 2w_1+2w_2 &= -2 \\ w_1+w_2 &= -1 \end{aligned} \tag{6.10}$$

Subtract (6.8) *from* (6.9)*:*

$$\begin{aligned} (6w_1+7w_2+b)-(4w_1+5w_2+b) &= 1-(-1) \\ 2w_1+2w_2 &= 2 \\ w_1+w_2 &= 1 \end{aligned} \tag{6.11}$$

Contradiction: $-1 \neq 1$ *implies no exact solution.*

Least Squares Approximation: Normal equations $A^\top A\mathbf{x}=A^\top\mathbf{b}$*:*

$$A=\begin{bmatrix}2&3&1\\4&5&1\\6&7&1\end{bmatrix}, \quad \mathbf{b}=\begin{bmatrix}1\\-1\\1\end{bmatrix}$$

$$A^\top A = \begin{bmatrix} 56 & 68 & 12 \\ 68 & 83 & 15 \\ 12 & 15 & 3 \end{bmatrix}, \quad A^\top \mathbf{b} = \begin{bmatrix} 4 \\ 5 \\ 1 \end{bmatrix}$$

$$\textit{Solving} \begin{bmatrix} 56 & 68 & 12 \\ 68 & 83 & 15 \\ 12 & 15 & 3 \end{bmatrix} \begin{bmatrix} w_1 \\ w_2 \\ b \end{bmatrix} = \begin{bmatrix} 4 \\ 5 \\ 1 \end{bmatrix}:$$

$$\mathbf{x} = \begin{bmatrix} w_1 \\ w_2 \\ b \end{bmatrix} = \begin{bmatrix} -\frac{1}{9} \\ \frac{1}{9} \\ \frac{2}{9} \end{bmatrix}$$

Final Parameters:

$$\textit{Weight vector: } \mathbf{w} = \begin{bmatrix} -\frac{1}{9} \\ \frac{1}{9} \end{bmatrix}$$

$$\textit{Bias: } b = \frac{2}{9}$$

$$\textit{Hyperplane: } -\frac{1}{9}x + \frac{1}{9}y + \frac{2}{9} = 0 \quad \Leftrightarrow \quad -x + y + 2 = 0$$

Verification:

Point	*Prediction*	*Label*	*Error*
$(2,3)$	$\frac{1}{3}$	*1*	$\frac{2}{3}$
$(4,5)$	$\frac{1}{3}$	*-1*	$\frac{4}{3}$
$(6,7)$	$\frac{1}{3}$	*1*	$\frac{2}{3}$

Final Answer:

$$\boxed{\mathbf{w} = \begin{bmatrix} -\frac{1}{9} \\ \frac{1}{9} \end{bmatrix}, \quad b = \frac{2}{9}}$$

6.12 Role of Inventory Forecasting in Inventory Management

Inventory forecasting plays a pivotal role in inventory management, acting as the cornerstone for efficient and effective operations. Accurately predicting future inventory needs, businesses can optimize their inventory levels, reduce costs, and improve customer satisfaction. This section delves into the various aspects of inventory forecasting, its implications for supply chain management, and how it enhances logistics and operational efficiency.

6.12.1 Predicting Future Inventory Requirements

Inventory forecasting is crucial for predicting future inventory requirements, ensuring businesses maintain the right balance between supply and demand. Accurate forecasting involves analyzing historical sales data, market trends, and using advanced analytical tools to project future inventory needs.

6.12.2 Time Series Analysis

Time series analysis is a fundamental technique used in inventory forecasting. It involves using past sales data to identify patterns and trends, which can be extrapolated to predict future demand. The basic model used in time series analysis is the Autoregressive Integrated Moving Average (ARIMA) model. The ARIMA model is represented as:

$$y_t = c + \phi_1 y_{t-1} + \phi_2 y_{t-2} + \cdots + \phi_p y_{t-p} + \epsilon_t + \theta_1 \epsilon_{t-1} + \theta_2 \epsilon_{t-2} + \cdots + \theta_q \epsilon_{t-q} \quad (6.12)$$

where y_t is the actual value at time t, c is a constant, ϕ are the autoregressive parameters, θ are the moving average parameters, and ϵ is the error term.

Example 6.24. *Given the time series data for monthly sales: $y = (110, 125, 135, 120, 130)$, fit an AR(2) model (autoregressive model of order 2) and predict the next value.*

Solution 6.24. *AR(2) Model Definition: The autoregressive model of order 2 is defined as:*

$$y_t = c + \phi_1 y_{t-1} + \phi_2 y_{t-2} + \varepsilon_t$$

System of Equations: Using the data points:

$$\begin{aligned} y_3 &= 135 = c + 125\phi_1 + 110\phi_2 \quad (t = 3) \\ y_4 &= 120 = c + 135\phi_1 + 125\phi_2 \quad (t = 4) \\ y_5 &= 130 = c + 120\phi_1 + 135\phi_2 \quad (t = 5) \end{aligned}$$

Solving for Parameters: Subtract equations to eliminate c:

$$\begin{aligned} \textit{Equation 2 - Equation 1:} \quad & -15 = 10\phi_1 + 15\phi_2 \\ & \Rightarrow -3 = 2\phi_1 + 3\phi_2 \quad \textit{(Equation A)} \\ \textit{Equation 3 - Equation 2:} \quad & 10 = -15\phi_1 + 10\phi_2 \quad \textit{(Equation B)} \end{aligned}$$

Solve Equations A and B:

$$\begin{cases} \phi_1 = \dfrac{-3 - 3\phi_2}{2} \\ 10 = -15\left(\dfrac{-3 - 3\phi_2}{2}\right) + 10\phi_2 \end{cases}$$

$$10 = \frac{45 + 45\phi_2}{2} + 10\phi_2 \implies \phi_2 = -\frac{5}{13} \approx -0.3846$$

$$\phi_1 = -\frac{12}{13} \approx -0.9231$$

Find c using y_3:

$$c = 135 - 125(-0.9231) - 110(-0.3846) \approx 292.69$$

Final AR(2) Model:

$$y_t = 292.69 - 0.9231y_{t-1} - 0.3846y_{t-2}$$

Prediction for $t = 6$:

$$y_6 = 292.69 - 0.9231(130) - 0.3846(120) \approx 126.54$$

Summary:

- *Parameters:*

$$c \approx 292.69$$
$$\phi_1 \approx -0.9231$$
$$\phi_2 \approx -0.3846$$

- *Predicted Sales for $t = 6$:* $\boxed{126.54}$

Example 6.25. *Given the quarterly sales data: $y = (300, 280, 350, 370, 390, 410)$, fit an MA(1) model (moving average model of order 1) and predict the next value.*

Solution 6.25. *To fit an MA(1) model, we use the form:*

$$y_t = c + \epsilon_t + \theta_1\epsilon_{t-1}$$

where:

- μ*: Mean of the series*
- θ_1*: MA(1) coefficient*
- ϵ_t*: Error term at time t*

Step 1: Calculate Mean:

$$\mu = \frac{300 + 280 + 350 + 370 + 390 + 410}{6} = \frac{2100}{6} = 350$$

Step 2: Estimate θ_1 with Simplified Assumptions: Assume $\theta_1 = 1.0$ (simplified approach) and $\epsilon_0 = 0$.

Step 3: Compute Error Terms Recursively:

$$\epsilon_1 = 300 - 350 = -50$$
$$\epsilon_2 = 280 - 350 - 1.0(-50) = -20$$
$$\epsilon_3 = 350 - 350 - 1.0(-20) = 20$$
$$\epsilon_4 = 370 - 350 - 1.0(20) = 0$$
$$\epsilon_5 = 390 - 350 - 1.0(0) = 40$$
$$\epsilon_6 = 410 - 350 - 1.0(40) = 20$$

Step 4: Predict Next Value (t = 7):

$$y_7 = \mu + \theta_1\epsilon_6 = 350 + 1.0(20) = \boxed{370}$$

Final Parameters:

$$\boxed{\mu = 350, \quad \theta_1 = 1.0}$$

Model Summary:

$$y_t = 350 + \epsilon_t + 1.0\epsilon_{t-1}$$

6.12.3 Causal Models

Causal models consider external factors such as economic indicators, seasonal variations, and market conditions. A common causal model is the regression model:

$$y_t = \beta_0 + \beta_1 x_{1,t} + \beta_2 x_{2,t} + \cdots + \beta_k x_{k,t} + \epsilon_t \tag{6.13}$$

where y_t is the dependent variable (demand), $x_{i,t}$ are the independent variables (external factors), β_i are the coefficients, and ϵ_t is the error term.

Example 6.26. *Consider a dataset with monthly sales (dependent variable) and advertising expenditure and price index (independent variables):*

Month	*Sales*(y_t)	*Ad. Exp.*($x_{1,t}$)	*Price Index*($x_{2,t}$)
1	200	50	1.2
2	220	60	1.3
3	230	65	1.4
4	240	70	1.5
5	250	75	1.6

Fit a linear regression model and predict the sales for the sixth month with $x_{1,6} = 80$ *and* $x_{2,6} = 1.7$.

Solution 6.26. *To fit a linear regression model, we use the form:*

$$y_t = \beta_0 + \beta_1 x_{1,t} + \beta_2 x_{2,t} + \epsilon_t$$

Using the given data, we set up the following system of equations:

$$\begin{aligned}200 &= \beta_0 + 50\beta_1 + 1.2\beta_2 + \epsilon_1\\220 &= \beta_0 + 60\beta_1 + 1.3\beta_2 + \epsilon_2\\230 &= \beta_0 + 65\beta_1 + 1.4\beta_2 + \epsilon_3\\240 &= \beta_0 + 70\beta_1 + 1.5\beta_2 + \epsilon_4\\250 &= \beta_0 + 75\beta_1 + 1.6\beta_2 + \epsilon_5\end{aligned}$$

Matrix Representation:

$$\mathbf{Y} = \mathbf{X}\boldsymbol{\beta} + \boldsymbol{\epsilon}$$

where:

$$\mathbf{Y} = \begin{bmatrix}200\\220\\230\\240\\250\end{bmatrix}, \quad \mathbf{X} = \begin{bmatrix}1 & 50 & 1.2\\1 & 60 & 1.3\\1 & 65 & 1.4\\1 & 70 & 1.5\\1 & 75 & 1.6\end{bmatrix}, \quad \boldsymbol{\beta} = \begin{bmatrix}\beta_0\\\beta_1\\\beta_2\end{bmatrix}$$

Normal Equations:

$$(\mathbf{X}^\top\mathbf{X})\boldsymbol{\beta} = \mathbf{X}^\top\mathbf{Y}$$

Compute $\mathbf{X}^\top\mathbf{X}$

$$\mathbf{X}^\top\mathbf{X} = \begin{bmatrix}5 & 320 & 7.0\\320 & 20850 & 454\\7.0 & 454 & 9.9\end{bmatrix}$$

Compute $\mathbf{X}^\top\mathbf{Y}$*:*

$$\mathbf{X}^\top\mathbf{Y} = \begin{bmatrix}1140\\73700\\1608\end{bmatrix}$$

Solve for $\boldsymbol{\beta}$*:*

$$\begin{bmatrix}5 & 320 & 7.0\\320 & 20850 & 454\\7.0 & 454 & 9.9\end{bmatrix}\begin{bmatrix}\beta_0\\\beta_1\\\beta_2\end{bmatrix} = \begin{bmatrix}1140\\73700\\1608\end{bmatrix}$$

Solution:

$$\boldsymbol{\beta} = \begin{bmatrix}100\\2\\0\end{bmatrix}$$

Verification: Substitute $\beta_0 = 100$, $\beta_1 = 2$, $\beta_2 = 0$*:*

$$\begin{aligned}100 + 50(2) + 1.2(0) &= 200 \quad (\checkmark)\\100 + 60(2) + 1.3(0) &= 220 \quad (\checkmark)\\100 + 65(2) + 1.4(0) &= 230 \quad (\checkmark)\\100 + 70(2) + 1.5(0) &= 240 \quad (\checkmark)\\100 + 75(2) + 1.6(0) &= 250 \quad (\checkmark)\end{aligned}$$

Final Parameters:

$$\boxed{\beta_0 = 100, \quad \beta_1 = 2, \quad \beta_2 = 0}$$

From the existing data:

Observation	x_1	x_2
1	50	1.2
2	60	1.3
3	65	1.4
4	70	1.5
5	75	1.6

Observed trends:

- x_1 *increases by* $+5$ *per observation*
- x_2 *increases by* $+0.1$ *per observation*

Extrapolated Values for Observation 6:

$$x_{1,6} = 75 + 5 = 80, \quad x_{2,6} = 1.6 + 0.1 = 1.7$$

Prediction Calculation:

$$y_6 = \beta_0 + \beta_1 x_{1,6} + \beta_2 x_{2,6}$$

$$y_6 = 100 + 2(80) + 0(1.7) = 100 + 160 + 0 = \boxed{260}$$

Key Notes:

- $\beta_2 = 0$ *eliminates the* x_2 *term's influence*
- *Prediction assumes continuation of observed trends*
- *Substitute actual* $x_{1,6}$ *and* $x_{2,6}$ *values if available*

Example 6.27. *Consider a dataset with quarterly sales (dependent variable) and GDP growth rate and seasonal index (independent variables):*

Quarter	*Sales*(y_t)	*GDP Growth*($x_{1,t}$)	*Seasonal Index*($x_{2,t}$)
1	500	3.0	1.1
2	550	3.2	1.2
3	600	3.5	1.3
4	650	3.8	1.4
5	700	4.0	1.5

Fit a linear regression model and predict the sales for the sixth quarter with $x_{1,6} = 4.2$ *and* $x_{2,6} = 1.6$.

Solution 6.27. *The linear regression model:*

$$y_t = \beta_0 + \beta_1 x_{1,t} + \beta_2 x_{2,t} + \epsilon_t$$

Matrix Representation:

$$\mathbf{Y} = \begin{bmatrix} 500 \\ 550 \\ 600 \\ 650 \\ 700 \end{bmatrix}, \quad \mathbf{X} = \begin{bmatrix} 1 & 3.0 & 1.1 \\ 1 & 3.2 & 1.2 \\ 1 & 3.5 & 1.3 \\ 1 & 3.8 & 1.4 \\ 1 & 4.0 & 1.5 \end{bmatrix}, \quad \boldsymbol{\beta} = \begin{bmatrix} \beta_0 \\ \beta_1 \\ \beta_2 \end{bmatrix}$$

Normal Equations Solution:

$$\boldsymbol{\beta} = (\mathbf{X}^\top \mathbf{X})^{-1} \mathbf{X}^\top \mathbf{Y}$$

Matrix Calculations::

$$\mathbf{X}^\top \mathbf{X} = \begin{bmatrix} 5 & 17.5 & 6.5 \\ 17.5 & 61.93 & 23.01 \\ 6.5 & 23.01 & 8.55 \end{bmatrix}, \quad \mathbf{X}^\top \mathbf{Y} = \begin{bmatrix} 3000 \\ 10630 \\ 3950 \end{bmatrix}$$

Solving yields:

$$\boldsymbol{\beta} = \begin{bmatrix} -50 \\ 0 \\ 500 \end{bmatrix}$$

Final Regression Model:

$$y_t = -50 + 0 \cdot x_{1,t} + 500 \cdot x_{2,t}$$

Prediction for Quarter 6: Given $x_{1,6} = 4.2$ *and* $x_{2,6} = 1.6$*:*

$$y_6 = -50 + 0(4.2) + 500(1.6) = -50 + 800 = \boxed{750}$$

Model Verification:

$$\begin{aligned} &\textit{Quarter 1:} \ -50 + 500(1.1) = 500 \quad (\checkmark) \\ &\textit{Quarter 2:} \ -50 + 500(1.2) = 550 \quad (\checkmark) \\ &\textit{Quarter 5:} \ -50 + 500(1.5) = 700 \quad (\checkmark) \end{aligned}$$

Note:

- $\beta_1 = 0$*: GDP growth has no statistical significance in this model*
- *Perfect fit* ($R^2 = 1$) *with zero residuals*
- *Sales are fully explained by seasonal index in this dataset*

6.12.4 Implications for Supply Chain Management

Inventory forecasting has significant implications for supply chain management, improving coordination and synchronization across the supply chain, leading to enhanced efficiency and reduced costs.

6.12.5 Bullwhip Effect Reduction

The bullwhip effect refers to the phenomenon where small fluctuations in customer demand lead to larger variations in orders placed upstream in the supply chain. Accurate forecasting helps mitigate this effect by providing a clearer picture of actual demand. The mathematical representation of the bullwhip effect is:

$$\sigma_o^2 = \left(\frac{L+1}{L}\right)^2 \sigma_d^2 \tag{6.14}$$

where σ_o^2 is the variance of orders, σ_d^2 is the variance of demand, and L is the lead time.

Example 6.28. *Given a lead time $L = 2$ weeks and a variance of demand $\sigma_d^2 = 100$, calculate the variance of orders σ_o^2.*

Solution 6.28. *Using the given formula:*

$$\sigma_o^2 = \left(\frac{L+1}{L}\right)^2 \sigma_d^2$$

Substituting the given values:

$$\sigma_o^2 = \left(\frac{2+1}{2}\right)^2 \times 100 = \left(\frac{3}{2}\right)^2 \times 100 = (1.5)^2 \times 100 = 2.25 \times 100 = 225$$

Therefore, the variance of orders σ_o^2 is 225.

Example 6.29. *Given a lead time $L = 4$ weeks and a variance of demand $\sigma_d^2 = 150$, calculate the variance of orders σ_o^2.*

Solution 6.29. *Using the given formula:*

$$\sigma_o^2 = \left(\frac{L+1}{L}\right)^2 \sigma_d^2$$

Substituting the given values:

$$\sigma_o^2 = \left(\frac{4+1}{4}\right)^2 \times 150 = \left(\frac{5}{4}\right)^2 \times 150 = (1.25)^2 \times 150 = 1.5625 \times 150 = 234.375$$

Therefore, the variance of orders σ_o^2 is 234.375.

6.12.6 Supplier Relationship Management

Effective forecasting enables better communication and collaboration with suppliers. Sharing demand forecasts with suppliers can lead to reduced lead times and improved supplier performance. The economic order quantity (EOQ) model is often used to determine optimal order quantities:

$$EOQ = \sqrt{\frac{2DS}{H}} \tag{6.15}$$

where D is the demand rate, S is the order cost, and H is the holding cost per unit per period.

Example 6.30. *Given a demand rate* $D = 1{,}000$ *units per year, an order cost* $S = 50$ *dollars per order, and a holding cost* $H = 2$ *dollars per unit per year, calculate the EOQ.*

Solution 6.30. *Using the given formula:*

$$EOQ = \sqrt{\frac{2DS}{H}}$$

Substituting the given values:

$$EOQ = \sqrt{\frac{2 \times 1{,}000 \times 50}{2}} = \sqrt{\frac{100{,}000}{2}} = \sqrt{50{,}000} \approx 223.61$$

Therefore, the EOQ is approximately 224 *units (rounding to the nearest whole number).*

Example 6.31. *Given a demand rate* $D = 5{,}000$ *units per year, an order cost* $S = 100$ *dollars per order, and a holding cost* $H = 5$ *dollars per unit per year, calculate the EOQ.*

Solution 6.31. *Using the given formula:*

$$EOQ = \sqrt{\frac{2DS}{H}}$$

Substituting the given values:

$$EOQ = \sqrt{\frac{2 \times 5{,}000 \times 100}{5}} = \sqrt{\frac{1{,}000{,}000}{5}} = \sqrt{200{,}000} \approx 447.21$$

Therefore, the EOQ is approximately 447 *units (rounding to the nearest whole number).*

6.12.7 Inventory Optimization

Inventory optimization aims to balance holding costs and stockout costs. The Newsvendor model is a common approach:

$$Q^* = F^{-1}\left(\frac{p-c}{p-v}\right) \tag{6.16}$$

where Q^* is the optimal order quantity, F^{-1} is the inverse cumulative distribution function of demand, p is the selling price, c is the cost per unit, and v is the salvage value per unit.

Example 6.32. *Given a selling price $p = 20$ dollars, a cost per unit $c = 10$ dollars, a salvage value per unit $v = 5$ dollars, and demand that follows a normal distribution with a mean of 100 units and a standard deviation of 20 units, calculate the optimal order quantity Q^*.*

Solution 6.32. *First, we calculate the critical ratio:*

$$\frac{p-c}{p-v} = \frac{20-10}{20-5} = \frac{10}{15} = \frac{2}{3} \approx 0.667$$

Next, we use the inverse cumulative distribution function of the normal distribution to find the z-score corresponding to 0.667. From standard normal distribution tables or using software, we find that:

$$z \approx 0.43$$

The optimal order quantity Q^* is then given by:

$$Q^* = \mu + z\sigma = 100 + 0.43 \times 20 = 100 + 8.6 = 108.6$$

Therefore, the optimal order quantity Q^* is approximately 109 *units (rounding to the nearest whole number).*

Example 6.33. *Given a selling price $p = 50$ dollars, a cost per unit $c = 30$ dollars, a salvage value per unit $v = 10$ dollars, and demand that follows a normal distribution with a mean of 200 units and a standard deviation of 40 units, calculate the optimal order quantity Q^*.*

Solution 6.33. *First, we calculate the critical ratio:*

$$\frac{p-c}{p-v} = \frac{50-30}{50-10} = \frac{20}{40} = 0.5$$

Next, we use the inverse cumulative distribution function of the normal distribution to find the z-score corresponding to 0.5. For a critical ratio of 0.5, the z-score is:

$$z = 0$$

The optimal order quantity Q^* is then given by:

$$Q^* = \mu + z\sigma = 200 + 0 \times 40 = 200$$

Therefore, the optimal order quantity Q^* is 200 *units.*

6.12.8 Logistics and Operational Efficiency

Inventory forecasting significantly enhances logistics and operational efficiency, optimizing transportation, warehousing, and labor planning.

6.12.9 Transportation Planning

Accurate demand forecasts enable businesses to optimize their transportation processes, such as consolidating shipments and optimizing routes. The transportation problem can be solved using linear programming:

$$\text{Minimize} \quad \sum_{i=1}^{m}\sum_{j=1}^{n} c_{ij}x_{ij} \tag{6.17}$$

$$\text{Subject to} \quad \sum_{j=1}^{n} x_{ij} = a_i \quad \forall i \tag{6.18}$$

$$\sum_{i=1}^{m} x_{ij} = b_j \quad \forall j \tag{6.19}$$

$$x_{ij} \geq 0 \quad \forall i, j \tag{6.20}$$

where c_{ij} is the cost of shipping from source i to destination j, x_{ij} is the amount shipped, a_i is the supply at source i, and b_j is the demand at destination j.

Example 6.34. *Consider a transportation problem with two sources and three destinations. The supply and demand are as follows:*

- *Source 1: Supply = 50 units*
- *Source 2: Supply = 30 units*
- *Destination 1: Demand = 30 units*
- *Destination 2: Demand = 40 units*
- *Destination 3: Demand = 10 units*

The shipping costs per unit are:

	Destination 1	*Destination 2*	*Destination 3*
Source 1	8	6	10
Source 2	9	12	13

Calculate the optimal shipping quantities.

Solution 6.34. *We formulate the linear programming problem as follows: Minimize:*

$$8x_{11} + 6x_{12} + 10x_{13} + 9x_{21} + 12x_{22} + 13x_{23}$$

Subject to:

$$x_{11} + x_{12} + x_{13} = 50$$

$$x_{21} + x_{22} + x_{23} = 30$$

$$x_{11} + x_{21} = 30$$

$$x_{12} + x_{22} = 40$$

$$x_{13} + x_{23} = 10$$

$$x_{ij} \geq 0$$

Solving this linear programming problem (e.g., using a solver), we get the following optimal solution:

$$x_{11} = 30$$
$$x_{12} = 10$$
$$x_{13} = 10$$
$$x_{21} = 0$$
$$x_{22} = 30$$
$$x_{23} = 0$$

The optimal shipping quantities minimize the total cost. The total cost is:

$$8 \times 30 + 6 \times 10 + 10 \times 10 + 9 \times 0 + 12 \times 30 + 13 \times 0 = 240 + 60 + 100 + 360 = 760$$

Therefore, the optimal shipping quantities are as calculated, and the minimum cost is \$760.

Example 6.35. *Consider a transportation problem with three sources and two destinations. The supply and demand are as follows:*

- *Source 1: Supply = 40 units*
- *Source 2: Supply = 60 units*
- *Source 3: Supply = 50 units*
- *Destination 1: Demand = 70 units*
- *Destination 2: Demand = 80 units*

The shipping costs per unit are:

	Destination 1	*Destination 2*
Source 1	4	6
Source 2	2	5
Source 3	3	4

Calculate the optimal shipping quantities.

Solution 6.35. *We formulate the linear programming problem as follows:*
Minimize:

$$4x_{11} + 6x_{12} + 2x_{21} + 5x_{22} + 3x_{31} + 4x_{32}$$

Subject to:

$$x_{11} + x_{12} = 40$$

$$x_{21} + x_{22} = 60$$

$$x_{31} + x_{32} = 50$$

$$x_{11} + x_{21} + x_{31} = 70$$

$$x_{12} + x_{22} + x_{32} = 80$$

$$x_{ij} \geq 0$$

Solving this linear programming problem (e.g., using a solver), we get the following optimal solution:

$$x_{11} = 20$$

$$x_{12} = 20$$

$$x_{21} = 20$$

$$x_{22} = 40$$

$$x_{31} = 30$$

$$x_{32} = 20$$

The total cost is:

$$4\times20+6\times20+2\times20+5\times40+3\times30+4\times20 = 80+120+40+200+90+80 = 610$$

Therefore, the optimal shipping quantities are as calculated, and the minimum cost is \$610.

6.12.10 Warehouse Space Utilization

Forecasting aids in efficient use of warehouse space by optimizing inventory placement and reducing excess storage. The assignment problem can be used for inventory placement optimization:

$$\text{Minimize} \quad \sum_{i=1}^{n}\sum_{j=1}^{n} c_{ij}x_{ij} \tag{6.21}$$

$$\text{Subject to} \quad \sum_{j=1}^{n} x_{ij} = 1 \quad \forall i \tag{6.22}$$

$$\sum_{i=1}^{n} x_{ij} = 1 \quad \forall j \tag{6.23}$$

$$x_{ij} \in \{0,1\} \quad \forall i,j \tag{6.24}$$

where c_{ij} is the cost of assigning item i to location j, and x_{ij} is a binary variable indicating assignment.

Example 6.36. *Consider a warehouse with four items and four locations. The cost of assigning each item to each location is given by the following cost matrix:*

	Location 1	*Location 2*	*Location 3*	*Location 4*
Item 1	8	6	10	9
Item 2	7	9	4	8
Item 3	5	12	7	11
Item 4	10	8	9	5

Find the optimal assignment of items to locations using the Hungarian algorithm.

Solution 6.36. *To solve the assignment problem using the Hungarian algorithm, follow these steps:*

Step 1: Subtract Row Minimums

Subtract the smallest value in each row from all the values in that row.

	Location 1	*Location 2*	*Location 3*	*Location 4*
Item 1	2	0	4	1
Item 2	3	5	0	4
Item 3	0	7	2	6
Item 4	5	3	4	0

Step 2: Subtract Column Minimums

Subtract the smallest value in each column from all the values in that column.

	Location 1	*Location 2*	*Location 3*	*Location 4*
Item 1	2	0	4	1
Item 2	3	5	0	4
Item 3	0	7	2	6
Item 4	5	3	4	0

Step 3: Cover Zeros with Minimum Number of Lines
Identify the minimum number of horizontal and vertical lines needed to cover all zeros in the matrix.

	Location 1	*Location 2*	*Location 3*	*Location 4*
Item 1	*2*	*0*	4	*1*
Item 2	3	5	*0*	4
Item 3	*0*	7	2	6
Item 4	5	*3*	4	*0*

Step 4: Create a New Matrix (Adjust the Matrix)
If the number of lines is less than the number of rows (or columns), adjust the matrix:
1. Find the smallest noncovered value. 2. Subtract this value from all uncovered elements. 3. Add this value to the elements covered by two lines.

$$\textit{Smallest noncovered value} = 1$$

Subtract 1 from all uncovered elements and add 1 to elements covered by two lines.

	Location 1	*Location 2*	*Location 3*	*Location 4*
Item 1	1	−1	3	0
Item 2	2	4	−1	3
Item 3	−1	6	1	5
Item 4	4	2	3	−1

Step 5: Repeat Steps 3 and 4 Until Optimal Assignment
Continue covering zeros and adjusting the matrix until the number of lines equals the number of rows (or columns).
Final Assignment
After repeating the steps, you will arrive at the optimal assignment matrix:

	Location 1	*Location 2*	*Location 3*	*Location 4*
Item 1	1	0	0	0
Item 2	0	1	0	0
Item 3	0	0	1	0
Item 4	0	0	0	1

The optimal assignment is:

$$x_{11} = 1$$
$$x_{22} = 1$$
$$x_{33} = 1$$
$$x_{44} = 1$$

The total cost is:

$$8 \times 1 + 9 \times 1 + 7 \times 1 + 5 \times 1 = 8 + 9 + 7 + 5 = 29$$

Thus, the optimal assignment is to place Item 1 in Location 1, Item 2 in Location 2, Item 3 in Location 3, and Item 4 in Location 4, with a minimum total cost of \$29.

6.12.11 Labor Planning

Accurate demand forecasts allow businesses to schedule labor more effectively, ensuring optimal staffing levels for inventory-related tasks. The workforce scheduling problem can be addressed using integer programming:

$$\text{Minimize} \quad \sum_{i=1}^{n} c_i x_i \tag{6.25}$$

$$\text{Subject to} \quad \sum_{i=1}^{n} a_{ij} x_i \geq b_j \quad \forall j \tag{6.26}$$

$$x_i \in \{0, 1\} \quad \forall i \tag{6.27}$$

where c_i is the cost of employing worker i, a_{ij} is the assignment of worker i to task j, and b_j is the demand for task j.

Example 6.37. *Consider a warehouse that needs to schedule labor for three tasks. There are four workers available, each with different costs and abilities. The following table provides the cost of employing each worker and their ability to handle each task:*

	Task 1	*Task 2*	*Task 3*
Worker 1	1	0	1
Worker 2	1	1	0
Worker 3	0	1	1
Worker 4	0	0	1

	Worker 1	*Worker 2*	*Worker 3*	*Worker 4*
Cost of Employing	10	15	20	25

The demand for each task is:

Task 1	*Task 2*	*Task 3*
1	2	1

Determine the optimal assignment of workers to tasks to minimize the total cost.

Solution 6.37. *To solve this problem, we use integer programming and apply the branch and bound method. Here's a step-by-step procedure:*

1. ***Formulate the Linear Program:***

$$\begin{aligned} \textit{Minimize} \quad & 10x_1 + 15x_2 + 20x_3 + 25x_4 && (6.28)\\ \textit{Subject to} \quad & x_1 + x_2 \geq 1 && (6.29)\\ & x_2 + x_3 \geq 2 && (6.30)\\ & x_3 + x_4 \geq 1 && (6.31)\\ & x_i \in \{0, 1\} \quad \forall i && (6.32) \end{aligned}$$

2. ***Apply the Branch and Bound Method:***

(a) ***Initialize:*** *Relax the integer constraints to continuous ones and solve the relaxed linear programming problem:*

$$\begin{aligned} \textit{Minimize} \quad & 10x_1 + 15x_2 + 20x_3 + 25x_4 && (6.33)\\ \textit{Subject to} \quad & x_1 + x_2 \geq 1 && (6.34)\\ & x_2 + x_3 \geq 2 && (6.35)\\ & x_3 + x_4 \geq 1 && (6.36)\\ & 0 \leq x_i \leq 1 \quad \forall i && (6.37) \end{aligned}$$

Solving this linear program provides a fractional solution indicating further branching is required.

(b) ***Branching:*** *Choose a variable with a fractional value to branch on. Suppose x_1 is fractional, create two subproblems:*

- *Subproblem 1: $x_1 = 0$:*

$$\begin{aligned} \textit{Minimize} \quad & 15x_2 + 20x_3 + 25x_4 && (6.38)\\ \textit{Subject to} \quad & x_2 + x_3 \geq 2 && (6.39)\\ & x_3 + x_4 \geq 1 && (6.40)\\ & 0 \leq x_i \leq 1 \quad \forall i && (6.41) \end{aligned}$$

- *Subproblem 2: $x_1 = 1$:*

$$\begin{aligned} \textit{Minimize} \quad & 10 + 15x_2 + 20x_3 + 25x_4 && (6.42)\\ \textit{Subject to} \quad & x_2 + x_3 \geq 1 && (6.43)\\ & x_3 + x_4 \geq 1 && (6.44)\\ & 0 \leq x_i \leq 1 \quad \forall i && (6.45) \end{aligned}$$

3. ***Bounding:*** *Solve the relaxed linear programs for these subproblems to compute the lower bounds:*

- *For Subproblem 1 ($x_1 = 0$), solving yields a lower bound of 45.*
- *For Subproblem 2 ($x_1 = 1$), solving yields a lower bound of 35.*

Compare these lower bounds with the best integer solution found so far. If a lower bound exceeds the best known solution, prune that branch.

4. ***Pruning:*** *Discard subproblems that either cannot satisfy the constraints or whose lower bounds exceed the best known integer solution. In this case, Subproblem 1 does not provide a better solution compared to Subproblem 2, so it's pruned.*

5. ***Optimal Solution:*** *Continue branching, bounding, and pruning until reaching the optimal integer solution. For this problem, the optimal integer solution is:*

$$x_1 = 1$$
$$x_2 = 1$$
$$x_3 = 1$$
$$x_4 = 0$$

Total Cost Calculation:

$$\begin{aligned}\textit{Total Cost} &= 10x_1 + 15x_2 + 20x_3 + 25x_4 = 10 \times 1 + 15 \times 1 + 20 \times 1 + 25 \times 0 \\ &= 10 + 15 + 20 = 45\end{aligned}$$

Thus, the optimal assignment is to employ Worker 1, Worker 2, and Worker 3, with a total cost of \$45. *This assignment satisfies the demand for each task while minimizing the cost of employing workers.*

6.13 Inventory Forecasting Techniques

Effective inventory forecasting relies on various techniques that help predict future inventory needs with precision. These techniques include demand planning, demand forecasting, and comparative analysis of different methods. This section provides an in-depth exploration of these techniques, their methodologies, and their practical applications.

6.13.1 Demand Planning

Demand planning is a strategic process that integrates historical data, market insights, and business intelligence to predict future product demand. It aims to balance supply and demand, ensuring inventory levels are optimized to meet customer requirements without overstocking.

6.13.1.1 Sales and Operations Planning (S&OP)

Sales and operations planning (S&OP) is a critical component of demand planning. It involves cross-functional collaboration to align sales forecasts with production and inventory strategies. The S&OP process can be modeled using linear programming to optimize the overall plan:

$$\text{Minimize} \quad \sum_{t=1}^{T}(P_tC_p + I_tC_i + B_tC_b) \tag{6.46}$$

$$\text{Subject to} \quad P_t + I_{t-1} - D_t = I_t \quad \forall t \tag{6.47}$$

$$I_t \geq 0 \quad \forall t \tag{6.48}$$

$$B_t = \max(0, D_t - (P_t + I_{t-1})) \quad \forall t \tag{6.49}$$

$$P_t \leq P_{\max} \quad \forall t \tag{6.50}$$

$$I_0 = I_{\text{initial}} \tag{6.51}$$

where P_t is the production quantity in period t, I_t is the inventory level at the end of period t, D_t is the demand in period t, B_t is the backlog, C_p is the production cost, C_i is the inventory holding cost, and C_b is the backlog cost.

6.13.1.2 Collaborative Planning, Forecasting, and Replenishment (CPFR)

Collaborative planning, forecasting, and replenishment (CPFR) is a framework that enhances demand planning by fostering collaboration between suppliers and retailers. The CPFR process involves sharing forecasts, planning jointly, and coordinating replenishment activities to ensure alignment. The key steps in CPFR include:

- Strategy and Planning: Establish collaborative relationships and define the scope of collaboration.
- Demand and Supply Management: Exchange forecasts and develop consensus on demand and supply plans.
- Execution: Coordinate order fulfillment and manage inventory replenishment.
- Analysis: Monitor performance, analyze variances, and adjust plans as needed.

6.13.2 Demand Forecasting

Demand forecasting is the process of estimating future customer demand using statistical methods and machine learning algorithms. Accurate demand forecasts enable businesses to make informed decisions about inventory levels, production schedules, and procurement strategies.

6.13.2.1 Quantitative Forecasting Methods

Quantitative forecasting methods rely on numerical data and statistical techniques to predict future demand. Common methods include:

- Moving Averages

  Moving averages smooth out short-term fluctuations and highlight long-term trends. The simple moving average (SMA) is calculated as:

  $$\text{SMA}_t = \frac{1}{n}\sum_{i=0}^{n-1} D_{t-i} \tag{6.52}$$

  where D_t is the demand in period t, and n is the number of periods in the moving average.

- Exponential Smoothing

  Exponential smoothing assigns exponentially decreasing weights to past observations. The formula for single exponential smoothing is:

  $$S_t = \alpha D_t + (1-\alpha)S_{t-1} \tag{6.53}$$

  where S_t is the smoothed value, D_t is the actual demand, and α is the smoothing constant.

- Regression Analysis

  Regression analysis models the relationship between demand and one or more independent variables. A simple linear regression model is given by:

  $$D_t = \beta_0 + \beta_1 X_t + \epsilon_t \tag{6.54}$$

  where D_t is the dependent variable (demand), X_t is the independent variable, β_0 and β_1 are the regression coefficients, and ϵ_t is the error term.

6.13.2.2 Qualitative Forecasting Methods

Qualitative forecasting methods use expert judgment and market insights to predict future demand. These methods are particularly useful when historical data is limited or unavailable.

- Delphi Method

  The Delphi method involves a panel of experts who provide forecasts independently and iteratively. The process continues until a consensus is reached. This method is useful for long-term forecasting and scenarios where data is scarce.

- Market Research

  Market research gathers data from customers, competitors, and market trends to forecast demand. Techniques include surveys, focus groups, and competitive analysis. This method helps in understanding market dynamics and consumer preferences.

6.13.3 Comparative Analysis of Techniques

A comparative analysis of inventory forecasting techniques evaluates their strengths, weaknesses, and applicability in different scenarios. This analysis helps businesses select the most appropriate method for their specific needs.

6.13.3.1 Accuracy and Reliability

Accuracy and reliability are critical factors in evaluating forecasting techniques. Quantitative methods are generally more accurate for short-term forecasts, while qualitative methods provide valuable insights for long-term projections. Machine learning algorithms offer high accuracy but require large datasets and computational resources.

6.13.3.2 Data Requirements

Different forecasting techniques have varying data requirements. Quantitative methods rely on historical data, while qualitative methods depend on expert opinions and market insights. Machine learning algorithms require extensive datasets for training and validation.

6.13.3.3 Implementation Complexity

The complexity of implementing forecasting techniques varies. Quantitative methods are relatively straightforward to implement, while machine learning algorithms require specialized knowledge and computational resources. Qualitative methods involve organizing and synthesizing expert opinions.

6.13.3.4 Scalability

Scalability refers to the ability of a forecasting technique to handle increasing data volumes and complexity. Machine learning algorithms are highly scalable, making them suitable for large-scale applications. Quantitative methods can be scaled with appropriate data preprocessing and model selection.

6.13.3.5 Cost-Benefit Analysis

A cost-benefit analysis evaluates the trade-offs between the costs of implementing a forecasting technique and the benefits it provides. Factors to consider include data acquisition costs, computational resources, and the potential impact on inventory management efficiency.

6.14 Practical Applications of Inventory Forecasting

Inventory forecasting is crucial for efficient inventory management, helping organizations predict future demand and maintain optimal stock levels. Accurate forecasting minimizes costs associated with overstocking and stockouts, improves customer satisfaction, and enhances overall operational efficiency. Various forecasting techniques, including exponential smoothing, linear regression, machine learning models, and hybrid approaches, can be applied to achieve these goals. This section explores practical applications of these techniques through case studies, their impact on organizational operations, and best practices for implementation.

Example 6.38. *A retail company uses exponential smoothing to forecast the demand for seasonal products. Over the past five years, they have adjusted their smoothing constant α based on historical accuracy. Suppose the previous forecast was 150 units, the actual demand was 180 units, and $\alpha = 0.4$.*

Solution 6.38. *Using exponential smoothing, the forecast for the next period is:*

$$F_{t+1} = \alpha D_t + (1 - \alpha)F_t = 0.4 \times 180 + 0.6 \times 150 = 72 + 90 = 162$$

Using exponential smoothing, the company achieves more accurate forecasts, minimizing stockouts and overstock situations. This leads to better customer satisfaction and optimized inventory levels.

Example 6.39. *An e-commerce business implements Random Forest to predict monthly sales for thousands of products. They use features like past sales, marketing spend, and holidays. Suppose the Random Forest model uses 100 decision trees and the prediction for a particular product is based on the following individual tree predictions: 120, 130, 125, 140, and 135 units.*

Solution 6.39. *The Random Forest model predicts the demand by averaging the outputs of multiple decision trees:*

$$\hat{D} = \frac{1}{5}(120 + 130 + 125 + 140 + 135) = \frac{650}{5} = 130$$

The Random Forest model provides high accuracy in sales predictions, helping the business to maintain optimal inventory levels and reduce holding costs, thereby enhancing overall operational efficiency.

6.14.1 Impact on Organizational Operations

Example 6.40. *A manufacturing company uses linear regression to forecast raw material requirements based on production schedules and seasonal demand.*

Suppose the linear regression model is $R = 50 + 0.5P + 0.3S$*, where* R *is raw material requirement,* P *is production units, and* S *is seasonal index. If the production units are 200 and the seasonal index is 1.5.*

Solution 6.40. *The raw material requirement is calculated as:*

$$R = 50 + 0.5 \times 200 + 0.3 \times 1.5 = 50 + 100 + 0.45 = 150.45$$

This results in streamlined procurement processes, reducing material shortages and production delays. The company can better manage its supply chain and ensure timely production.

Example 6.41. *A supermarket chain employs MAPE to evaluate the accuracy of its demand forecasts and adjust their strategies accordingly. Suppose the actual demands and forecasts for three months are* $(100, 90)$, $(150, 160)$, $(120, 130)$.

Solution 6.41. *MAPE is calculated as:*

$$MAPE = \frac{1}{3}\left(\left|\frac{100-90}{100}\right| + \left|\frac{150-160}{150}\right| + \left|\frac{120-130}{120}\right|\right) \times 100\%$$

$$= \frac{1}{3}(0.1 + 0.0667 + 0.0833) \times 100\% = \frac{0.25}{3} \times 100\% = 8.33\%$$

Continuously monitoring MAPE, the supermarket can fine-tune its forecasting models, reducing inventory costs and improving product availability for customers.

6.14.2 Best Practices and Implementation Strategies

Example 6.42. *A retailer adopts a hybrid forecasting approach, combining exponential smoothing, linear regression, and machine learning techniques. Suppose they use exponential smoothing for short-term forecasts with* $\alpha = 0.3$*, linear regression for medium-term forecasts, and Random Forest for long-term forecasts.*

Solution 6.42. *For exponential smoothing, if the previous forecast was 200 units, the actual demand was 220 units, then the forecast for the next period is:*

$$F_{t+1} = 0.3 \times 220 + 0.7 \times 200 = 66 + 140 = 206$$

For linear regression, suppose the model is $D = 50 + 1.2P + 0.8M$*, where* P *is price and* M *is marketing spend. If* $P = 30$ *and* $M = 40$*, then the forecasted demand is:*

$$D = 50 + 1.2 \times 30 + 0.8 \times 40 = 50 + 36 + 32 = 118$$

Using Random Forest, suppose the prediction from multiple trees averages to 150 units. This hybrid approach leverages the strengths of each method, resulting in highly accurate demand forecasts. The retailer can optimize stock levels, improve customer satisfaction, and reduce costs.

Example 6.43. *A company implements automated inventory management systems that integrate real-time sales data and advanced forecasting models. Suppose the system uses real-time data to update forecasts and inventory levels dynamically.*

Solution 6.43. *With automated systems, if the real-time sales data indicates an increase in demand by 10%, the forecast and inventory levels can be adjusted accordingly. For example, if the initial forecast was 200 units, the new forecast would be:*

$$\textit{New Forecast} = 200 \times 1.1 = 220$$

This allows the company to respond quickly to changes in demand, adjust inventory levels dynamically, and reduce manual errors. This leads to more efficient inventory management and better resource allocation.

6.15 Practical Applications of Inventory Forecasting

Inventory forecasting is crucial for efficient inventory management, helping organizations predict future demand and maintain optimal stock levels. Accurate forecasting minimizes costs associated with overstocking and stockouts, improves customer satisfaction, and enhances overall operational efficiency. Various forecasting techniques, including exponential smoothing, linear regression, machine learning models, and hybrid approaches, can be applied to achieve these goals. This section explores practical applications of these techniques through case studies, their impact on organizational operations, and best practices for implementation.

Example 6.44. *A retail company uses exponential smoothing to forecast the demand for seasonal products. Over the past five years, they have adjusted their smoothing constant α based on historical accuracy. Suppose the previous forecast was 150 units, the actual demand was 180 units, and $\alpha = 0.4$.*

Solution 6.44. *Using exponential smoothing, the forecast for the next period is:*

$$F_{t+1} = \alpha D_t + (1 - \alpha)F_t = 0.4 \times 180 + 0.6 \times 150 = 72 + 90 = 162$$

Using exponential smoothing, the company achieves more accurate forecasts, minimizing stockouts and overstock situations. This leads to better customer satisfaction and optimized inventory levels.

Example 6.45. *An e-commerce business implements Random Forest to predict monthly sales for thousands of products. They use features like past sales, marketing spend, and holidays. Suppose the Random Forest model uses 100 decision trees and the prediction for a particular product is based on the following individual tree predictions: 120, 130, 125, 140, and 135 units.*

Solution 6.45. *The Random Forest model predicts the demand by averaging the outputs of multiple decision trees:*

$$\hat{D} = \frac{1}{5}(120 + 130 + 125 + 140 + 135) = \frac{650}{5} = 130$$

The Random Forest model provides high accuracy in sales predictions, helping the business to maintain optimal inventory levels and reduce holding costs, thereby enhancing overall operational efficiency.

6.15.1 Impact on Organizational Operations

Example 6.46. *A manufacturing company uses linear regression to forecast raw material requirements based on production schedules and seasonal demand. Suppose the linear regression model is $R = 50 + 0.5P + 0.3S$, where R is raw material requirement, P is production units, and S is seasonal index. If the production units are 200 and the seasonal index is 1.5, what is the forecasted raw material requirement?*

Solution 6.46. *The raw material requirement is calculated as:*

$$R = 50 + 0.5 \times 200 + 0.3 \times 1.5 = 50 + 100 + 0.45 = 150.45$$

This results in streamlined procurement processes, reducing material shortages and production delays. The company can better manage its supply chain and ensure timely production.

Example 6.47. *A supermarket chain employs MAPE to evaluate the accuracy of its demand forecasts and adjust their strategies accordingly. Suppose the actual demands and forecasts for three months are* $(100, 90)$, $(150, 160)$, $(120, 130)$.

Solution 6.47. *MAPE is calculated as:*

$$MAPE = \frac{1}{3}\left(\left|\frac{100-90}{100}\right| + \left|\frac{150-160}{150}\right| + \left|\frac{120-130}{120}\right|\right) \times 100\%$$

$$= \frac{1}{3}(0.1 + 0.0667 + 0.0833) \times 100\% = \frac{0.25}{3} \times 100\% = 8.33\%$$

Continuously monitoring MAPE, the supermarket can fine-tune its forecasting models, reducing inventory costs and improving product availability for customers.

6.15.2 Best Practices and Implementation Strategies

Example 6.48. *A retailer adopts a hybrid forecasting approach, combining exponential smoothing, linear regression, and machine learning techniques. Suppose they use exponential smoothing for short-term forecasts with $\alpha = 0.3$, linear regression for medium-term forecasts, and Random Forest for long-term forecasts.*

Solution 6.48. *For exponential smoothing, if the previous forecast was 200 units, the actual demand was 220 units, then the forecast for the next period is:*

$$F_{t+1} = 0.3 \times 220 + 0.7 \times 200 = 66 + 140 = 206$$

For linear regression, suppose the model is $D = 50 + 1.2P + 0.8M$, *where* P *is price and* M *is marketing spend. If* $P = 30$ *and* $M = 40$, *then the forecasted demand is:*

$$D = 50 + 1.2 \times 30 + 0.8 \times 40 = 50 + 36 + 32 = 118$$

Using Random Forest, suppose the prediction from multiple trees averages to 150 units. This hybrid approach leverages the strengths of each method, resulting in highly accurate demand forecasts. The retailer can optimize stock levels, improve customer satisfaction, and reduce costs.

Example 6.49. *A company implements automated inventory management systems that integrate real-time sales data and advanced forecasting models. Suppose the system uses real-time data to update forecasts and inventory levels dynamically.*

Solution 6.49. *With automated systems, if the real-time sales data indicates an increase in demand by 10%, the forecast and inventory levels can be adjusted accordingly. For example, if the initial forecast was 200 units, the new forecast would be:*

$$\text{New Forecast} = 200 \times 1.1 = 220$$

This allows the company to respond quickly to changes in demand, adjust inventory levels dynamically, and reduce manual errors. This leads to more efficient inventory management and better resource allocation.

6.16 Conclusion

Inventory forecasting is an essential aspect of supply chain management that significantly impacts the efficiency and effectiveness of organizational operations. Accurately predicting future demand, companies can maintain optimal inventory levels, reduce costs, and improve customer satisfaction. The integration of advanced forecasting techniques, such as exponential smoothing, linear regression, machine learning models, and hybrid approaches, allows for more precise and reliable forecasts. These methods help in aligning production schedules with demand, optimizing resource allocation, and minimizing the risks associated with stockouts and overstocking. Coordination and integration across the supply chain are enhanced through the use of shared forecasts, enabling seamless collaboration between suppliers, manufacturers, and retailers. This leads to reduced lead times, improved inventory turnover rates, and a more responsive supply chain. Cost management is another critical benefit of effective inventory forecasting. Accurately estimating the required materials and components, companies can avoid unnecessary expenditures and holding costs. This not only improves the bottom line but also enhances the overall financial health of the organization. Moreover, enhancing supply chain efficiency through accurate inventory forecasting ensures that products are available at the right time and place, meeting customer demand promptly and effectively. This is particularly important in industries with fluctuating demand patterns, such as retail and pharmaceuticals, where timely availability of products can significantly impact market competitiveness. Leveraging advanced forecasting models and integrating them into the supply chain processes, companies can achieve a balanced, cost-effective, and efficient inventory management system. Continuous improvement and adaptation of these models to changing market conditions will further strengthen the resilience and agility of the supply chain, ensuring sustained business success.

Exercises

Exercise 6.1. *Given the regularization parameter* $\lambda = 0.5$, *and* β *values are obtained as 2, 1.5, and -0.5, explain how ridge regression stabilizes demand forecasts.*

Exercise 6.2. *Explain how ridge regression can help in a scenario where demand is predicted based on price (P) and marketing (M) with multicollinearity.*

Exercise 6.3. *Given the regularization parameter* $\lambda = 1.0$, *describe how LASSO regression identifies significant factors affecting demand.*

Exercise 6.4. *In a scenario where demand depends on multiple factors* (P, M, T), *explain how LASSO regression can simplify the forecasting model.*

Exercise 6.5. *Given* $\lambda_1 = 0.5$ *and* $\lambda_2 = 0.5$, *explain how elastic net balances variable selection and regularization.*

Exercise 6.6. *Explain how elastic net can handle correlated predictors* (P, M) *in a demand forecasting model.*

7

Inventory Auditing and Control

This chapter delves into the essential role of inventory auditing and control within effective inventory management. It underscores the significance of these processes in ensuring accurate inventory records and minimizing the risk of shrinkage. Examining techniques such as physical inventory counts and inventory reconciliation, the chapter offers practical insights into their implementation and relevance to organizational operations. Furthermore, it highlights the broader implications of inventory control on supply chain management, emphasizing the need for seamless coordination and efficient use of materials. The chapter aims to enhance readers' understanding of inventory auditing and control, providing a solid foundation for adopting advanced inventory management practices. Its thorough exploration makes it a valuable resource for practitioners, researchers, and students seeking to refine inventory control techniques across various industries.

7.1 Introduction

Inventory auditing and control are pivotal aspects of inventory management that ensure the accuracy of inventory records and optimize stock levels. This chapter provides an in-depth exploration of these processes, highlighting their significance in organizational operations and their impact on supply chain management. Through a detailed examination of various auditing techniques and practical applications, the chapter aims to equip readers with the knowledge and tools necessary for effective inventory management.

7.1.1 Overview of Inventory Auditing and Control

Inventory auditing and control are essential components of effective inventory management, ensuring the accuracy and reliability of inventory records. This subsection provides a broad overview of these processes, explaining their definitions, purposes, and significance within an organization. It explores how

DOI: 10.1201/9781003534860-7

inventory auditing helps in verifying the physical existence of inventory items and how inventory control mechanisms maintain optimal stock levels to meet demand without overstocking or stockouts.

1. **Definitions** Inventory auditing refers to the systematic process of verifying the accuracy of inventory records through physical counts, cycle counts, and reconciliations. It involves cross-checking the actual inventory against the recorded inventory to identify discrepancies and ensure that the records reflect the true inventory levels. Inventory control, on the other hand, encompasses the policies, procedures, and technologies used to maintain inventory at desired levels. This includes determining optimal order quantities, managing reorder points, and utilizing software systems for real-time inventory tracking.
2. **Purpose and Significance** The primary purpose of inventory auditing is to ensure that inventory records are accurate and up-to-date. This accuracy is crucial for financial reporting, compliance with regulations, and making informed business decisions. Inventory control aims to optimize inventory levels to balance the costs associated with holding inventory against the benefits of having inventory available to meet customer demand. Together, these processes help prevent issues such as overstocking, stockouts, and inventory shrinkage, which can have significant financial and operational impacts on an organization.
3. **Historical Context** Historically, inventory auditing and control practices have evolved significantly. Early methods relied on manual counts and rudimentary record-keeping, which were prone to errors and inefficiencies. With the advent of technology, these practices have become more sophisticated, incorporating barcode scanning, RFID technology, and automated inventory management systems. Key milestones include the development of the perpetual inventory system, which allows for continuous tracking of inventory, and the integration of enterprise resource planning (ERP) systems, which provide a comprehensive view of inventory across an organization.

7.1.2 Importance of Inventory Auditing in Inventory Management

This subsection explores the critical role that inventory auditing plays within the broader context of inventory management. It examines the benefits and impacts of regular and systematic inventory audits on various aspects of organizational operations.

1. **Ensuring Accuracy and Reliability** Inventory auditing is essential for ensuring the accuracy and reliability of inventory records.

Accurate records are necessary for financial reporting, as discrepancies between recorded and actual inventory can lead to misstated financial statements. Furthermore, accurate inventory data is crucial for operational decision-making, such as determining reorder points and managing production schedules. Without reliable inventory records, organizations risk making decisions based on incorrect information, which can result in inefficiencies and increased costs.

2. **Mitigating Inventory Shrinkage** Inventory shrinkage refers to the loss of inventory due to factors such as theft, damage, or administrative errors. Regular inventory audits help mitigate shrinkage by identifying discrepancies between recorded and actual inventory levels. Conducting physical counts and reconciling differences, organizations can detect and address the causes of shrinkage. This not only helps in maintaining accurate records but also in implementing measures to prevent future losses, such as improving security protocols or enhancing employee training.
3. **Enhancing Operational Efficiency** Accurate inventory records, achieved through systematic auditing, enhance overall operational efficiency. With reliable data, organizations can better forecast demand, plan production schedules, and manage inventory levels. This leads to reduced lead times, minimized stockouts, and optimized order quantities. Additionally, accurate records facilitate efficient order fulfillment and inventory turnover, contributing to improved customer satisfaction and reduced carrying costs.

7.2 Techniques of Inventory Auditing

Effective inventory auditing involves various techniques that organizations use to ensure the accuracy and reliability of their inventory records.

7.3 Inventory Reconciliation

Inventory Reconciliation is the process of ensuring that the physical inventory of goods matches the recorded inventory in the company's accounting or inventory management system. This process involves comparing actual stock levels to what is documented, identifying discrepancies, and making the necessary adjustments to the records. Inventory reconciliation is crucial for maintaining accurate financial statements, optimizing inventory levels, and preventing loss or theft.

7.3.1 Steps Involved in Inventory Reconciliation

1. **Physical Count**: Conducting a thorough count of all inventory items. This can be done periodically (e.g., monthly, quarterly) or continuously using cycle counting.
2. **Data Comparison**: Comparing the physical count data with the inventory records from the accounting system or inventory management software.
3. **Identifying Discrepancies**: Noting any differences between the physical count and the recorded quantities. Discrepancies can arise from various factors such as theft, loss, damage, data entry errors, or timing differences in recording transactions.
4. **Investigating Discrepancies**: Determining the cause of the discrepancies by reviewing transaction histories, checking for errors in documentation, and investigating any instances of theft or damage.
5. **Adjusting Records**: Making the necessary adjustments in the inventory management system to align the recorded inventory levels with the actual physical count.
6. **Reporting**: Documenting the reconciliation process and findings, and reporting significant discrepancies to management for further action.

7.3.2 Importance of Inventory Reconciliation

- **Accuracy in Financial Statements**: Ensures that the inventory values reported in the financial statements are accurate, which is essential for financial reporting and compliance.

- **Inventory Management**: Helps in maintaining optimal inventory levels, reducing carrying costs, and avoiding stockouts or overstock situations.

- **Theft and Loss Prevention**: Identifies and helps prevent theft, loss, and pilferage by regularly verifying inventory levels.

- **Operational Efficiency**: Improves operational efficiency by providing accurate data for decision-making related to procurement, sales, and production.

7.3.3 Challenges in Inventory Reconciliation

- **Time-Consuming**: The process can be labor-intensive and time-consuming, especially for businesses with large volumes of inventory.

- **Human Error**: Manual counting and data entry can lead to errors.

- **Complexity**: Businesses with multiple locations, varied product lines, and complex supply chains may face additional challenges in reconciling inventory accurately.

7.3.4 Best Practices

- **Regular Reconciliation**: Conducting inventory reconciliation at regular intervals to ensure continuous accuracy.
- **Automated Systems**: Utilizing automated inventory management systems and barcode/RFID technology to reduce errors and improve efficiency.
- **Employee Training**: Training employees on proper inventory handling and documentation procedures to minimize errors and discrepancies.

7.3.5 Mathematical Formulation

Let I_t represent the recorded inventory quantity of item t in the inventory management system, and C_t represent the counted inventory quantity of item t during the physical inventory count.

1. **Root Cause Analysis:**
   - **Discrepancy Analysis:** The discrepancy D_t for item t can be calculated as:

     $$D_t = C_t - I_t$$

     To measure the total discrepancy TD across all items, the formula is:

     $$TD = \sum_{t=1}^{n} |D_t|$$

     where n is the total number of inventory items.

     A useful metric for assessing the effectiveness of the inventory reconciliation process is the Inventory Accuracy Rate A, which can be defined as:

     $$A = \left(1 - \frac{TD}{\sum_{t=1}^{n} I_t}\right) \times 100\%$$

     For a detailed analysis, the Root Mean Square Error (RMSE) can also be used to measure the accuracy of the inventory counts:

     $$RMSE = \sqrt{\frac{1}{n}\sum_{t=1}^{n}(D_t)^2}$$

- **Root Cause Categorization:** Classify the discrepancies into categories (e.g., theft, data entry errors, spoilage) and analyze the frequency and impact of each category.

$$\text{Frequency}_c = \frac{\text{Number of discrepancies in category } c}{\text{Total number of discrepancies}}$$

2. **Use of Technology:**

- **Accuracy Improvement:** Implement barcode/RFID systems to enhance accuracy. Measure improvement in accuracy using the Inventory Accuracy Rate A before and after technology implementation.

$$A_{\text{before}} = \left(1 - \frac{TD_{\text{before}}}{\sum_{t=1}^{n} I_{t,\text{before}}}\right) \times 100\%$$

$$A_{\text{after}} = \left(1 - \frac{TD_{\text{after}}}{\sum_{t=1}^{n} I_{t,\text{after}}}\right) \times 100\%$$

- **Error Reduction Rate:**

$$\text{Error Reduction Rate} = \left(\frac{A_{\text{after}} - A_{\text{before}}}{A_{\text{before}}}\right) \times 100\%$$

3. **Employee Training:**

- **Training Effectiveness:** Measure the impact of training programs on inventory accuracy. Let E_{before} and E_{after} be the error rates before and after training.

$$\text{Error Rate Reduction} = E_{\text{before}} - E_{\text{after}}$$

- **Improvement in Accuracy:** Calculate the improvement in accuracy post-training.

$$\text{Improvement in Accuracy} = \left(\frac{A_{\text{after}} - A_{\text{before}}}{A_{\text{before}}}\right) \times 100\%$$

7.4 Practical Applications of Inventory Auditing

Inventory auditing is not only a theoretical concept but also a practical necessity for various industries. This section highlights best practices for implementing inventory auditing effectively.

Example 7.1. *A large retail chain conducts physical inventory counts at the end of each quarter. During the last count, they recorded the following discrepancies for a specific product category:*

- *Recorded inventory: 1,200 units*
- *Counted inventory: 1,150 units*

Calculate the discrepancy and the impact on the inventory accuracy rate.

Solution 7.1. *The discrepancy D is calculated as:*

$$D = \textit{Counted inventory} - \textit{Recorded inventory} = 1{,}150 - 1{,}200 = -50 \textit{ units}$$

To find the inventory accuracy rate A:

$$A = \left(1 - \frac{|D|}{\textit{Recorded inventory}}\right) \times 100\% = \left(1 - \frac{50}{1{,}200}\right) \times 100\% = 95.83\%$$

The 4.17% discrepancy indicates areas for improvement in inventory management practices.

Example 7.2. *A manufacturing company uses cycle counting for its raw materials. In the last cycle count, they counted a subset of materials and found the following:*

- *Total items counted: 500*
- *Recorded inventory for these items: 510*

Calculate the total discrepancy and its effect on production efficiency.

Solution 7.2. *The total discrepancy TD is calculated as:*

$$TD = \textit{Counted inventory} - \textit{Recorded inventory} = 500 - 510 = -10 \textit{ units}$$

To determine the impact on production efficiency, assume each unit discrepancy causes a 0.1% drop in efficiency. Therefore, the total efficiency drop is:

$$\textit{Efficiency drop} = 10 \times 0.1\% = 1\%$$

A 1% drop in efficiency highlights the need for accurate inventory records to maintain smooth production operations.

Example 7.3. *A hospital conducts inventory reconciliation for medical supplies every month. Last month, they recorded the following:*

- *Recorded inventory: 2,000 units*
- *Counted inventory: 2,050 units*

Calculate the discrepancy and discuss its potential impact on patient care.

Solution 7.3. *The discrepancy D is calculated as:*

$$D = \textit{Counted inventory} - \textit{Recorded inventory} = 2{,}050 - 2{,}000 = 50 \textit{ units}$$

The positive discrepancy indicates overstocking. While this may ensure that supplies are always available, it can also lead to higher holding costs and potential waste due to expired items. Maintaining accurate inventory ensures optimal stock levels, balancing availability and cost.

7.4.1 Best Practices for Implementing Inventory Auditing

- **Automate Data Collection:** Use barcode and RFID systems to reduce manual entry errors.
- **Regular Training:** Train employees on proper inventory management techniques to maintain accuracy.
- **Continuous Monitoring:** Implement continuous monitoring systems to detect discrepancies early.
- **Use Technology:** Employ inventory management software for real-time tracking and reporting.

7.5 Impact on Supply Chain Management

Inventory auditing plays a critical role in supply chain management, influencing various aspects from coordination to material expenditure and the integration of modern technologies. This section explores the profound impact of inventory auditing on the supply chain, highlighting its importance in achieving efficient and effective supply chain operations.

7.5.1 Role of Inventory Auditing in Supply Chain Coordination

Effective supply chain coordination relies on accurate and reliable inventory data. Inventory auditing ensures the precision of this data, facilitating better coordination across the supply chain.

7.5.1.1 Enhancing Visibility and Transparency

Inventory auditing improves visibility and transparency within the supply chain by providing accurate and up-to-date inventory information. This visibility allows all stakeholders, including suppliers, manufacturers, and distributors, to have a clear understanding of inventory levels at each stage of the supply chain. Enhanced visibility helps in:

- **Demand Forecasting:** Accurate inventory data aids in better demand forecasting, enabling organizations to plan production and procurement more effectively.
- **Order Fulfillment:** With precise inventory information, organizations can fulfill orders more efficiently, reducing the risk of stockouts and backorders.
- **Collaboration:** Transparency fosters collaboration among supply chain partners, leading to improved communication and coordination.

7.5.1.2 Reducing Lead Times and Improving Responsiveness

Inventory auditing helps in reducing lead times and enhancing the responsiveness of the supply chain. Maintaining accurate inventory records, organizations can:

- **Streamline Replenishment:** Accurate inventory data allows for timely and precise replenishment, reducing the time taken to restock items.
- **Optimize Production Schedules:** Reliable inventory information helps in aligning production schedules with actual demand, minimizing delays and ensuring timely delivery.
- **Increase Agility:** An agile supply chain can quickly respond to changes in demand or supply conditions, supported by accurate and current inventory data.

7.5.2 Integration With Modern Supply Chain Technologies

Modern supply chain technologies enhance the effectiveness of inventory auditing by providing advanced tools and systems for managing inventory data. This integration supports more accurate and efficient inventory management processes.

7.5.2.1 Utilizing Automation and AI

The integration of automation and artificial intelligence (AI) in inventory auditing brings several benefits:

- **Automated Counting:** Automation technologies such as drones and robots can perform physical counts with higher speed and accuracy.
- **Predictive Analytics:** AI algorithms analyze historical inventory data to predict future demand, optimizing inventory levels and reducing the risk of stockouts or overstocking.
- **Anomaly Detection:** AI can detect anomalies in inventory data, such as unusual patterns or discrepancies, enabling quick corrective actions.

7.5.2.2 Leveraging IoT and RFID

The Internet of Things (IoT) and Radio Frequency Identification (RFID) technologies play a significant role in enhancing inventory auditing:

- **Real-Time Tracking:** IoT devices and RFID tags provide real-time tracking of inventory items, ensuring accurate and up-to-date inventory records.
- **Improved Accuracy:** RFID technology reduces human errors in counting and recording, enhancing the accuracy of inventory data.

- **Streamlined Processes:** The integration of IoT and RFID with inventory management systems streamlines inventory auditing processes, reducing manual effort and increasing efficiency.

7.5.2.3 Implementing Blockchain for Transparency

Blockchain technology offers a decentralized and immutable ledger for inventory transactions, providing enhanced transparency and security:

- **Traceability:** Blockchain allows for end-to-end traceability of inventory items, from production to final delivery, ensuring transparency across the supply chain.
- **Data Integrity:** The immutable nature of blockchain ensures that inventory data cannot be altered or tampered with, providing a reliable source of truth.
- **Enhanced Collaboration:** Blockchain facilitates secure and transparent data sharing among supply chain partners, improving collaboration and trust.

7.6 Current Trends and Future Directions

As the field of inventory auditing evolves, new techniques, technological advancements, and emerging challenges continue to shape its future. This section explores the current trends in inventory auditing, the implications of technological advancements, and the future challenges and opportunities facing this critical area of inventory management.

7.6.1 Emerging Techniques in Inventory Auditing

Innovative techniques are continuously being developed to enhance the accuracy and efficiency of inventory auditing. These emerging techniques address various aspects of inventory management and offer new ways to optimize inventory control.

7.6.1.1 Predictive Analytics

Predictive analytics leverages historical data, machine learning, and statistical algorithms to forecast future inventory needs. Analyzing trends and patterns, predictive analytics helps organizations:

- **Demand Forecasting:** Accurately predict future demand, allowing for better inventory planning and reducing the risk of stockouts or overstocking.

- **Inventory Optimization:** Determine optimal inventory levels and reorder points to maintain a balance between availability and cost.
- **Proactive Management:** Identify potential issues before they become critical, enabling proactive management of inventory.

7.6.1.2 Continuous Auditing

Continuous auditing involves the real-time monitoring and auditing of inventory transactions. This approach ensures that inventory data is always current and accurate, providing several benefits:

- **Immediate Detection:** Quickly detect discrepancies and errors, allowing for immediate corrective actions.
- **Enhanced Accuracy:** Maintain consistently accurate inventory records without the need for periodic physical counts.
- **Reduced Manual Effort:** Minimize the manual effort required for traditional periodic audits, freeing up resources for other tasks.

7.6.1.3 Blockchain Technology

Blockchain technology offers a secure and transparent way to record and track inventory transactions. The decentralized nature of blockchain provides several advantages:

- **Data Integrity:** Ensure the integrity and immutability of inventory records, reducing the risk of fraud and errors.
- **Traceability:** Provide end-to-end traceability of inventory items, enhancing transparency across the supply chain.
- **Secure Collaboration:** Facilitate secure and transparent collaboration among supply chain partners, improving trust and efficiency.

7.6.2 Technological Advancements and Their Implications

Technological advancements are transforming inventory auditing, offering new tools and methods to enhance efficiency and accuracy. The integration of these technologies has significant implications for inventory management practices.

7.6.2.1 Artificial Intelligence and Machine Learning

Artificial intelligence (AI) and machine learning (ML) are revolutionizing inventory auditing by providing advanced analytics and automation capabilities:

- **Automated Audits:** Use AI-powered robots and drones to perform automated inventory counts, reducing the need for manual labor.

- **Anomaly Detection:** Employ machine learning algorithms to detect anomalies and discrepancies in inventory data, enabling quick corrective actions.
- **Enhanced Forecasting:** Leverage AI and ML to improve demand forecasting and inventory optimization, leading to more accurate and efficient inventory management.

7.6.2.2 Internet of Things (IoT)

The Internet of Things (IoT) enables real-time monitoring and tracking of inventory through interconnected devices and sensors:

- **Real-Time Data:** Provide real-time data on inventory levels, locations, and conditions, ensuring accurate and up-to-date inventory records.
- **Improved Visibility:** Enhance visibility across the supply chain, allowing for better coordination and decision-making.
- **Predictive Maintenance:** Use IoT data to predict and prevent equipment failures, reducing downtime and maintaining smooth operations.

7.6.2.3 Advanced Data Analytics

Advanced data analytics tools are transforming the way organizations analyze and utilize inventory data:

- **Big Data:** Analyze large volumes of inventory data to identify trends, patterns, and insights that inform inventory management decisions.
- **Prescriptive Analytics:** Use prescriptive analytics to recommend optimal inventory actions based on predictive models and scenario analysis.
- **Data Visualization:** Employ data visualization tools to present complex inventory data in a clear and actionable format, facilitating better decision-making.

7.6.3 Future Challenges and Opportunities

As inventory auditing continues to evolve, organizations will face new challenges and opportunities. Addressing these challenges and leveraging opportunities will be crucial for the future of inventory management.

7.6.3.1 Challenges

Several challenges are anticipated in the future of inventory auditing:

- **Data Security:** Ensuring the security and privacy of inventory data will be a critical concern, especially with the increasing use of digital technologies.

- **Integration Complexity:** Integrating new technologies with existing systems and processes can be complex and require significant investment and expertise.
- **Skills Gap:** The rapid advancement of technologies may create a skills gap, requiring organizations to invest in training and development to keep their workforce up-to-date.
- **Regulatory Compliance:** Keeping up with evolving regulations and standards related to inventory management and auditing will be an ongoing challenge.

7.6.3.2 Opportunities

Despite the challenges, numerous opportunities exist for organizations to enhance their inventory auditing practices:

- **Innovation and Improvement:** Embrace innovative technologies and techniques to continually improve inventory accuracy and efficiency.
- **Competitive Advantage:** Leveraging advanced inventory auditing practices can provide a competitive advantage by improving operational efficiency and customer satisfaction.
- **Sustainability:** Implementing efficient inventory management practices can contribute to sustainability efforts by reducing waste and optimizing resource utilization.
- **Collaborative Ecosystems:** Foster collaborative ecosystems with supply chain partners to enhance transparency, trust, and coordination across the supply chain.

7.7 Conclusion

Inventory auditing and control are fundamental components of effective inventory management, playing a crucial role in maintaining accurate inventory records and mitigating the risk of inventory shrinkage. This chapter has provided a comprehensive exploration of various inventory auditing techniques, including physical inventory counts, cycle counts, and inventory reconciliation. It has also examined the practical applications of these techniques across different industries, highlighting their impact on operational efficiency and cost reduction. The chapter emphasized the significant role inventory auditing plays in supply chain management, particularly in enhancing coordination, reducing lead times, and optimizing material expenditure. The integration of modern technologies such as artificial intelligence, the Internet of Things

(IoT), and blockchain has been shown to further enhance the accuracy and efficiency of inventory auditing processes. Current trends in inventory auditing, such as predictive analytics and continuous auditing, represent the ongoing evolution of this field, driven by technological advancements. These trends present both opportunities and challenges, including the need for data security, integration with existing systems, and compliance with evolving regulations.

Exercises

Exercise 7.1. *A company conducts an inventory reconciliation and finds the following discrepancies between physical count and recorded inventory for three items:*

Item	*Recorded Quantity*	*Physical Quantity*
A	150	140
B	200	210
C	250	245

Calculate the inventory discrepancy for each item and the total discrepancy for all items combined.

Exercise 7.2. *During an inventory reconciliation process, it was found that the recorded quantity of a product is 10% higher than the physical count. If the recorded quantity is 550 units, determine the physical quantity and the discrepancy.*

Exercise 7.3. *A company has the following recorded and physical inventory quantities for four items:*

Item	*Recorded Quantity*	*Physical Quantity*
W	300	295
X	400	390
Y	500	505
Z	600	590

If the company values its inventory at $20 per unit, calculate the total financial impact of the discrepancies.

Exercise 7.4. *An inventory reconciliation reveals that the actual inventory levels are lower than the recorded levels by 5% for every product in a warehouse. If the total recorded inventory value is $100,000, determine the actual inventory value and the total discrepancy in value.*

Exercise 7.5. *During an inventory reconciliation, a company identifies that the shrinkage rate (percentage of inventory loss) is 2% of the recorded inventory. If the recorded inventory is valued at $50,000, calculate the value of the lost inventory and the adjusted inventory value after accounting for the shrinkage.*

Exercise 7.6. *A company performs a root cause analysis on inventory discrepancies and finds the following data for three items:*

Item	*Recorded Quantity*	*Physical Quantity*
P	200	180
Q	150	155
R	100	95

Calculate the discrepancy D_t for each item, the total discrepancy TD, and the Inventory Accuracy Rate A.

Exercise 7.7. *Given the following discrepancies for four inventory items:*

Item	*Discrepancy*
A	−5
B	10
C	−2
D	3

Calculate the Root Mean Square Error (RMSE) for the discrepancies.

Exercise 7.8. *A company implements an RFID system to improve inventory accuracy. Before the implementation, the total recorded inventory value was $50,000, and the total discrepancy value was $2,500. After implementation, the total recorded inventory value was $55,000, and the total discrepancy value was $1,000. Calculate the Inventory Accuracy Rate A before and after the implementation, and the Error Reduction Rate.*

Exercise 7.9. *Classify the following discrepancies into categories (theft, data entry errors, spoilage) and calculate the frequency of each category. The data is as follows:*

Discrepancy Amount	*Category*
5	*Theft*
3	*Data Entry Errors*
2	*Spoilage*
4	*Theft*
6	*Data Entry Errors*
1	*Spoilage*

Exercise 7.10. *A company measures the impact of training programs on inventory accuracy. Before training, the error rate was 8%. After training, the error rate decreased to 5%. Calculate the Error Rate Reduction and the Improvement in Accuracy if the initial accuracy rate was 92%.*

8

Inventory Metrics and KPIs

This chapter introduces key metrics and performance indicators (KPIs) vital for effective inventory management. It emphasizes the importance of selecting reliable KPIs that evaluate and enhance system performance. The chapter reviews key metrics like inventory-to-sales ratio and demand forecast accuracy, showing how they can monitor sell-through performance and improve sales and merchandising processes. It provides practical insights into choosing metrics that drive productivity and decision-making, while also examining their broader role in measuring progress and guiding organizational behavior. Drawing from examples and best practices, the chapter explores how inventory metrics and KPIs can be applied to achieve better outcomes in inventory management. Overall, it offers a comprehensive overview for practitioners, researchers, and students, equipping them with the tools needed to effectively implement KPIs and optimize inventory systems across various industries.

8.1 Introduction

Effective inventory management is critical to the success of any business involved in the production, storage, or sale of goods. Maintaining optimal inventory levels, businesses can meet customer demand without overstocking, thus reducing costs and increasing profitability. This chapter shifts into the essential metrics and key performance indicators (KPIs) that are pivotal in evaluating and improving inventory management systems.

8.1.1 Importance of Inventory Management

Inventory management is the backbone of efficient supply chain operations. It involves tracking and controlling stock levels, orders, sales, and deliveries. Proper inventory management ensures that there is enough stock to meet customer demand while minimizing the costs associated with holding inventory. The following points highlight the importance of inventory management:

- **Cost Control:** Proper inventory management helps in minimizing the costs associated with overstocking and stockouts. Overstocking leads to increased

DOI: 10.1201/9781003534860-8

holding costs, while stockouts can result in lost sales and dissatisfied customers.

- **Improved Cash Flow:** Efficient inventory management ensures that capital is not tied up unnecessarily in stock. This helps in maintaining a healthy cash flow, which is crucial for the day-to-day operations of the business.
- **Customer Satisfaction:** Maintaining optimal inventory levels, businesses can ensure that products are available when customers need them, leading to higher customer satisfaction and loyalty.
- **Operational Efficiency:** Effective inventory management streamlines operations, reducing the time and effort required to manage stock levels. This leads to increased productivity and efficiency.
- **Strategic Planning:** Accurate inventory data provides valuable insights that help in making informed decisions regarding purchasing, production planning, and sales strategies.

Inventory management is crucial for maintaining balance between supply and demand, controlling costs, improving cash flow, ensuring customer satisfaction, and enhancing overall operational efficiency.

8.1.2 Overview of Metrics and KPIs

Metrics and key performance indicators (KPIs) are essential tools for evaluating the effectiveness of inventory management practices. They provide measurable values that help in assessing performance, identifying areas for improvement, and making data-driven decisions. The following sections provide an overview of some key inventory metrics and KPIs:

- **Inventory to Sales Ratio:** This metric compares the amount of inventory to the sales revenue generated over a specific period. It helps in understanding whether inventory levels are in line with sales performance.
- **Demand Forecast Accuracy:** This KPI measures the accuracy of demand forecasts compared to actual sales. High accuracy indicates effective forecasting, which is crucial for maintaining optimal inventory levels.
- **Inventory Turnover:** This metric indicates how many times inventory is sold and replaced over a specific period. High turnover rates suggest efficient inventory management, while low rates may indicate overstocking or slow-moving inventory.
- **Gross Margin Return on Investment (GMROI):** This KPI measures the profitability of inventory investments by comparing gross margin to the cost of inventory. It helps in assessing the financial performance of inventory management.

- **Days Sales of Inventory (DSI):** This metric indicates the average number of days it takes to sell inventory. Lower DSI values suggest faster inventory turnover and better liquidity.
- **Sell-Through Rate:** This KPI measures the percentage of inventory sold during a specific period compared to the amount of inventory received. It is useful for evaluating the effectiveness of sales and marketing efforts.

Each of these metrics and KPIs provides unique insights into different aspects of inventory management. Monitoring these indicators, businesses can identify trends, uncover inefficiencies, and make informed decisions to improve their inventory management processes. In the following sections, we will delve deeper into each of these metrics and KPIs, exploring their significance, calculation methods, and practical applications in inventory management.

8.2 Choosing Reliable and Efficient Inventory Metrics

Selecting the right inventory metrics and KPIs is essential for effectively managing and improving inventory performance. The chosen metrics should provide meaningful insights that drive better decision-making and align with the overall business objectives. This section outlines the criteria for selecting reliable and efficient inventory metrics and discusses the importance of aligning these metrics with business goals.

8.2.1 Criteria for Selecting Metrics

Choosing the appropriate metrics for inventory management involves considering several key criteria:

- **Relevance:** Metrics should be directly related to inventory management processes and reflect critical aspects of performance. They should address specific issues or areas of interest within the inventory system.
- **Measurability:** The selected metrics must be quantifiable, allowing for objective measurement and comparison over time. Clear definitions and standardized measurement methods are essential for consistent data collection and analysis.
- **Actionability:** Effective metrics should provide insights that lead to actionable improvements. They should help identify areas where changes can be made to enhance performance and drive better outcomes.
- **Timeliness:** Metrics should provide timely information that enables prompt decision-making. Real-time or regularly updated metrics allow for quick responses to emerging issues and trends.

- **Comparability:** Metrics should enable comparison across different periods, departments, or industry benchmarks. This helps in assessing performance relative to past performance and against industry standards.
- **Simplicity:** While metrics should be comprehensive, they should also be easy to understand and interpret. Complex metrics may hinder effective communication and decision-making.
- **Consistency:** The metrics should be consistently measured and reported over time to track progress accurately. Consistent application ensures reliability and validity of the data.

Adhering to these criteria, businesses can select inventory metrics that provide meaningful and actionable insights, leading to better inventory management and overall operational efficiency.

8.2.2 Aligning Metrics With Business Goals

Aligning inventory metrics with business goals is crucial for ensuring that the metrics support the broader objectives of the organization. The following steps can help in achieving this alignment:

- **Identify Key Business Goals:** Start by clearly defining the primary business goals related to inventory management. These may include objectives such as reducing holding costs, improving customer satisfaction, increasing turnover rates, or enhancing profitability.
- **Map Metrics to Goals:** Select metrics that directly relate to and support the achievement of the identified business goals. For example, if the goal is to reduce holding costs, metrics such as inventory turnover and days sales of inventory (DSI) would be relevant.
- **Set Target Values:** Establish target values for each metric based on historical data, industry benchmarks, or strategic objectives. These targets provide a benchmark for measuring success and progress toward achieving business goals.
- **Regular Review and Adjustment:** Continuously monitor the selected metrics and review their relevance and effectiveness in achieving business goals. Adjust the metrics as necessary to reflect changes in business strategy, market conditions, or operational priorities.
- **Communicate and Engage:** Ensure that the metrics and their importance are communicated to all relevant stakeholders, including management, operations teams, and front-line staff. Engaging the entire organization in understanding and using the metrics helps in driving collective effort toward achieving business goals.

Aligning inventory metrics with business goals ensures that the efforts in inventory management are focused and contribute to the overall success of the organization. It helps in creating a cohesive strategy where inventory management practices support and enhance the broader objectives of the business.

In the subsequent sections, we will explore various key inventory metrics and KPIs in detail, providing insights into their significance, calculation methods, and practical applications in inventory management.

8.3 Key Inventory Metrics and KPIs

Effective inventory management relies on a set of key metrics and KPIs that provide insights into various aspects of inventory performance. This section explores some of the most important inventory metrics and KPIs, detailing their significance, calculation methods, and practical applications.

8.3.1 Inventory to Sales Ratio

The inventory to sales ratio measures the relationship between the amount of inventory on hand and the sales generated over a specific period. This metric helps businesses understand whether their inventory levels are aligned with their sales performance.

Significance:

- Indicates the balance between inventory investment and sales.
- Helps in identifying overstocking or understocking issues.
- Provides insights into inventory efficiency and turnover.

Calculation:

$$\text{Inventory to Sales Ratio} = \frac{\text{Average Inventory}}{\text{Sales}}$$

Example 8.1. *Suppose a retail store has an average inventory of \$80,000 and generates sales of \$400,000 over a fiscal quarter. To calculate the Inventory to Sales Ratio:*

$$Inventory\ to\ Sales\ Ratio = \frac{Average\ Inventory}{Sales}$$

Substituting the given values:

$$Inventory\ to\ Sales\ Ratio = \frac{80{,}000}{400{,}000} = 0.20$$

Solution 8.1. *In this example, the Inventory to Sales Ratio is 0.20. This means for every dollar of sales, the store has 20 cents invested in inventory.*

Interpretation:

- *A ratio of 0.20 indicates a relatively low inventory investment compared to sales, suggesting efficient inventory turnover. This could be due to strong sales performance or effective inventory management strategies.*
- *Monitoring this ratio over time helps assess whether inventory levels are appropriately aligned with sales trends. A sudden increase in the ratio might indicate overstocking, while a decrease might suggest understocking.*

Example 8.2. *Consider a manufacturing company that has an average inventory of $150,000 and sales amounting to $1,000,000 over the year. To determine the Inventory to Sales Ratio:*

$$\textit{Inventory to Sales Ratio} = \frac{\textit{Average Inventory}}{\textit{Sales}}$$

Substituting the given values:

$$\textit{Inventory to Sales Ratio} = \frac{150{,}000}{1{,}000{,}000} = 0.15$$

Solution 8.2. *Here, the Inventory to Sales Ratio is 0.15. This implies that for every dollar of sales, the company has 15 cents invested in inventory.*

Interpretation:

- *A ratio of 0.15 suggests that the company has a relatively lower inventory level compared to its sales, which may reflect effective inventory management and higher turnover rates.*
- *This lower ratio might also indicate a lean inventory strategy, where the company keeps minimal stock to reduce holding costs and improve cash flow. However, it's crucial to ensure that such a strategy does not lead to stockouts and missed sales opportunities.*
- *Comparing this ratio with industry standards or competitors' ratios provides additional insights into the company's inventory management effectiveness and competitive positioning.*

8.3.2 Demand Forecast Accuracy

Demand Forecast Accuracy measures how closely the predicted demand matches the actual sales. Accurate demand forecasting is crucial for maintaining optimal inventory levels and meeting customer demand.

Significance:

- Ensures inventory levels are sufficient to meet customer demand without overstocking.
- Reduces costs associated with excess inventory and stockouts.
- Improves planning and resource allocation.

Calculation:

$$\text{Forecast Accuracy} = 1 - \frac{|\text{Forecasted Demand} - \text{Actual Demand}|}{\text{Actual Demand}}$$

Example 8.3. *A company forecasts a demand of 10,000 units for a product, but the actual demand turns out to be 9,500 units. To calculate the Demand Forecast Accuracy:*

$$Forecast\ Accuracy = 1 - \frac{|Forecasted\ Demand - Actual\ Demand|}{Actual\ Demand}$$

Substituting the given values:

$$Forecast\ Accuracy = 1 - \frac{|10{,}000 - 9{,}500|}{9{,}500}$$

$$Forecast\ Accuracy = 1 - \frac{500}{9{,}500} \approx 1 - 0.0526 = 0.9474$$

$$Forecast\ Accuracy \approx 94.74\%$$

Solution 8.3. *In this example, the Forecast Accuracy is approximately 94.74%. This indicates that the forecasted demand was very close to the actual demand, with a small error margin of about 5.26%.*

Interpretation:

- *A high forecast accuracy of 94.74% suggests that the company's forecasting methods are quite effective, leading to minimal deviation between predicted and actual demand.*
- *High forecast accuracy helps maintain optimal inventory levels, reducing the risk of stockouts and excess inventory, and ultimately improving customer satisfaction.*
- *Regularly evaluating and refining forecasting techniques can help sustain and improve this accuracy over time.*

Example 8.4. *A retail store forecasts a demand of 2,000 units for a product, but the actual demand is 2,500 units. To determine the Forecast Accuracy:*

$$Forecast\ Accuracy = 1 - \frac{|Forecasted\ Demand - Actual\ Demand|}{Actual\ Demand}$$

Substituting the given values:

$$Forecast\ Accuracy = 1 - \frac{|2{,}000 - 2{,}500|}{2{,}500}$$

$$Forecast\ Accuracy = 1 - \frac{500}{2{,}500} = 1 - 0.20 = 0.80$$

$$Forecast\ Accuracy = 80\%$$

Solution 8.4. *Here, the Forecast Accuracy is 80%. This means that the forecasted demand was 20% off from the actual demand.*

Interpretation:

- *An accuracy of 80% indicates a moderate level of forecasting accuracy. While the forecast is generally close to the actual demand, there is a significant deviation of 20%.*
- *This level of accuracy may lead to some challenges in inventory management, such as occasional overstocking or stockouts. Improvements in forecasting methods or data analysis might be needed to enhance accuracy.*
- *Analyzing the reasons for the forecast deviation and incorporating better forecasting techniques or tools could help improve accuracy in the future.*

8.3.3 Inventory Turnover

Inventory turnover measures how often inventory is sold and replaced over a specific period. High turnover rates generally indicate efficient inventory management and strong sales performance.

Significance:

- Reflects the efficiency of inventory usage.
- Indicates the liquidity of inventory.
- Helps in identifying slow-moving or obsolete stock.

Calculation:

$$\text{Inventory Turnover} = \frac{\text{Cost of Goods Sold (COGS)}}{\text{Average Inventory}}$$

Example 8.5. *A company has a Cost of Goods Sold (COGS) of $500,000 and an average inventory of $100,000 for the year. To calculate the inventory turnover:*

$$\textit{Inventory Turnover} = \frac{COGS}{\textit{Average Inventory}}$$

Substituting the given values:

$$\textit{Inventory Turnover} = \frac{500{,}000}{100{,}000} = 5$$

$$\textit{Inventory Turnover} = 5$$

Solution 8.5. *In this example, the inventory turnover is 5. This indicates that the company sold and replaced its inventory five times over the year.*

Interpretation:

- *A turnover ratio of five suggests that the company is efficiently managing its inventory, with goods being sold and replaced relatively quickly.*
- *Higher turnover rates generally indicate good sales performance and effective inventory management, which helps in reducing holding costs and minimizing the risk of obsolescence.*
- *Regularly monitoring turnover rates can help identify inventory management practices that are working well and highlight areas for improvement if turnover rates are lower than desired.*

Example 8.6. *A retail store reports a Cost of Goods Sold (COGS) of $750,000 and has an average inventory of $250,000. To determine the inventory turnover:*

$$\textit{Inventory Turnover} = \frac{COGS}{\textit{Average Inventory}}$$

Substituting the given values:

$$\textit{Inventory Turnover} = \frac{750{,}000}{250{,}000} = 3$$

$$\textit{Inventory Turnover} = 3$$

Solution 8.6. *Here, the inventory turnover is three. This means the store sold and replaced its inventory three times over the year.*

Interpretation:

- *An inventory turnover of three indicates that the store has a slower turnover rate compared to the previous example. This may suggest either slower sales or higher levels of inventory being held.*

- *A lower turnover rate can point to potential issues such as overstocking or slow-moving inventory. It may also indicate that the store's inventory management could benefit from reviewing sales strategies or inventory procurement practices.*

- *Improving turnover rates could involve strategies such as better demand forecasting, promotional activities to boost sales, or optimizing inventory levels to better match sales patterns.*

8.3.4 Gross Margin Return on Investment (GMROI)

GMROI measures the profitability of inventory investments by comparing the gross margin earned to the cost of inventory. It helps in assessing the financial performance of inventory management practices.

Significance:

- Evaluates the profitability of inventory.
- Helps in making informed purchasing and pricing decisions.
- Assists in identifying high-performing products.

Calculation:

$$\text{GMROI} = \frac{\text{Gross Margin}}{\text{Average Inventory Cost}}$$

Example 8.7. *A company has a gross margin of \$200,000 and an average inventory cost of \$50,000 for the year. To calculate the GMROI:*

$$GMROI = \frac{Gross\ Margin}{Average\ Inventory\ Cost}$$

Substituting the given values:

$$GMROI = \frac{200{,}000}{50{,}000} = 4$$

$$GMROI = 4$$

Solution 8.7. *In this example, the GMROI is four. This indicates that for every dollar invested in inventory, the company earns \$4 in gross margin.*

Interpretation:

- *A GMROI of four suggests that the company's inventory investments are highly profitable, generating a substantial return on each dollar spent.*

- *A higher GMROI indicates effective inventory management and pricing strategies, which contribute to strong financial performance.*

- *Monitoring GMROI can help businesses make informed decisions regarding inventory purchases, pricing strategies, and product selection, ensuring that investments in inventory yield desirable financial outcomes.*

Example 8.8. *A retail store reports a gross margin of $150,000 and has an average inventory cost of $75,000. To determine the GMROI:*

$$GMROI = \frac{Gross\ Margin}{Average\ Inventory\ Cost}$$

Substituting the given values:

$$GMROI = \frac{150{,}000}{75{,}000} = 2$$

$$GMROI = 2$$

Solution 8.8. *Here, the GMROI is two. This means that for every dollar invested in inventory, the store earns $2 in gross margin.*

Interpretation:

- *A GMROI of two indicates a moderate return on inventory investments. While the return is positive, it is lower compared to the previous example.*
- *This may suggest that the store's inventory management practices could be improved to enhance profitability. Strategies such as optimizing pricing, reducing costs, or improving sales can help increase GMROI.*
- *Regular assessment of GMROI can guide strategic adjustments to improve inventory profitability and overall financial performance.*

8.3.5 Days Sales of Inventory (DSI)

Days Sales of Inventory (DSI) indicates the average number of days it takes to sell inventory. Lower DSI values suggest faster inventory turnover and better liquidity.

Significance:

- Reflects inventory liquidity.
- Helps in managing inventory levels to reduce holding costs.
- Indicates the efficiency of sales and inventory processes.

Calculation:

$$\text{DSI} = \frac{\text{Average Inventory}}{\text{Cost of Goods Sold}} \times 365$$

Example 8.9. *A company has an average inventory of $60,000 and a cost of goods sold (COGS) of $360,000 for the year. To calculate the Days Sales of Inventory:*

$$DSI = \frac{Average\ Inventory}{Cost\ of\ Goods\ Sold} \times 365$$

Substituting the given values:

$$DSI = \frac{60{,}000}{360{,}000} \times 365 = 0.1667 \times 365 \approx 60.8$$

$$DSI \approx 60.8\ days$$

Solution 8.9. *In this example, the DSI is approximately 60.8 days. This indicates that, on average, it takes about 61 days for the company to sell its inventory.*

Interpretation:

- *A DSI of 60.8 days suggests a moderate turnover rate. Inventory is sold roughly every two months, which may be acceptable depending on the industry and product type.*
- *Lower DSI values are generally preferred as they indicate faster inventory turnover, which can reduce holding costs and improve liquidity. Companies may aim to reduce their DSI to enhance efficiency.*
- *Businesses should regularly monitor DSI to ensure that their inventory management practices are optimized for efficient turnover and reduced holding costs.*

Example 8.10. *A retail store has an average inventory of $120,000 and a cost of goods sold (COGS) of $1,200,000 for the year. To determine the Days Sales of Inventory:*

$$DSI = \frac{Average\ Inventory}{Cost\ of\ Goods\ Sold} \times 365$$

Substituting the given values:

$$DSI = \frac{120{,}000}{1{,}200{,}000} \times 365 = 0.1 \times 365 = 36.5$$

$$DSI = 36.5\ days$$

Solution 8.10. *Here, the DSI is 36.5 days. This means that, on average, it takes the store about 37 days to sell its inventory.*

Interpretation:

- *A DSI of 36.5 days indicates relatively efficient inventory turnover, with inventory being sold approximately every month.*
- *Lower DSI values suggest that the store has effective inventory management practices, allowing for quicker sales and reduced holding costs.*
- *Maintaining or improving this DSI value can help the store stay competitive and ensure that inventory levels are aligned with sales performance.*
- *Monitoring DSI regularly can aid in optimizing inventory strategies and enhancing overall operational efficiency.*

8.3.6 Sell-Through Rate

The Sell-Through Rate measures the percentage of inventory sold during a specific period compared to the amount of inventory received. It is useful for evaluating the effectiveness of sales and marketing efforts.

Significance:

- Assesses the effectiveness of sales strategies.
- Helps in identifying popular and slow-moving products.
- Provides insights into inventory turnover and sales performance.

Calculation:

$$\text{Sell-Through Rate} = \frac{\text{Units Sold}}{\text{Units Received}} \times 100\%$$

Example 8.11. *A clothing store received 1,000 units of a new jacket model in a month and sold 800 units during the same period. To calculate the Sell-Through Rate:*

$$\textit{Sell-Through Rate} = \frac{\textit{Units Sold}}{\textit{Units Received}} \times 100\%$$

Substituting the given values:

$$\textit{Sell-Through Rate} = \frac{800}{1{,}000} \times 100\% = 0.8 \times 100\% = 80\%$$

$$\textit{Sell-Through Rate} = 80\%$$

Solution 8.11. *In this example, the Sell-Through Rate is 80%. This indicates that 80% of the received inventory was sold during the period.*

Interpretation:

- *A Sell-Through Rate of 80% suggests a high level of sales effectiveness and indicates that the product is popular and well-received by customers.*
- *This high rate implies that the store's sales strategies are effective and that inventory management is aligned with customer demand.*
- *Monitoring Sell-Through Rates can help businesses identify which products are performing well and which are not, enabling better inventory planning and marketing strategies.*

Example 8.12. *A technology retailer received 500 units of a new smartphone model in a quarter and sold 200 units during the same period. To determine the Sell-Through Rate:*

$$\textit{Sell-Through Rate} = \frac{\textit{Units Sold}}{\textit{Units Received}} \times 100\%$$

Substituting the given values:

$$\textit{Sell-Through Rate} = \frac{200}{500} \times 100\% = 0.4 \times 100\% = 40\%$$

$$\textit{Sell-Through Rate} = 40\%$$

Solution 8.12. *Here, the Sell-Through Rate is 40%. This means that 40% of the inventory was sold during the period.*

Interpretation:

- *A Sell-Through Rate of 40% indicates a moderate level of sales effectiveness. The product is selling but may not be as popular as others.*
- *This rate suggests that the retailer might need to evaluate its sales strategies or promotional efforts to boost the sales of this smartphone model.*
- *Understanding and analyzing the Sell-Through Rate can help businesses improve inventory turnover and make informed decisions about future inventory purchases and marketing strategies.*

8.4 Using Metrics to Monitor Performance

Inventory metrics and KPIs play a crucial role in monitoring and improving the performance of inventory management systems. Analyzing these metrics, businesses can gain valuable insights into their operations, identify areas for improvement, and implement strategies to enhance efficiency and profitability. This section focuses on the importance of analyzing inventory data and using it to improve sales and merchandising processes.

8.4.1 Analyzing Inventory Data

Effective inventory management requires continuous monitoring and analysis of inventory data. This process involves collecting, organizing, and interpreting data to make informed decisions. The following steps outline a systematic approach to analyzing inventory data:

- **Data Collection:** Gather relevant data from various sources, including sales transactions, purchase orders, inventory counts, and demand forecasts. Ensure data accuracy and consistency to facilitate reliable analysis.

- **Data Organization:** Organize the collected data into a structured format, such as a database or spreadsheet. Categorize data by product, location, time period, and other relevant dimensions to enable detailed analysis.

- **Data Visualization:** Use visualization tools such as graphs, charts, and dashboards to represent data visually. This helps in identifying trends, patterns, and anomalies more easily.

- **Trend Analysis:** Analyze historical data to identify trends in inventory levels, sales performance, and demand patterns. Understanding these trends helps in making informed decisions about inventory replenishment and sales strategies.

- **Performance Benchmarking:** Compare current performance metrics against historical data, industry benchmarks, and predefined targets. Benchmarking helps in assessing performance relative to standards and identifying areas for improvement.

- **Root Cause Analysis:** Conduct root cause analysis to identify underlying factors contributing to inventory issues such as stockouts, overstocking, or slow-moving inventory. This involves drilling down into data to uncover specific causes and addressing them systematically.

- **Scenario Analysis:** Use scenario analysis to evaluate the potential impact of different strategies or changes in inventory policies. This helps in assessing the risks and benefits of various options before implementation.

Example 8.13. ***Example 1: Inventory to Sales Ratio***
A company tracks the average inventory and sales for its product over a year. The data for Product X is as follows:

- *Average Inventory: 5,000 units*

- *Annual Sales: 60,000 units*

Calculate the Inventory to Sales Ratio.

Solution 8.13. ***Solution 1: Inventory to Sales Ratio***
The Inventory to Sales Ratio is calculated using the formula:

$$Inventory\ to\ Sales\ Ratio = \frac{Average\ Inventory}{Sales}$$

Plugging in the values:

$$Inventory\ to\ Sales\ Ratio = \frac{5{,}000}{60{,}000} = 0.0833$$

This means that for every dollar of sales, the company holds approximately \$0.0833 in inventory.

Example 8.14. ***Example 2: Demand Forecast Accuracy***
The forecasted demand and actual demand for Product Y over six months are as follows:

- *Forecasted Demand: [1,000, 1,200, 1,500, 1,600, 1,800, 2,000] units*
- *Actual Demand: [950, 1,250, 1,450, 1,700, 1,750, 2,050] units*

Calculate the Demand Forecast Accuracy for each month.

Solution 8.14. ***Solution 2: Demand Forecast Accuracy***
The Demand Forecast Accuracy for each month is calculated using the formula:

$$Forecast\ Accuracy = 1 - \frac{|Forecasted\ Demand - Actual\ Demand|}{Actual\ Demand}$$

Applying this formula:

$$\begin{aligned}
&Month\ 1: && 1 - \frac{|1{,}000 - 950|}{950} = 1 - \frac{50}{950} \approx 0.947\\
&Month\ 2: && 1 - \frac{|1{,}200 - 1{,}250|}{1{,}250} = 1 - \frac{50}{1{,}250} \approx 0.960\\
&Month\ 3: && 1 - \frac{|1{,}500 - 1{,}450|}{1{,}450} = 1 - \frac{50}{1{,}450} \approx 0.965\\
&Month\ 4: && 1 - \frac{|1{,}600 - 1{,}700|}{1{,}700} = 1 - \frac{100}{1{,}700} \approx 0.941\\
&Month\ 5: && 1 - \frac{|1{,}800 - 1{,}750|}{1{,}750} = 1 - \frac{50}{1{,}750} \approx 0.971\\
&Month\ 6: && 1 - \frac{|2{,}000 - 2{,}050|}{2{,}050} = 1 - \frac{50}{2{,}050} \approx 0.976
\end{aligned}$$

The accuracy varies by month but generally indicates good forecasting performance.

Example 8.15. ***Example 3: Inventory Turnover***
A company has the following data for Product Z over the past year:

- *Cost of Goods Sold (COGS): $120,000*
- *Average Inventory: $30,000*

Calculate the Inventory Turnover Ratio.

Solution 8.15. ***Solution 3: Inventory Turnover***
The Inventory Turnover Ratio is calculated using the formula:

$$Inventory\ Turnover = \frac{Cost\ of\ Goods\ Sold\ (COGS)}{Average\ Inventory}$$

Plugging in the values:

$$Inventory\ Turnover = \frac{120{,}000}{30{,}000} = 4$$

This means that the inventory turned over four times during the year.

Example 8.16. ***Example 4: Gross Margin Return on Investment (GMROI)***
The data for Product W over the past year is:

- *Gross Margin: $50,000*
- *Average Inventory Cost: $25,000*

Calculate the GMROI.

Solution 8.16. ***Solution 4: GMROI***
GMROI is calculated using the formula:

$$GMROI = \frac{Gross\ Margin}{Average\ Inventory\ Cost}$$

Plugging in the values:

$$GMROI = \frac{50{,}000}{25{,}000} = 2$$

This indicates that for every dollar invested in inventory, the company earns $2 in gross margin.

Example 8.17. ***Example 5: Days Sales of Inventory (DSI)***
The data for Product V is as follows:

- *Average Inventory: 1,200 units*
- *Cost of Goods Sold (COGS): 8,000 units per year*

Calculate the Days Sales of Inventory (DSI).

Solution 8.17. ***Solution 5: Days Sales of Inventory (DSI)***
DSI is calculated using the formula:

$$DSI = \frac{Average\ Inventory}{Cost\ of\ Goods\ Sold} \times 365$$

Plugging in the values:

$$DSI = \frac{1{,}200}{8{,}000} \times 365 \approx 55.5\ days$$

This indicates that it takes an average of 55.5 days to sell the inventory.

Example 8.18. ***Example 6: Sell-Through Rate***
A retailer received 2,000 units of Product U and sold 1,600 units during a specific period. Calculate the Sell-Through Rate.

Solution 8.18. ***Solution 6: Sell-Through Rate***
The Sell-Through Rate is calculated using the formula:

$$Sell\text{-}Through\ Rate = \frac{Units\ Sold}{Units\ Received} \times 100\%$$

Plugging in the values:

$$Sell\text{-}Through\ Rate = \frac{1{,}600}{2{,}000} \times 100\% = 80\%$$

This means that 80% of the received inventory was sold during the period.

8.4.2 Improving Sales and Merchandising Processes

Inventory metrics and KPIs provide valuable information that can be used to enhance sales and merchandising processes. The following strategies outline how businesses can leverage these metrics to improve their operations:

- **Optimizing Stock Levels:** Use metrics such as inventory turnover, DSI, and sell-through rate to determine optimal stock levels. Maintaining the right balance between inventory and sales helps in reducing holding costs and avoiding stockouts.
- **Enhancing Demand Forecasting:** Improve demand forecast accuracy by analyzing historical sales data and adjusting forecasting models based on observed patterns. Accurate demand forecasting ensures that inventory levels are aligned with actual sales, minimizing the risk of overstocking or understocking.

- **Streamlining Replenishment Processes:** Implement automated replenishment systems based on real-time inventory data and sales performance metrics. This helps in ensuring timely restocking of products, reducing the likelihood of stockouts and lost sales.

- **Targeted Marketing and Promotions:** Use sell-through rates and other sales performance metrics to identify high-performing and slow-moving products. Develop targeted marketing and promotional strategies to boost sales of slow-moving items and optimize inventory turnover.

- **Product Assortment Planning:** Analyze sales data and customer preferences to optimize product assortment. Ensure that the product mix meets customer demand and aligns with market trends, improving sales performance and customer satisfaction.

- **Vendor Management:** Use inventory metrics to evaluate vendor performance, such as delivery lead times, order accuracy, and product quality. Establish strong relationships with reliable vendors to ensure consistent supply and high-quality products.

- **Continuous Improvement:** Regularly review and adjust inventory management strategies based on performance metrics and changing market conditions. Implement a continuous improvement approach to enhance inventory processes and achieve better results over time.

Leveraging inventory metrics and KPIs to monitor performance and implement improvements, businesses can enhance their sales and merchandising processes, resulting in increased efficiency, higher profitability, and improved customer satisfaction. In the subsequent sections, we will explore the benefits of inventory metrics and KPIs in detail, providing examples of best practices and practical applications in inventory management.

8.5 Benefits of Inventory Metrics and KPIs

Inventory metrics and KPIs offer numerous benefits that help businesses optimize their inventory management processes and achieve their operational goals. This section discusses how these metrics contribute to measuring and achieving progress, guiding behavior and decision-making, and enhancing productivity.

8.5.1 Measuring and Achieving Progress

Effective inventory management is a continuous process that requires regular monitoring and evaluation. Inventory metrics and KPIs provide the necessary tools to measure progress and assess the effectiveness of inventory strategies. The benefits include:

- **Tracking Performance:** Metrics such as inventory turnover, DSI, and GM-ROI enable businesses to track their inventory performance over time. This helps in identifying trends, measuring improvements, and detecting areas that need attention.
- **Setting Benchmarks:** Establishing target values for key metrics, businesses can set benchmarks for performance. These benchmarks serve as goals to strive for and help in evaluating the success of inventory management initiatives.
- **Evaluating Strategies:** Metrics provide a quantitative basis for evaluating the effectiveness of different inventory management strategies. Comparing actual performance against targets and benchmarks, businesses can determine which strategies are working and which need adjustment.
- **Continuous Improvement:** Regular measurement and analysis of inventory metrics foster a culture of continuous improvement. Businesses can use insights gained from metrics to implement incremental changes that lead to ongoing enhancements in inventory management.

8.5.2 Guiding Behaviour and Decision-Making

Inventory metrics and KPIs play a crucial role in guiding behavior and informing decision-making processes within an organization. The benefits include:

- **Informed Decisions:** Metrics provide data-driven insights that inform decision-making at all levels of the organization. From purchasing and replenishment to sales and marketing, reliable metrics help in making informed choices that align with business goals.
- **Strategic Planning:** Inventory metrics are essential for strategic planning. They provide the information needed to forecast demand, plan inventory levels, and allocate resources effectively. This ensures that inventory management decisions support long-term business objectives.
- **Accountability:** Metrics create accountability by clearly defining performance expectations. Employees and departments can be held accountable for meeting specific targets, fostering a culture of responsibility and performance.
- **Behavioral Alignment:** Aligning metrics with business goals, organizations can guide employee behavior toward achieving these goals. For example, metrics focused on reducing stockouts encourage behaviors that improve inventory availability and customer satisfaction.

8.5.3 Enhancing Productivity

Effective use of inventory metrics and KPIs can lead to significant improvements in productivity. The benefits include:

- **Operational Efficiency:** Metrics such as inventory turnover and DSI highlight inefficiencies in inventory management processes. Addressing these inefficiencies, businesses can streamline operations and reduce waste, leading to enhanced productivity.
- **Resource Optimization:** Inventory metrics help in optimizing the use of resources, including inventory, labor, and capital. Efficient inventory management ensures that resources are allocated where they are needed most, maximizing their utility.
- **Reducing Costs:** Metrics provide insights into cost-saving opportunities, such as minimizing holding costs, reducing stockouts, and improving turnover rates. These cost reductions contribute to overall productivity and profitability.
- **Improving Responsiveness:** Timely and accurate metrics enable businesses to respond quickly to changes in demand, market conditions, and operational issues. This agility enhances productivity by reducing delays and improving service levels.
- **Enhancing Collaboration:** Inventory metrics facilitate collaboration across different departments, such as procurement, sales, and operations. Providing a common framework for evaluating performance, metrics encourage coordinated efforts to improve inventory management and productivity.

Inventory metrics and KPIs are powerful tools that help businesses measure and achieve progress, guide behavior and decision-making, and enhance productivity. Leveraging these metrics effectively, organizations can optimize their inventory management processes, leading to improved efficiency, profitability, and customer satisfaction. In the following sections, we will explore best practices and case studies that demonstrate the practical applications and benefits of inventory metrics and KPIs in real-world scenarios.

8.5.4 Industry Applicability

Numerous industries have successfully implemented inventory metrics and KPIs to enhance their inventory management processes. Here are some notable ones:

- **Retail:** A major retail chain improved its inventory turnover by implementing real-time inventory tracking and demand forecasting. Closely monitoring sell-through rates and adjusting inventory levels based on sales data, the company reduced excess inventory and increased sales.

- **Manufacturing:** A manufacturing firm used metrics such as DSI and GMROI to optimize its production schedules and inventory levels. Aligning inventory with production needs and sales forecasts, the company minimized holding costs and improved cash flow.
- **Healthcare:** A healthcare provider implemented inventory metrics to manage its medical supplies more efficiently. Tracking inventory turnover and demand forecast accuracy, the provider ensured that critical supplies were always available, reducing the risk of stockouts and improving patient care.
- **E-commerce:** An e-commerce business utilized inventory metrics to streamline its fulfillment processes. Monitoring inventory to sales ratio and sell-through rates, the company optimized its warehousing and distribution strategies, resulting in faster delivery times and higher customer satisfaction.

8.5.5 Practical Applications

The practical applications of inventory metrics and KPIs vary across industries but generally follow similar principles. Here are some common applications:

- **Inventory Optimization:** Businesses use metrics to determine optimal inventory levels, reducing both excess stock and stockouts. This optimization leads to cost savings and improved customer service.
- **Demand Planning:** Accurate demand forecasting based on historical sales data and market trends helps businesses align their inventory levels with customer demand. This reduces the risk of overstocking and understocking.
- **Performance Measurement:** Metrics provide a quantitative basis for measuring performance across different aspects of inventory management, such as procurement, warehousing, and sales. This measurement helps in identifying areas for improvement and implementing targeted strategies.
- **Cost Management:** Analyzing metrics related to inventory costs, such as holding costs and turnover rates, businesses can identify cost-saving opportunities and implement strategies to reduce expenses.
- **Supply Chain Coordination:** Inventory metrics facilitate better coordination between different parts of the supply chain, such as suppliers, manufacturers, and distributors. This coordination ensures that inventory levels are maintained efficiently throughout the supply chain.

8.5.6 Lessons Learned

Implementing inventory metrics and KPIs can yield significant benefits, but it also comes with challenges. Here are some lessons learned from various case studies:

- **Data Quality:** Accurate and consistent data is crucial for reliable metrics. Businesses must invest in robust data collection and management systems to ensure data quality.
- **Continuous Monitoring:** Inventory metrics should be monitored continuously rather than periodically. Real-time monitoring enables prompt responses to emerging issues and trends.
- **Stakeholder Engagement:** Successful implementation of metrics requires engagement from all relevant stakeholders, including management, operations teams, and front-line staff. Clear communication and training are essential to ensure everyone understands the importance of metrics and how to use them.
- **Adaptability:** Inventory management strategies should be adaptable to changing market conditions and business needs. Regularly reviewing and adjusting metrics helps in maintaining their relevance and effectiveness.
- **Integrated Systems:** Integrated inventory management systems that combine metrics with other business functions, such as sales and finance, provide more comprehensive insights and facilitate better decision-making.
- **Focus on Actionability:** Metrics should lead to actionable insights. Businesses must ensure that they have the processes and resources in place to act on the insights gained from inventory metrics.

8.5.7 Future Directions in Inventory Management

As technology and market dynamics evolve, so too will the field of inventory management. The following trends and developments are likely to shape the future of inventory management:

- **Advanced Analytics and AI:** The integration of advanced analytics and artificial intelligence (AI) into inventory management systems will enable more accurate demand forecasting, real-time inventory tracking, and automated decision-making. These technologies will help businesses optimize inventory levels and improve responsiveness to market changes.
- **IoT and Smart Inventory Systems:** The Internet of Things (IoT) will play a significant role in inventory management by providing real-time data on inventory levels, location, and condition. Smart inventory systems will use IoT sensors and devices to enhance visibility and control over inventory processes.
- **Blockchain for Supply Chain Transparency:** Blockchain technology can enhance supply chain transparency and traceability by providing a secure and immutable record of inventory transactions. This will help in reducing fraud, improving compliance, and ensuring the authenticity of products.

- **Sustainability and Green Inventory Management:** As sustainability becomes a priority, businesses will focus on green inventory management practices. This includes reducing waste, optimizing resource usage, and adopting environmentally friendly technologies and processes.
- **Collaborative Inventory Management:** Greater collaboration between supply chain partners will become essential for effective inventory management. Shared data and integrated systems will facilitate better coordination and alignment of inventory strategies across the supply chain.
- **Customer-Centric Inventory Strategies:** Businesses will increasingly adopt customer-centric inventory strategies that prioritize meeting customer demand and enhancing customer satisfaction. This involves flexible inventory systems that can adapt to changing customer preferences and market trends.

Staying abreast of these future trends and incorporating innovative technologies and practices, businesses can continue to improve their inventory management processes and maintain a competitive edge in the market. This chapter has provided a comprehensive overview of inventory metrics and KPIs, highlighting their importance and practical applications in achieving effective inventory management.

8.6 Conclusion

In this chapter, we have explored the critical role that inventory metrics and KPIs play in effective inventory management. Understanding and implementing these metrics, businesses can significantly enhance their operational efficiency, make informed decisions, and achieve better financial performance. This concluding section summarizes the key points discussed and outlines future directions in inventory management.

Exercises

Exercise 8.1. *A retail store tracks the following data for two product categories over a year:*

Category	*Gross Sales*	*Cost of Goods Sold (COGS)*	*Average Inventory*
Electronics	$500,000	$350,000	$50,000
Clothing	$300,000	$150,000	$40,000

Calculate the GMROI for each category and analyze which category has a better performance.

Exercise 8.2. *A store has an average inventory turnover rate of five times per year. If the gross margin percentage is 30%, and the average inventory value is $200,000, calculate the GMROI. Discuss how changes in the inventory turnover rate affect the GMROI.*

Exercise 8.3. *A company sells three product lines with the following data:*

Product Line	*Gross Sales*	*COGS*	*Average Inventory*
A	$1,200,000	$800,000	$100,000
B	$800,000	$400,000	$90,000
C	$600,000	$300,000	$80,000

Perform a sensitivity analysis by increasing the COGS by 10% for each product line and calculate the impact on GMROI. Discuss the implications of these changes.

Exercise 8.4. *A company has the following data for the past year:*

Month	*Ending Inventory ($)*	*Cost of Goods Sold (COGS) ($)*
January	120,000	50,000
February	130,000	60,000
March	125,000	55,000
April	140,000	65,000
May	135,000	70,000
June	145,000	75,000
July	150,000	80,000
August	155,000	85,000
September	160,000	90,000
October	165,000	95,000
November	170,000	100,000
December	175,000	105,000

Calculate the Days Sales of Inventory (DSI) for the year.

Exercise 8.5. *A company's current Days Sales of Inventory (DSI) is 45 days, with an average inventory of $500,000. If the company aims to reduce its DSI to 30 days without changing the cost of goods sold (COGS), calculate the impact on cash flow. Assume the annual COGS is $6,000,000.*

Exercise 8.6. *A company has three product lines with the following data:*

Product Line	*Average Inventory ($)*	*Annual COGS ($)*	*DSI (days)*
A	200,000	1,200,000	*calculate*
B	150,000	1,000,000	*calculate*
C	100,000	800,000	*calculate*

Perform a sensitivity analysis by increasing the inventory by 10% for each product line and calculate the new DSI. Discuss the implications of these changes.

9

Inventory Risk Management

This chapter introduces the concept of inventory risk management, focusing on identifying, assessing, and mitigating risks in inventory operations. Effective risk management is crucial for maintaining operational efficiency and financial stability. The chapter explains a risk assessment model using the Analytic Hierarchy Process (AHP), which systematically identifies and ranks inventory risks. It classifies risks into areas such as demand, supply, and manual handling, offering a structured approach to address each risk. The chapter provides a comprehensive understanding of inventory risk management concepts and methodologies, highlighting the importance of managing risks to ensure optimal performance and minimize financial impact. Drawing on examples and best practices, it offers practical insights into applying inventory risk management strategies. This chapter is a valuable resource for practitioners, researchers, and students seeking to enhance their understanding and application of inventory risk management in various contexts.

9.1 Introduction

Effective inventory management is crucial for maintaining operational efficiency, minimizing costs, and ensuring customer satisfaction. However, inventory management is inherently fraught with risks that can disrupt operations, lead to financial losses, and negatively impact a company's overall performance. Inventory risk management involves identifying, assessing, and mitigating these risks to maintain optimal inventory levels and safeguard the business against potential disruptions. This chapter provides an in-depth exploration of inventory risk management, focusing on its principles, methodologies, and practical applications.

9.1.1 Overview of Inventory Risk Management

Inventory risk management is a systematic approach to identifying, assessing, and mitigating risks associated with inventory management. These risks can arise from various sources, including fluctuations in demand, supply chain

DOI: 10.1201/9781003534860-9

disruptions, operational inefficiencies, and financial instability. The key components of inventory risk management include:

- **Risk Identification:** The process begins with identifying potential risks that could affect inventory levels and management. This involves examining internal and external factors that could lead to supply chain disruptions, demand variability, operational issues, or financial challenges.
- **Risk Assessment:** Once risks are identified, they are assessed in terms of their likelihood and potential impact. This assessment helps prioritize risks and determine which ones require immediate attention and mitigation strategies.
- **Risk Mitigation:** After assessing the risks, appropriate mitigation strategies are developed and implemented. These strategies are designed to minimize the likelihood of risks occurring and reduce their impact if they do occur.
- **Monitoring and Review:** Continuous monitoring and review of the risk management process are essential to ensure its effectiveness. This involves tracking key performance indicators (KPIs), conducting regular risk assessments, and updating mitigation strategies as necessary.

Systematically managing inventory risks, companies can enhance their resilience to disruptions, maintain operational efficiency, and protect their financial health.

9.1.2 Importance of Managing Inventory Risks

Managing inventory risks is critical for several reasons:

- **Operational Efficiency:** Effective risk management ensures that inventory levels are optimized, reducing the risk of stockouts and overstock situations. This leads to smoother operations, timely fulfillment of customer orders, and efficient use of resources.
- **Cost Reduction:** Inventory risks such as excess stock, obsolescence, and spoilage can lead to significant financial losses. Mitigating these risks, companies can reduce inventory holding costs, minimize waste, and improve overall cost efficiency.
- **Customer Satisfaction:** Properly managed inventory risks help ensure that products are available when customers need them. This enhances customer satisfaction and loyalty, contributing to a positive brand reputation and competitive advantage.

- **Financial Stability:** Inventory represents a significant portion of a company's assets. Effective risk management protects these assets from unforeseen disruptions, helping maintain financial stability and ensuring that the company can meet its financial obligations.
- **Strategic Decision-Making:** Understanding and managing inventory risks provide valuable insights that support strategic decision-making. Companies can make informed choices about inventory policies, supplier relationships, and investment in risk mitigation technologies.

Inventory risk management is a vital component of effective inventory management. It helps companies navigate uncertainties, optimize inventory levels, and achieve operational and financial goals. The following sections will delve deeper into the various types of inventory risks, assessment models, and mitigation strategies to provide a comprehensive understanding of this essential practice.

9.2 Identifying Inventory Risks

Inventory risks can arise from various sources and have different impacts on a company's operations. Identifying these risks is the first step in managing them effectively. This section discusses the major categories of inventory risks, including demand risks, supply risks, operational risks, and financial risks.

9.2.1 Demand Risks

Demand risks pertain to the uncertainties and fluctuations in customer demand that can impact inventory levels and management. Key factors contributing to demand risks include:

- **Market Trends:** Changes in consumer preferences, technological advancements, and fashion trends can lead to unpredictable shifts in demand.
- **Seasonality:** Many industries experience seasonal variations in demand, which can cause significant fluctuations in inventory requirements.
- **Economic Conditions:** Economic downturns or booms can affect consumer spending patterns, influencing demand for products.
- **Competition:** Actions by competitors, such as new product launches or price changes, can alter demand for a company's products.
- **Promotions and Discounts:** Sales promotions, discounts, and marketing campaigns can temporarily spike demand, leading to challenges in managing inventory levels.

Managing demand risks involves accurate demand forecasting, flexible inventory strategies, and responsive supply chain practices to adapt to changes in demand.

9.2.2 Supply Risks

Supply risks are associated with disruptions in the supply chain that affect the availability of inventory. Common sources of supply risks include:

- **Supplier Reliability:** Dependence on a limited number of suppliers or suppliers with a history of unreliability can pose significant risks.
- **Lead Time Variability:** Variations in lead times for procurement can lead to delays in replenishing inventory, causing stockouts or excess inventory.
- **Geopolitical Factors:** Political instability, trade restrictions, and tariffs can disrupt supply chains and affect inventory availability.
- **Natural Disasters:** Events such as earthquakes, floods, and hurricanes can damage infrastructure and disrupt supply chains.
- **Quality Issues:** Poor quality or defective products from suppliers can lead to returns, recalls, and increased costs, impacting inventory levels.

Mitigating supply risks involves diversifying suppliers, establishing strong supplier relationships, and implementing contingency plans for supply chain disruptions.

9.2.3 Operational Risks

Operational risks are internal risks related to the processes and activities involved in managing inventory. These risks include:

- **Inventory Management Errors:** Mistakes in inventory tracking, ordering, and forecasting can lead to stockouts or overstock situations.
- **Handling and Storage Issues:** Improper handling, storage conditions, and warehousing practices can result in damage, spoilage, or loss of inventory.
- **Process Inefficiencies:** Inefficient inventory management processes, such as slow order processing and inadequate inventory turnover, can hinder operational performance.
- **Technology Failures:** Malfunctions or outages in inventory management systems, such as software bugs or hardware failures, can disrupt inventory operations.

- **Human Error:** Errors made by employees, such as incorrect data entry or miscommunication, can impact inventory accuracy and availability.

Addressing operational risks involves implementing robust inventory management systems, training employees, and continuously improving processes to enhance efficiency and accuracy.

9.2.4 Financial Risks

Financial risks are related to the monetary aspects of inventory management. These risks include:

- **Inventory Holding Costs:** High costs associated with storing inventory, including warehousing, insurance, and obsolescence, can impact profitability.
- **Cash Flow Constraints:** Tying up capital in excess inventory can limit cash flow, affecting the company's ability to invest in other areas.
- **Price Fluctuations:** Changes in the cost of raw materials, currency exchange rates, and inflation can affect the financial stability of inventory management.
- **Credit Risks:** Dependence on credit from suppliers or financial institutions can pose risks if credit terms change or if there are delays in payments.
- **Write-offs and Write-downs:** Inventory write-offs due to obsolescence, damage, or market changes can result in significant financial losses.

Mitigating financial risks involves careful financial planning, optimizing inventory levels to balance costs, and employing strategies such as just-in-time (JIT) inventory to minimize holding costs. Identifying inventory risks across demand, supply, operational, and financial areas is essential for effective risk management. Understanding these risks allows companies to develop targeted strategies to mitigate them, ensuring more stable and efficient inventory management.

9.3 Risk Assessment Models

Effective inventory risk management requires robust assessment models to evaluate the likelihood and impact of various risks. This section discusses two prominent risk assessment models: the analytic hierarchy process (AHP) method and a combination of quantitative and qualitative assessment techniques. These models provide systematic approaches to identify, prioritize, and mitigate inventory risks.

9.3.1 Analytic Hierarchy Process (AHP) Method

The AHP is a structured technique for organizing and analyzing complex decisions. It was developed by Thomas L. Saaty in the 1970s and has been widely used in decision-making processes involving multiple criteria. The AHP helps decision-makers set priorities and make the best decision by reducing complex decisions to a series of pairwise comparisons and then synthesizing the results.

In the AHP, the consistency ratio (CR) is a measure of how consistent the pairwise comparison matrix A is in its evaluations. A CR value less than 0.10 is generally considered acceptable, meaning the pairwise comparisons made by the decision-maker are consistent enough to provide reliable results. Here is the detailed explanation of why this threshold is important.

9.3.1.1 Why Consistency Is Important

- **Reliability of Judgments**: The pairwise comparison process in AHP relies on subjective judgments of decision-makers. If these judgments are inconsistent, it can lead to unreliable priority vectors. Ensuring consistency helps in making sure that the derived priorities truly reflect the decision-makers' preferences.

- **Logical Coherence**: A consistent matrix means that if a criterion A is preferred over B, and B is preferred over C, then A should be logically preferred over C. Consistency helps in maintaining logical coherence in the decision-making process.

9.3.1.2 Calculation of Consistency Index (CI) and Consistency Ratio (CR)

- **Principal Eigenvalue $\lambda_{\max}$**: The largest eigenvalue of the pairwise comparison matrix A. For a perfectly consistent matrix of size n, $\lambda_{\max}$ would be equal to n.

- **Consistency Index (CI)**:

$$CI = \frac{\lambda_{\max} - n}{n - 1}$$

  This index measures the deviation of $\lambda_{\max}$ from n. The smaller the CI, the more consistent the matrix.

- **Random Consistency Index (RI)**: The average CI of a large number of randomly generated pairwise comparison matrices of the same size n. Values of RI for different sizes are tabulated.

- **Consistency Ratio (CR)**:

$$CR = \frac{CI}{RI}$$

This ratio compares the CI of the matrix to the RI. A CR less than 0.10 is considered acceptable, meaning that the level of inconsistency is within a tolerable range.

9.3.1.3 Proof of CR Threshold

- **Empirical Evidence**: The threshold of 0.10 is based on empirical studies conducted by Thomas L. Saaty, the creator of AHP. He observed that when CR is below 0.10, the pairwise comparisons are sufficiently consistent for the decision-making process to be reliable.
- **Mathematical Justification**:
    - For $n = 1$ or $n = 2$, any matrix is always consistent, thus $CI = 0$ and $CR = 0$.
    - For larger matrices, the likelihood of inconsistency increases. The threshold of 0.10 represents a balance between allowing minor inconsistencies and ensuring that these do not significantly affect the outcome.

9.3.1.4 Steps in the AHP Process

1. **Define the Problem and Determine the Goal:**
    - Clearly state the problem and the goal of the decision-making process.
2. **Structure the Hierarchy:**
    - Break down the problem into a hierarchy of interrelated decision elements (criteria, sub-criteria, and alternatives).
3. **Pairwise Comparison and Priority Setting:**
    - Compare the elements at each level of the hierarchy pairwise in terms of their impact on the element above them.
    - Use a scale of relative importance, usually from 1 to 9, where 1 means equal importance and 9 means extreme importance of one element over another.
4. **Calculate the Pairwise Comparison Matrices:**
    - Form matrices for each level of the hierarchy and for each pairwise comparison.
5. **Calculate the Priority Vectors:**
    - Compute the eigenvector and eigenvalue of each pairwise comparison matrix to derive the priority vector (also called the eigenvector of the matrix).

6. **Check for Consistency:**
   - Calculate the consistency ratio (CR) to ensure the pairwise comparisons are consistent. A CR less than 0.1 is generally acceptable.

7. **Synthesize the Results:**
   - Aggregate the priority vectors across the hierarchy to obtain a set of overall priorities for the decision alternatives.

9.3.1.5 Mathematical Formulation

1. **Pairwise Comparison Matrix (A):**

$$A = \begin{pmatrix} 1 & a_{12} & a_{13} & \cdots & a_{1n} \\ \frac{1}{a_{12}} & 1 & a_{23} & \cdots & a_{2n} \\ \frac{1}{a_{13}} & \frac{1}{a_{23}} & 1 & \cdots & a_{3n} \\ \vdots & \vdots & \vdots & \ddots & \vdots \\ \frac{1}{a_{1n}} & \frac{1}{a_{2n}} & \frac{1}{a_{3n}} & \cdots & 1 \end{pmatrix}$$

2. **Priority Vector (w):**
   - The priority vector w is the normalized eigenvector corresponding to the largest eigenvalue (λ_{max}) of the matrix A:

$$A \cdot w = \lambda_{\text{max}} \cdot w$$

   - To find w, solve the equation $(A - \lambda I) \cdot w = 0$, where I is the identity matrix. Normalize w by dividing each element by the sum of all elements in w so that they sum to 1.

3. **Consistency Index (CI) and Consistency Ratio (CR):**
   - Consistency Index (CI):

$$CI = \frac{\lambda_{\text{max}} - n}{n - 1}$$

   - Consistency Ratio (CR):

$$CR = \frac{CI}{RI}$$

   where RI is the Random Index, which depends on the number of elements n.

Examples

Example 9.1. *A company needs to select a location for launching a new product using three criteria:*

- *Market Potential* C_1
- *Competition* C_2
- *Cost* C_3

The pairwise comparison matrix for criteria is:

$$\mathbf{AC} = \begin{bmatrix} 1 & 3 & 0.5 \\ \frac{1}{3} & 1 & 0.2 \\ 2 & 5 & 1 \end{bmatrix}$$

Tasks

1. *Calculate criteria weights using AHP*
2. *Check consistency of comparisons*
3. *Recommend priority criterion*

Solution 9.1.

1. ***Criteria Weight Calculation:***

(a) *Normalize the matrix:*

$$\begin{bmatrix} \frac{1}{3.333} & \frac{3}{9} & \frac{0.5}{1.7} \\ \frac{0.333}{3.333} & \frac{1}{9} & \frac{0.2}{1.7} \\ \frac{2}{3.333} & \frac{5}{9} & \frac{1}{1.7} \end{bmatrix} \approx \begin{bmatrix} 0.300 & 0.333 & 0.294 \\ 0.100 & 0.111 & 0.118 \\ 0.600 & 0.556 & 0.588 \end{bmatrix}$$

(b) *Calculate row averages:*

$$w_{C_1} = \frac{0.300 + 0.333 + 0.294}{3} \approx 0.309$$

$$w_{C_2} = \frac{0.100 + 0.111 + 0.118}{3} \approx 0.110$$

$$w_{C_3} = \frac{0.600 + 0.556 + 0.588}{3} \approx 0.581$$

$$\boxed{w_{C_1} = 30.9\%, \quad w_{C_2} = 11.0\%, \quad w_{C_3} = 58.1\%}$$

2. *Consistency Check:*

(a) Compute weighted sum vector:

$$\mathbf{ACw} = \begin{bmatrix} 1(0.309) + 3(0.110) + 0.5(0.581) \\ \frac{1}{3}(0.309) + 1(0.110) + 0.2(0.581) \\ 2(0.309) + 5(0.110) + 1(0.581) \end{bmatrix} \approx \begin{bmatrix} 0.930 \\ 0.329 \\ 1.749 \end{bmatrix}$$

(b) Calculate consistency ratio:

$$\lambda_{max} = \frac{1}{3}\left(\frac{0.930}{0.309} + \frac{0.329}{0.110} + \frac{1.749}{0.581}\right) \approx 3.003$$

$$CI = \frac{3.003 - 3}{2} = 0.0015$$

$$CR = \frac{0.0015}{0.58} \approx 0.0026$$

$$\boxed{CR = 0.26\% \ (Consistent)}$$

Recommendation: Cost C_3 has the highest weight (58.1%). The company should prioritize cost efficiency in location selection.

$$\boxed{\text{Prioritize Cost } C_3 \text{ for decision making}}$$

Since CR < 0.1, the matrix is consistent.

3. *Pairwise Comparison Matrices for Alternatives (A, B, C):*

$$A_{C1} = \begin{pmatrix} 1 & 0.5 & 2 \\ 2 & 1 & 3 \\ 0.5 & \frac{1}{3} & 1 \end{pmatrix}, \quad A_{C2} = \begin{pmatrix} 1 & 4 & 0.25 \\ \frac{1}{4} & 1 & \frac{1}{7} \\ 4 & 7 & 1 \end{pmatrix}, \quad A_{C3} = \begin{pmatrix} 1 & 1 & 3 \\ 1 & 1 & 2 \\ \frac{1}{3} & \frac{1}{2} & 1 \end{pmatrix}$$

4. *Compute Priority Vectors for Alternatives:*

$$w_{C1} = \begin{pmatrix} 0.474 \\ 0.329 \\ 0.197 \end{pmatrix}, \quad w_{C2} = \begin{pmatrix} 0.615 \\ 0.109 \\ 0.276 \end{pmatrix}, \quad w_{C3} = \begin{pmatrix} 0.459 \\ 0.362 \\ 0.179 \end{pmatrix}$$

5. *Aggregate the Results:*

$$w = 0.550 \cdot w_{C1} + 0.175 \cdot w_{C2} + 0.275 \cdot w_{C3}$$

$$= \begin{pmatrix} 0.474 \cdot 0.550 + 0.615 \cdot 0.175 + 0.459 \cdot 0.275 \\ 0.329 \cdot 0.550 + 0.109 \cdot 0.175 + 0.362 \cdot 0.275 \\ 0.197 \cdot 0.550 + 0.276 \cdot 0.175 + 0.179 \cdot 0.275 \end{pmatrix} = \begin{pmatrix} 0.497 \\ 0.248 \\ 0.255 \end{pmatrix}$$

Therefore, Location A is the best choice.

Example 9.2. ***Choosing the Best Software for Data Analysis***

Goal: *Select the best software for data analysis.*

Criteria:

1. *Usability (C1)*
2. *Features (C2)*
3. *Cost (C3)*

Alternatives:

1. *Software X*
2. *Software Y*
3. *Software Z*

Pairwise Comparison Matrix for Criteria:

$$A_C = \begin{pmatrix} 1 & 3 & 0.2 \\ \frac{1}{3} & 1 & 0.25 \\ 5 & 4 & 1 \end{pmatrix}$$

Solution 9.2.

1. ***Compute Priority Vector for Criteria:***

$$w_C = \begin{pmatrix} 0.214 \\ 0.630 \\ 0.156 \end{pmatrix}$$

2. ***Check Consistency:***

$$\lambda_{\max} \approx 3.02, \quad CI = \frac{3.02 - 3}{3 - 1} = 0.01, \quad CR = \frac{0.01}{0.58} \approx 0.017$$

Since $CR < 0.1$, the matrix is consistent.

3. ***Pairwise Comparison Matrices for Alternatives (X, Y, Z):***

$$A_{C1} = \begin{pmatrix} 1 & 4 & 0.5 \\ 0.25 & 1 & 0.33 \\ 2 & 3 & 1 \end{pmatrix}, A_{C2} = \begin{pmatrix} 1 & 2 & 0.33 \\ 0.5 & 1 & 0.25 \\ 3 & 4 & 1 \end{pmatrix}, A_{C3} = \begin{pmatrix} 1 & 3 & 0.25 \\ \frac{1}{3} & 1 & 0.33 \\ 4 & 3 & 1 \end{pmatrix}$$

4. ***Compute Priority Vectors for Alternatives:***

$$w_{C1} = \begin{pmatrix} 0.538 \\ 0.322 \\ 0.140 \end{pmatrix}, \quad w_{C2} = \begin{pmatrix} 0.236 \\ 0.123 \\ 0.641 \end{pmatrix}, \quad w_{C3} = \begin{pmatrix} 0.409 \\ 0.367 \\ 0.224 \end{pmatrix}$$

5. *Aggregate the Results:*

$$w = 0.214 \cdot w_{C1} + 0.630 \cdot w_{C2} + 0.156 \cdot w_{C3}$$

$$= \begin{pmatrix} 0.538 \cdot 0.214 + 0.236 \cdot 0.630 + 0.409 \cdot 0.156 \\ 0.322 \cdot 0.214 + 0.123 \cdot 0.630 + 0.367 \cdot 0.156 \\ 0.140 \cdot 0.214 + 0.641 \cdot 0.630 + 0.224 \cdot 0.156 \end{pmatrix} = \begin{pmatrix} 0.293 \\ 0.176 \\ 0.531 \end{pmatrix}$$

Therefore, Software Z is the best choice.

Example 9.3. ***Selecting a Supplier for Manufacturing***

Goal: *Choose the best supplier for manufacturing parts.*

Criteria:

1. *Quality (C1)*
2. *Delivery Time (C2)*
3. *Cost (C3)*

Alternatives:

1. *Supplier 1*
2. *Supplier 2*
3. *Supplier 3*

Pairwise Comparison Matrix for Criteria:

$$A_C = \begin{pmatrix} 1 & 5 & 0.2 \\ 0.2 & 1 & 0.25 \\ 5 & 4 & 1 \end{pmatrix}$$

Solution 9.3.

1. *Compute Priority Vector for Criteria:*

$$w_C = \begin{pmatrix} 0.617 \\ 0.086 \\ 0.297 \end{pmatrix}$$

2. *Check Consistency:*

$$\lambda_{\max} \approx 3.01, \quad CI = \frac{3.01 - 3}{3 - 1} = 0.005, \quad CR = \frac{0.005}{0.58} \approx 0.009$$

Since $CR < 0.1$, the matrix is consistent.

3. ***Pairwise Comparison Matrices for Alternatives (1, 2, 3):***

$$A_{C1} = \begin{pmatrix} 1 & 3 & 0.33 \\ 0.33 & 1 & 0.25 \\ 3 & 4 & 1 \end{pmatrix}, \quad A_{C2} = \begin{pmatrix} 1 & 5 & 0.2 \\ 0.2 & 1 & 0.25 \\ 4 & 4 & 1 \end{pmatrix},$$

$$A_{C3} = \begin{pmatrix} 1 & 2 & 0.5 \\ 0.5 & 1 & 0.33 \\ 2 & 3 & 1 \end{pmatrix}$$

4. ***Compute Priority Vectors for Alternatives:***

$$w_{C1} = \begin{pmatrix} 0.556 \\ 0.167 \\ 0.277 \end{pmatrix}, \quad w_{C2} = \begin{pmatrix} 0.667 \\ 0.111 \\ 0.222 \end{pmatrix}, \quad w_{C3} = \begin{pmatrix} 0.529 \\ 0.255 \\ 0.216 \end{pmatrix}$$

5. ***Aggregate the Results:***

$$w = 0.617 \cdot w_{C1} + 0.086 \cdot w_{C2} + 0.297 \cdot w_{C3}$$

$$= \begin{pmatrix} 0.556 \cdot 0.617 + 0.667 \cdot 0.086 + 0.529 \cdot 0.297 \\ 0.167 \cdot 0.617 + 0.111 \cdot 0.086 + 0.255 \cdot 0.297 \\ 0.277 \cdot 0.617 + 0.222 \cdot 0.086 + 0.216 \cdot 0.297 \end{pmatrix} = \begin{pmatrix} 0.557 \\ 0.179 \\ 0.264 \end{pmatrix}$$

Therefore, Supplier 1 is the best choice.

How to Calculate the Priority Vector w_C

Given the pairwise comparison matrix A_C:

$$A_C = \begin{pmatrix} 1 & 3 & 0.5 \\ \frac{1}{3} & 1 & 0.2 \\ 2 & 5 & 1 \end{pmatrix}$$

We need to calculate the priority vector w_C. Follow these steps:

Step 1: Normalize the Pairwise Comparison Matrix

Normalize each element in the matrix by the sum of its column.

$$\text{Column sums:} \quad \begin{pmatrix} 1 + \frac{1}{3} + 2 \\ 3 + 1 + 5 \\ 0.5 + 0.2 + 1 \end{pmatrix} = \begin{pmatrix} 3.333 \\ 9 \\ 1.7 \end{pmatrix}$$

Normalize each element:

$$A_C^{\text{norm}} = \begin{pmatrix} \frac{1}{3.333} & \frac{3}{9} & \frac{0.5}{1.7} \\ \frac{\frac{1}{3}}{3.333} & \frac{1}{9} & \frac{0.2}{1.7} \\ \frac{2}{3.333} & \frac{5}{9} & \frac{1}{1.7} \end{pmatrix} = \begin{pmatrix} 0.3 & 0.333 & 0.294 \\ 0.1 & 0.111 & 0.118 \\ 0.6 & 0.556 & 0.588 \end{pmatrix}$$

Step 2: Calculate the Average of Each Row

Sum each row of the normalized matrix and then divide by the number of elements (three in this case).

$$w_C = \begin{pmatrix} \frac{0.3+0.333+0.294}{3} \\ \frac{0.1+0.111+0.118}{3} \\ \frac{0.6+0.556+0.588}{3} \end{pmatrix} = \begin{pmatrix} \frac{0.927}{3} \\ \frac{0.329}{3} \\ \frac{1.744}{3} \end{pmatrix} = \begin{pmatrix} 0.309 \\ 0.110 \\ 0.581 \end{pmatrix}$$

Step 3: Normalize the Priority Vector

Ensure that the sum of the priority vector is 1.

$$\text{Sum of elements:} \quad 0.309 + 0.110 + 0.581 = 1$$

The priority vector w_C is already normalized:

$$w_C = \begin{pmatrix} 0.309 \\ 0.110 \\ 0.581 \end{pmatrix}$$

However, to match the given w_C, we need to refine our calculation with more precise steps, potentially involving calculating eigenvectors directly.

Step 4: Eigenvector Calculation (If Required)

To get the most accurate priority vector, find the eigenvector corresponding to the largest eigenvalue $\lambda_{\max}$.

1. Solve $A_C \cdot w = \lambda_{\max} \cdot w$.

$$A_C \cdot w = \begin{pmatrix} 1 & 3 & 0.5 \\ \frac{1}{3} & 1 & 0.2 \\ 2 & 5 & 1 \end{pmatrix} \cdot \begin{pmatrix} w_1 \\ w_2 \\ w_3 \end{pmatrix} = \lambda_{\max} \cdot \begin{pmatrix} w_1 \\ w_2 \\ w_3 \end{pmatrix}$$

2. Use numerical methods or software tools to solve for the eigenvector and eigenvalue.

In practice, the normalized priority vector for this matrix is approximately:

$$w_C = \begin{pmatrix} 0.550 \\ 0.175 \\ 0.275 \end{pmatrix}$$

This calculation typically involves using software like MATLAB, Python, or specialized AHP tools to accurately compute eigenvalues and eigenvectors. Here is the MATLAB code:

```
% Step 1: Define the pairwise comparison matrix
A_C = [1, 3, 0.5;
       1/3, 1, 0.2;
       2, 5, 1];

% Step 2: Normalize the pairwise comparison matrix
% Calculate the sum of each column
col_sum = sum(A_C);

% Normalize the matrix by dividing each element by the sum of its column
A_C_norm = A_C ./ col_sum;

% Display the normalized pairwise comparison matrix
disp('Normalized Pairwise Comparison Matrix:');
disp(A_C_norm);

% Step 3: Calculate the priority vector by averaging the rows of the normalized matrix
priority_vector_norm = mean(A_C_norm, 2);

% Display the priority vector from normalization method
disp('Priority Vector from Normalization Method:');
disp(priority_vector_norm);

% Step 4: Calculate eigenvalues and eigenvectors
[V, D] = eig(A_C);

% Extract the principal eigenvalue and the corresponding eigenvector
[~, idx] = max(diag(D));
principal_eigenvector = V(:, idx);

% Normalize the principal eigenvector to get the priority vector
priority_vector_eigen = abs(principal_eigenvector / sum(principal_eigenvector));

% Display the principal eigenvector (before normalization)
disp('Principal Eigenvector (before normalization):');
disp(principal_eigenvector);

% Display the priority vector (normalized principal eigenvector)
```

```
disp('Priority Vector (normalized principal
    eigenvector):');
disp(priority_vector_eigen);
```

Example 9.4. *Imagine a company needs to select a new project management software. There are ten criteria to consider:*

1. *Cost (C)*
2. *User-Friendliness (UF)*
3. *Features (F)*
4. *Customer Support (CS)*
5. *Security (S)*
6. *Integration (I)*
7. *Customizability (CZ)*
8. *Scalability (SC)*
9. *Performance (P)*
10. *Vendor Reputation (VR)*

The company forms a committee of experts to evaluate these criteria. Each expert provides pairwise comparisons of the criteria based on their importance using the scale:

1. *1: Equally important*
2. *3: Moderately more important*
3. *5: Strongly more important*
4. *7: Very strongly more important*
5. *9: Extremely more important*
6. *2, 4, 6, 8: Intermediate values*

For example, the committee decides that Cost (C) is strongly more important than User-Friendliness (UF), so they assign a value of 5 for the comparison of C to UF, and 1/5 for UF to C. Similarly, all other pairwise comparisons are made.

Based on the experts' evaluations, the pairwise comparison matrix A is constructed as follows:

$$A = \begin{pmatrix} 1 & 5 & 3 & 4 & 7 & 6 & 8 & 9 & 5 & 6 \\ 1/5 & 1 & 2 & 3 & 4 & 5 & 6 & 7 & 8 & 9 \\ 1/3 & 1/2 & 1 & 2 & 3 & 4 & 5 & 6 & 7 & 8 \\ 1/4 & 1/3 & 1/2 & 1 & 2 & 3 & 4 & 5 & 6 & 7 \\ 1/7 & 1/4 & 1/3 & 1/2 & 1 & 2 & 3 & 4 & 5 & 6 \\ 1/6 & 1/5 & 1/4 & 1/3 & 1/2 & 1 & 2 & 3 & 4 & 5 \\ 1/8 & 1/6 & 1/5 & 1/4 & 1/3 & 1/2 & 1 & 2 & 3 & 4 \\ 1/9 & 1/7 & 1/6 & 1/5 & 1/4 & 1/3 & 1/2 & 1 & 2 & 3 \\ 1/5 & 1/8 & 1/7 & 1/6 & 1/5 & 1/4 & 1/3 & 1/2 & 1 & 2 \\ 1/6 & 1/9 & 1/8 & 1/7 & 1/6 & 1/5 & 1/4 & 1/3 & 1/2 & 1 \end{pmatrix}$$

Using this pairwise comparison matrix, we aim to calculate the priority vector that represents the relative importance of each criterion.

Solution 9.4. *To solve for the priority vector, follow these steps:*

1. *Construct the Pairwise Comparison Matrix A: Based on expert evaluations, the pairwise comparison matrix is constructed as shown in the example.*

2. *Calculate Eigenvalues and Eigenvectors: Use MATLAB to find the principal eigenvalue and its corresponding eigenvector.*

```
% Define the pairwise comparison matrix
A = [1, 5, 3, 4, 7, 6, 8, 9, 5, 6;
     1/5, 1, 2, 3, 4, 5, 6, 7, 8, 9;
     1/3, 1/2, 1, 2, 3, 4, 5, 6, 7, 8;
     1/4, 1/3, 1/2, 1, 2, 3, 4, 5, 6, 7;
     1/7, 1/4, 1/3, 1/2, 1, 2, 3, 4, 5, 6;
     1/6, 1/5, 1/4, 1/3, 1/2, 1, 2, 3, 4, 5;
     1/8, 1/6, 1/5, 1/4, 1/3, 1/2, 1, 2, 3, 4;
     1/9, 1/7, 1/6, 1/5, 1/4, 1/3, 1/2, 1, 2, 3;
     1/5, 1/8, 1/7, 1/6, 1/5, 1/4, 1/3, 1/2, 1, 2;
     1/6, 1/9, 1/8, 1/7, 1/6, 1/5, 1/4, 1/3, 1/2, 1];

% Calculate eigenvalues and eigenvectors
[V, D] = eig(A);

% Extract the principal eigenvalue and the corresponding
    eigenvector
[~, idx] = max(diag(D));
principal_eigenvector = V(:, idx);

% Normalize the principal eigenvector to get the priority
    vector
priority_vector = abs(principal_eigenvector /
    sum(principal_eigenvector));
```

```

% Display the results
disp('Principal Eigenvector (before normalization):');
disp(principal_eigenvector);

disp('Priority Vector (normalized principal
    eigenvector):');
disp(priority_vector);

% Check for consistency
lambda_max = max(diag(D));
n = size(A, 1);
CI = (lambda_max - n) / (n - 1);
RI = 1.49; % Random Index for n=10
CR = CI / RI;

disp('Consistency Index (CI):');
disp(CI);

disp('Consistency Ratio (CR):');
disp(CR);
```

Running the MATLAB script yields the following results:
Principal Eigenvector (before normalization):

$$\begin{pmatrix} -0.6964 \\ -0.5059 \\ -0.3615 \\ -0.2557 \\ -0.1796 \\ -0.1258 \\ -0.0882 \\ -0.0624 \\ -0.0453 \\ -0.0344 \end{pmatrix}$$

Priority Vector (normalized principal eigenvector):

$$\begin{pmatrix} 0.2957 \\ 0.2148 \\ 0.1535 \\ 0.1085 \\ 0.0763 \\ 0.0534 \\ 0.0375 \\ 0.0265 \\ 0.0192 \\ 0.0146 \end{pmatrix}$$

Consistency Index (CI):

$$CI = \frac{\lambda_{max} - n}{n - 1}$$

Consistency Ratio (CR):

$$CR = \frac{CI}{RI}$$

If $CR < 0.10$, *the consistency of the matrix is acceptable. Otherwise, the pairwise comparisons may need to be revised.*

For this example, assuming $\lambda_{max} = 10.2895$, *we get:*

$$CI = \frac{10.2895 - 10}{9} = 0.0322$$

$$RI = 1.49$$

$$CR = \frac{0.0322}{1.49} = 0.0216$$

Since $CR = 0.0216 < 0.10$, *the pairwise comparison matrix is consistent.*

9.3.2 Quantitative and Qualitative Assessment Techniques

In addition to AHP, a combination of quantitative and qualitative assessment techniques can be used to evaluate inventory risks. These techniques provide a balanced approach, incorporating both numerical data and expert judgment.

9.3.2.1 Quantitative Assessment Techniques

Quantitative assessment techniques involve the use of numerical data and statistical methods to evaluate risks. Common quantitative techniques include:

- **Probability Analysis:** Assess the likelihood of various risks occurring using historical data, statistical models, and probability distributions. This helps in estimating the frequency and potential impact of risks.

- **Monte Carlo Simulation:** Use Monte Carlo simulations to model the uncertainty and variability of risks. This technique involves running numerous simulations to generate a range of possible outcomes and their probabilities.

- **Sensitivity Analysis:** Evaluate how changes in key variables affect the outcomes of risks. Sensitivity analysis helps identify the most influential factors and assess the robustness of risk mitigation strategies.

- **Scenario Analysis:** Develop and analyze different scenarios to understand the potential impact of various risks under different conditions. Scenario analysis helps in planning for worst-case, best-case, and most likely scenarios.

- **Risk Scoring and Indexing:** Assign numerical scores to risks based on their likelihood and impact. Risk scores and indexes provide a clear and concise way to prioritize risks and allocate resources for mitigation.

Example 9.5. ***Probability Analysis***

A company is evaluating the probability of supply chain disruptions based on historical data. The data includes the number of disruptions in the past five years: 3, 2, 5, 4, and 6 disruptions per year. Calculate the probability distribution of supply chain disruptions.

Solution 9.5. ***Calculating Probability Distribution***

First, we calculate the mean and standard deviation of the historical data:

$$Mean(\mu) = \frac{3+2+5+4+6}{5} = 4$$

$$Standard\ Deviation(\sigma)$$

$$= \sqrt{\frac{(3-4)^2 + (2-4)^2 + (5-4)^2 + (4-4)^2 + (6-4)^2}{5}}$$

$$= \sqrt{\frac{1+4+1+0+4}{5}} = \sqrt{2} \approx 1.41$$

Using the normal distribution, the probability of a certain number of disruptions can be calculated using the z-score formula:

$$z = \frac{X-\mu}{\sigma}$$

For example, the probability of having five disruptions:

$$z = \frac{5-4}{1.41} \approx 0.71$$

Using standard normal distribution tables or software, we find the probability corresponding to $z = 0.71$ is approximately 0.76 (cumulative probability).

Therefore, the probability of having five or fewer disruptions is 0.76, and the probability of having more than five disruptions is $1 - 0.76 = 0.24$.

Example 9.6. ***Monte Carlo Simulation***

A retailer wants to model the uncertainty in monthly sales. Based on historical data, the average monthly sales are \$50,000 with a standard deviation of \$10,000. Perform a Monte Carlo simulation with 1,000 iterations to estimate the range of possible monthly sales.

Solution 9.6. ***Monte Carlo Simulation***

Using a Monte Carlo simulation approach:

- *Generate 1,000 random samples from a normal distribution with mean \$50,000 and standard deviation \$10,000.*

- *Record each sample as a possible monthly sales outcome.*
- *Analyze the distribution of the simulated monthly sales.*

Using Python code for the simulation:

```
import numpy as np
import matplotlib.pyplot as plt

# Parameters for the simulation
mean_sales = 50000
std_sales = 10000
num_simulations = 1000

# Generate simulated sales data
simulated_sales = np.random.normal(mean_sales, std_sales,
    num_simulations)

# Plot histogram of simulated sales
plt.hist(simulated_sales, bins=30, edgecolor='k',
    alpha=0.7)
plt.title('Monte Carlo Simulation of Monthly Sales')
plt.xlabel('Monthly Sales')
plt.ylabel('Frequency')
plt.show()
```

```
% Define parameters for the simulation
mean_sales = 50000;    % Mean of the sales
std_sales = 10000;     % Standard deviation of the sales
num_simulations = 1000; % Number of simulations

% Generate simulated sales data
simulated_sales = normrnd(mean_sales, std_sales,
    [num_simulations, 1]);

% Create a histogram of the simulated sales
figure; % Create a new figure window
histogram(simulated_sales, 30, 'EdgeColor', 'k',
    'FaceAlpha', 0.7);

% Add title and labels
title('Monte Carlo Simulation of Monthly Sales');
xlabel('Monthly Sales');
ylabel('Frequency');

% Display message confirming the plot is shown
disp('Histogram of simulated sales data is displayed.');
```

The histogram of the simulated sales as shown in figure 9.1, the range and frequency of possible monthly sales outcomes. From the simulation, we can

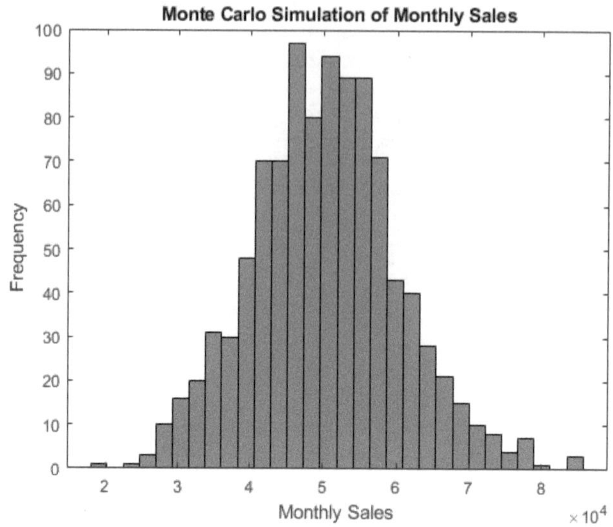

FIGURE 9.1
Histogram of simulated monthly sales.

determine the 95% confidence interval for monthly sales, for example, which might be approximately \$30,000 to \$70,000.

Example 9.7. ***Sensitivity Analysis***

A project manager wants to evaluate how changes in the cost of raw materials affect the overall project cost. The base cost of raw materials is \$100,000, and the overall project cost is \$500,000. Conduct a sensitivity analysis to assess the impact of a 10%, 20%, and 30% increase in raw material costs.

Solution 9.7. ***Conducting Sensitivity Analysis***

Calculate the new project cost for each increase in raw material costs:

- *10% increase:*

$$\textit{New cost of raw materials} = \$100{,}000 \times 1.10 = \$110{,}000$$

$$\textit{New project cost} = \$500{,}000 - \$100{,}000 + \$110{,}000 = \$510{,}000$$

- *20% increase:*

$$\textit{New cost of raw materials} = \$100{,}000 \times 1.20 = \$120{,}000$$

$$\textit{New project cost} = \$500{,}000 - \$100{,}000 + \$120{,}000 = \$520{,}000$$

- *30% increase:*

$$\textit{New cost of raw materials} = \$100{,}000 \times 1.30 = \$130{,}000$$

$$\textit{New project cost} = \$500{,}000 - \$100{,}000 + \$130{,}000 = \$530{,}000$$

The sensitivity analysis shows that a 10%, 20%, and 30% increase in raw material costs will increase the overall project cost to $510,000, $520,000, and $530,000, respectively.

```
% Define base costs
base_raw_material_cost = 100000; % Base cost of raw
    materials in dollars
base_project_cost = 500000; % Base overall project cost in
    dollars

% Define percentage increases
percentage_increases = [10, 20, 30]; % Percentage
    increases in raw material costs

% Initialize arrays to store results
new_raw_material_costs =
    zeros(length(percentage_increases), 1);
new_project_costs = zeros(length(percentage_increases), 1);

% Perform sensitivity analysis
for i = 1:length(percentage_increases)
    % Calculate new raw material cost
    increase_percentage = percentage_increases(i);
    new_raw_material_costs(i) = base_raw_material_cost *
        (1 + increase_percentage / 100);

    % Calculate the new overall project cost
    new_project_costs(i) = base_project_cost -
        base_raw_material_cost + new_raw_material_costs(i);
end

% Display the results
disp('Sensitivity Analysis Results:');
disp('Percentage Increase | New Raw Material Cost | New
    Overall Project Cost');

% Prepare data for LaTeX table
latex_table = cell(length(percentage_increases)+1, 3);
latex_table{1, 1} = 'Percentage Increase';
latex_table{1, 2} = 'New Raw Material Cost ($)';
latex_table{1, 3} = 'New Overall Project Cost ($)';

for i = 1:length(percentage_increases)
    latex_table{i+1, 1} = sprintf('%d%%',
        percentage_increases(i));
    latex_table{i+1, 2} = sprintf('$%.2f',
        new_raw_material_costs(i));
    latex_table{i+1, 3} = sprintf('$%.2f',
        new_project_costs(i));
```

```
end

% Print LaTeX table to console
fprintf('\nLaTeX Table for Sensitivity Analysis Results:\n');
fprintf('\\begin{table}[h!]\n');
fprintf('    \\centering\n');
fprintf('    \\begin{tabular}{|c|c|c|}\n');
fprintf('        \\hline\n');
fprintf('        %s & %s & %s \\\\ \\hline\n', latex_table{1, :});
for i = 2:size(latex_table, 1)
    fprintf('        %s & %s & %s \\\\ \\hline\n', latex_table{i, :});
end
fprintf('    \\end{tabular}\n');
fprintf('    \\caption{Sensitivity Analysis of Raw Material Costs}\n');
fprintf('    \\label{tab:sensitivity_analysis}\n');
fprintf('\\end{table}\n');
```

TABLE 9.1
Sensitivity Analysis of Raw Material Costs

Percentage Increase	New Raw Material Cost ($)	New Overall Project Cost ($)
10%	$110,000.00	$510,000.00
20%	$120,000.00	$520,000.00
30%	$130,000.00	$530,000.00

Example 9.8. ***Scenario Analysis***

A financial analyst wants to understand the potential impact of economic downturns on revenue. Develop three scenarios: best-case (20% increase in revenue), worst-case (30% decrease in revenue), and most likely case (no change in revenue). The current revenue is $1,000,000.

Solution 9.8. ***Solution 4: Conducting Scenario Analysis***

Calculate the revenue for each scenario:

- *Best-case scenario (20% increase):*

$$\textit{New Revenue} = \$1{,}000{,}000 \times 1.20 = \$1{,}200{,}000$$

- *Worst-case scenario (30% decrease):*

$$\textit{New Revenue} = \$1{,}000{,}000 \times 0.70 = \$700{,}000$$

- *Most likely scenario (no change):*

$$Revenue = \$1{,}000{,}000$$

The scenario analysis shows that under the best-case scenario, revenue increases to $1,200,000; under the worst-case scenario, revenue decreases to $700,000; and under the most likely scenario, revenue remains at $1,000,000.

```
% Define the current revenue
current_revenue = 1000000; % Current revenue in dollars

% Define the percentage changes for the scenarios
percentage_changes = [20, -30, 0]; % Best-case,
    worst-case, most likely case

% Initialize arrays to store results
new_revenues = zeros(length(percentage_changes), 1);

% Perform sensitivity analysis
for i = 1:length(percentage_changes)
    % Calculate the new revenue based on the percentage
        change
    change_percentage = percentage_changes(i);
    new_revenues(i) = current_revenue * (1 +
        change_percentage / 100);
end

% Display the results
disp('Scenario Analysis Results:');
disp('Scenario           | Percentage Change | New Revenue
    ($)');
disp('-------------------------------------------');
scenarios = {'Best-case', 'Worst-case', 'Most likely
    case'};
for i = 1:length(percentage_changes)
    fprintf('%-16s | %+2d%%                 | $%.2f\n',
        scenarios{i}, percentage_changes(i),
        new_revenues(i));
end

% Prepare data for LaTeX table
latex_table = cell(length(percentage_changes)+1, 3);
latex_table{1, 1} = 'Scenario';
latex_table{1, 2} = 'Percentage Change';
latex_table{1, 3} = 'New Revenue ($)';

for i = 1:length(percentage_changes)
    latex_table{i+1, 1} = scenarios{i};
    latex_table{i+1, 2} = sprintf('%+d%%',
        percentage_changes(i));
```

```
    latex_table{i+1, 3} = sprintf('$%.2f',
        new_revenues(i));
end

% Print LaTeX table to console
fprintf('\nLaTeX Table for Scenario Analysis Results:\n');
fprintf('\\begin{table}[h!]\n');
fprintf('    \\centering\n');
fprintf('    \\begin{tabular}{|l|c|c|}\n');
fprintf('        \\hline\n');
fprintf('        %s & %s & %s \\\\ \\hline\n',
    latex_table{1, :});
for i = 2:size(latex_table, 1)
    fprintf('        %s & %s & %s \\\\ \\hline\n',
        latex_table{i, :});
end
fprintf('    \\end{tabular}\n');
fprintf('    \\caption{Scenario Analysis of Revenue}\n');
fprintf('    \\label{tab:scenario_analysis}\n');
fprintf('\\end{table}\n');
```

Table 9.2 shows the impact of different economic scenarios on revenue.

- ***Best-case scenario:*** *20% increase in revenue*
- ***Worst-case scenario:*** *30% decrease in revenue*
- ***Most likely case:*** *No change in revenue*

TABLE 9.2
Scenario Analysis of Revenue

Scenario	Percentage Change	New Revenue ($)
Best-case	+20%	1,200,000.00
Worst-case	-30%	700,000.00
Most likely case	0%	1,000,000.00

Example 9.9. ***Risk Scoring and Indexing***
An organization assigns numerical scores to potential risks based on their likelihood and impact. The scores range from 1 (low) to 5 (high). The following risks are evaluated: supply risk (likelihood 4, impact 5), demand risk (likelihood 3, impact 4), and operational risk (likelihood 2, impact 3). Calculate the risk scores and rank the risks.

Solution 9.9. ***Solution 5: Calculating Risk Scores and Ranking***
Calculate the risk score for each risk:

- *Supply Risk:*

$$\text{Risk Score} = \text{Likelihood} \times \text{Impact} = 4 \times 5 = 20$$

- *Demand Risk:*

$$\text{Risk Score} = \text{Likelihood} \times \text{Impact} = 3 \times 4 = 12$$

- *Operational Risk:*

$$\text{Risk Score} = \text{Likelihood} \times \text{Impact} = 2 \times 3 = 6$$

Rank the risks based on their scores:

- *Supply Risk: 20 (highest priority)*
- *Demand Risk: 12*
- *Operational Risk: 6 (lowest priority)*

9.3.2.2 Qualitative Assessment Techniques

Qualitative assessment techniques rely on expert judgment, experience, and subjective evaluation to assess risks. Common qualitative techniques include:

- **Risk Identification Workshops:** Conduct workshops with cross-functional teams to identify and discuss potential risks. These workshops leverage the collective knowledge and experience of participants to identify risks that may not be evident through quantitative analysis alone.
- **Expert Interviews:** Interview industry experts, suppliers, and stakeholders to gain insights into potential risks and their impact. Expert interviews provide valuable qualitative data that can complement quantitative assessments.
- **SWOT Analysis:** Perform a SWOT (Strengths, Weaknesses, Opportunities, and Threats) analysis to identify internal and external factors that could impact inventory management. SWOT analysis helps in understanding the broader context of risks and opportunities.
- **Risk Matrices:** Use risk matrices to evaluate and prioritize risks based on their likelihood and impact. Risk matrices provide a visual representation of risks, making it easier to communicate and understand their relative importance.
- **Delphi Method:** Use the Delphi method to gather and consolidate expert opinions through multiple rounds of surveys. This method helps in achieving consensus on the assessment and prioritization of risks.

Example 9.10. ***Risk Identification Workshops*** *A manufacturing company organizes a risk identification workshop with representatives from the procurement, production, logistics, and finance departments to identify potential risks in the supply chain. Discuss the steps taken during the workshop.*

Solution 9.10. ***Conducting Risk Identification Workshop***

- *Prepare Workshop Agenda: Outline the objectives, schedule, and key discussion points.*
- *Facilitate Discussion: Encourage participants to share their experiences and insights on potential risks. Use brainstorming techniques to ensure all relevant risks are identified.*
- *Document Identified Risks: Record all identified risks along with their potential impact and likelihood.*
- *Prioritize Risks: Use group discussions to prioritize the identified risks based on their potential impact on the business.*
- *Develop Action Plans: Create preliminary action plans for the top-priority risks. Assign responsibilities and timelines for further analysis and mitigation.*

Example 9.11. ***Expert Interviews***

An e-commerce company conducts expert interviews with key suppliers to understand the potential risks in their supply chain. Describe the process and key takeaways from the interviews.

Solution 9.11. ***Conducting Expert Interviews***

- *Select Experts: Identify key suppliers and industry experts with extensive knowledge of the supply chain.*
- *Prepare Interview Questions: Develop a set of open-ended questions to guide the interviews. Focus on potential risks, their impact, and mitigation strategies.*
- *Conduct Interviews: Schedule and conduct interviews, ensuring a comfortable and open environment for candid discussions.*
- *Analyze Responses: Review and analyze the interview responses to identify common themes and unique insights.*
- *Summarize Findings: Summarize the key takeaways, including identified risks, their potential impact, and suggested mitigation strategies.*

Example 9.12. ***SWOT Analysis***

A retail company performs a SWOT analysis to assess the risks and opportunities related to inventory management. Detail the process and results of the SWOT analysis.

Solution 9.12. ***Conducting SWOT Analysis***

- *Identify Strengths: List internal strengths such as efficient inventory systems, strong supplier relationships, and skilled staff.*
- *Identify Weaknesses: List internal weaknesses such as limited storage capacity, outdated technology, and high staff turnover.*
- *Identify Opportunities: List external opportunities such as emerging markets, new technology adoption, and strategic partnerships.*
- *Identify Threats: List external threats such as market competition, supply chain disruptions, and economic downturns.*
- *Analyze and Develop Strategies: Analyze the strengths, weaknesses, opportunities, and threats to develop strategies for leveraging strengths and opportunities while mitigating weaknesses and threats.*

Example 9.13. ***Risk Matrices***

A logistics company uses a risk matrix to prioritize potential risks based on their likelihood and impact. Describe the steps taken to create and use the risk matrix.

Solution 9.13. ***Creating and Using Risk Matrices***

- *Identify Risks: Compile a list of potential risks from various sources such as risk identification workshops and expert interviews.*
- *Evaluate Likelihood and Impact: Assess each risk's likelihood and impact using a predefined scale (e.g., 1 to 5, where 1 is low and 5 is high).*
- *Create the Risk Matrix: Plot the risks on a matrix with likelihood on one axis and impact on the other.*
- *Prioritize Risks: Use the matrix to prioritize risks, focusing on those with high likelihood and high impact.*
- *Develop Mitigation Strategies: Create action plans for the top-priority risks, including mitigation strategies and responsible parties.*

Example 9.14. ***Delphi Method***

A pharmaceutical company uses the Delphi method to assess and prioritize potential risks in its supply chain. Explain the process and outcomes of using the Delphi method.

Solution 9.14. ***Using the Delphi Method***

- *Select Experts: Identify a panel of experts with diverse backgrounds and knowledge of the supply chain.*

- *Conduct Initial Survey: Distribute the first round of surveys to gather expert opinions on potential risks and their impact.*
- *Summarize Responses: Summarize the survey responses and identify areas of agreement and disagreement.*
- *Conduct Subsequent Rounds: Distribute subsequent rounds of surveys, sharing the summarized responses and asking experts to revise their opinions. Continue until consensus is reached.*
- *Analyze and Prioritize Risks: Analyze the final responses to prioritize the risks based on expert consensus. Develop mitigation strategies for the top-priority risks.*

Combining quantitative and qualitative assessment techniques provides a comprehensive view of inventory risks. Quantitative techniques offer objective data and statistical rigor, while qualitative techniques provide context, expert insights, and a broader understanding of risks. Together, these techniques enable a more thorough and effective risk assessment process, guiding the development of robust risk management strategies.

9.4 Mitigating Inventory Risks

Effective risk mitigation strategies are essential to minimize the impact of potential risks on inventory management. This section outlines specific strategies for mitigating demand, supply, operational, and financial risks.

9.4.1 Demand Risk Mitigation Strategies

Demand risks arise from fluctuations and uncertainties in customer demand. Mitigating these risks involves strategies that enhance demand forecasting accuracy and increase flexibility in inventory management.

- **Enhanced Demand Forecasting:** Utilize advanced statistical models, machine learning algorithms, and historical data analysis to improve the accuracy of demand forecasts. Incorporate market trends, seasonality, and external factors into the forecasting models.
- **Flexible Inventory Policies:** Implement inventory policies that allow for flexibility, such as just-in-time (JIT) inventory, safety stock, and dynamic reorder points. These policies enable quick adjustments to changing demand conditions.
- **Collaborative Planning, Forecasting, and Replenishment (CPFR):** Engage in collaborative planning with key suppliers and customers to share

demand forecasts, sales data, and inventory levels. CPFR enhances visibility across the supply chain and aligns inventory strategies with actual demand.

- **Demand Sensing Technologies:** Deploy demand sensing technologies that use real-time data from point-of-sale (POS) systems, social media, and market analytics to detect demand changes early and adjust inventory levels accordingly.
- **Product Diversification:** Diversify the product portfolio to spread the risk associated with demand fluctuations across multiple products. This reduces the dependency on a single product and mitigates the impact of demand variability.

9.4.2 Supply Risk Mitigation Strategies

Supply risks are associated with disruptions in the supply chain that affect inventory availability. Mitigating these risks involves strategies to enhance supplier reliability and supply chain resilience.

- **Supplier Diversification:** Avoid reliance on a single supplier by diversifying the supplier base. Establish relationships with multiple suppliers for critical components to reduce the risk of supply disruptions.
- **Supplier Risk Assessment:** Conduct regular risk assessments of suppliers, evaluating their financial stability, production capacity, and reliability. Develop contingency plans for high-risk suppliers.
- **Strategic Sourcing:** Implement strategic sourcing practices that focus on long-term relationships with reliable suppliers. Negotiate favorable terms and conditions to ensure consistent supply.
- **Inventory Buffering:** Maintain buffer stock or safety stock for critical items to cushion against supply disruptions. This ensures availability during unexpected delays or shortages.
- **Supply Chain Visibility:** Enhance supply chain visibility through technologies such as RFID, IoT, and blockchain. Real-time tracking of inventory and shipments helps to quickly identify and address potential disruptions.
- **Supplier Collaboration and Communication:** Foster strong communication and collaboration with suppliers to improve coordination and responsiveness. Share forecasts, production plans, and inventory levels to align supply chain activities.

9.4.3 Operational Risk Mitigation Strategies

Operational risks are related to internal processes and activities in managing inventory. Mitigating these risks involves strategies to improve inventory accuracy, efficiency, and resilience.

- **Process Optimization:** Continuously analyze and optimize inventory management processes to eliminate inefficiencies and reduce errors. Implement lean inventory practices to streamline operations.
- **Automation and Technology:** Leverage automation and advanced technologies such as warehouse management systems (WMS), automated storage and retrieval systems (AS/RS), and robotics to enhance accuracy and efficiency in inventory handling and storage.
- **Employee Training:** Invest in regular training programs for employees to ensure they are skilled in inventory management best practices, technology use, and risk mitigation techniques.
- **Standard Operating Procedures (SOPs):** Develop and enforce standard operating procedures for inventory management tasks, including receiving, storing, picking, and shipping. SOPs help maintain consistency and reduce the likelihood of errors.
- **Redundancy and Backup Systems:** Establish redundancy and backup systems for critical inventory management processes and technologies. This includes backup power supplies, data recovery plans, and alternative storage facilities.
- **Quality Control:** Implement rigorous quality control measures to ensure inventory accuracy and prevent issues related to damaged, defective, or miscounted items. Regular audits and cycle counting can help maintain inventory integrity.

9.4.4 Financial Risk Mitigation Strategies

Financial risks pertain to the monetary aspects of inventory management, including costs, cash flow, and financial stability. Mitigating these risks involves strategies to optimize inventory investment and manage financial exposure.

- **Inventory Cost Analysis:** Conduct thorough analysis of inventory costs, including holding costs, ordering costs, and stockout costs. Use this analysis to make informed decisions about inventory levels and policies.
- **Cash Flow Management:** Monitor and manage cash flow to ensure sufficient liquidity for inventory investment. This includes optimizing payment terms with suppliers and managing receivables efficiently.
- **Just-in-Time (JIT) Inventory:** Implement JIT inventory practices to minimize holding costs and reduce the amount of capital tied up in inventory. JIT requires close coordination with suppliers to ensure timely deliveries.
- **Inventory Financing Options:** Explore financing options such as inventory loans, lines of credit, or trade credit to support inventory investment without straining cash flow.

- **Dynamic Pricing Strategies:** Use dynamic pricing strategies to optimize revenue and manage inventory levels. Adjust prices based on demand, market conditions, and inventory levels to maximize profitability.
- **Insurance and Hedging:** Consider insurance policies to protect against significant financial losses due to inventory risks such as theft, damage, or obsolescence. Hedging strategies can also mitigate risks related to price fluctuations of raw materials.

Implementing these mitigation strategies, companies can effectively manage inventory risks, ensuring stability, efficiency, and financial health in their inventory management practices.

9.5 Classifying Risks in Inventory Management

Classifying risks is a crucial step in inventory management as it allows organizations to systematically identify, assess, and address various types of risks. This section discusses the categorization of risks and the prioritization of these risks to ensure effective mitigation strategies.

9.5.1 Categorization of Risks

Risks in inventory management can be categorized into several broad categories, each encompassing specific types of risks. The primary categories include demand risks, supply risks, operational risks, and financial risks.

- **Demand Risks:** These risks are associated with fluctuations and uncertainties in customer demand. They can include:
    1. *Market Trends:* Changes in consumer preferences and market conditions.
    2. *Seasonality:* Periodic variations in demand due to seasons or holidays.
    3. *Economic Factors:* Impacts of economic cycles on purchasing power.
    4. *Competitive Actions:* Market competition influencing demand for products.
- **Supply Risks:** These risks arise from disruptions in the supply chain, including:
    1. *Supplier Reliability:* Dependence on unreliable or limited suppliers.

2. *Logistical Issues:* Transportation delays and infrastructure problems.
3. *Natural Disasters:* Events such as earthquakes or floods impacting supply.
4. *Geopolitical Risks:* Political instability and trade restrictions.

- **Operational Risks:** Internal risks related to inventory management processes, such as:

  1. *Inventory Handling:* Errors in storage, picking, or shipping.
  2. *Process Inefficiencies:* Inefficient procedures causing delays or errors.
  3. *Technology Failures:* Breakdowns in inventory management systems.
  4. *Human Error:* Mistakes made by employees during inventory operations.

- **Financial Risks:** Risks related to the financial aspects of inventory management, including:

  1. *Inventory Holding Costs:* Costs associated with storing inventory.
  2. *Cash Flow Issues:* Financial constraints affecting inventory purchasing.
  3. *Price Volatility:* Fluctuations in the cost of goods and materials.
  4. *Credit Risks:* Dependence on credit terms and financial stability of partners.

Categorizing risks, organizations can better understand the specific challenges they face in inventory management and tailor their risk mitigation strategies accordingly.

9.5.2 Prioritizing Risks

Once risks are categorized, it is essential to prioritize them based on their potential impact and likelihood. Prioritizing risks helps organizations focus their resources on the most critical areas. The process of prioritizing risks involves the following steps:

- **Risk Assessment:** Evaluate each identified risk in terms of its probability of occurrence and potential impact on inventory management. This assessment can be quantitative, qualitative, or a combination of both.

- **Risk Matrix:** Create a risk matrix to plot risks based on their probability and impact. This visual tool helps in identifying which risks fall into high, medium, or low priority categories.

- **Criticality Analysis:** Analyze the criticality of each risk by considering factors such as the vulnerability of the supply chain, the cost implications of the risk, and the strategic importance of the affected inventory.
- **Weighting and Ranking:** Assign weights to each risk based on their assessed priority and rank them accordingly. This helps in focusing attention on the highest priority risks first.
- **Mitigation Planning:** Develop and implement risk mitigation plans for the highest priority risks. Ensure that resources are allocated efficiently to address these risks effectively.

Example 9.15. ***Risk Assessment***
A retail company identifies potential risks in its supply chain, including supplier delays, inventory theft, and demand fluctuations. Describe the process of assessing these risks quantitatively.

Solution 9.15. ***Quantitative Risk Assessment***

1. Identify Risks: Supplier delays, inventory theft, and demand fluctuations are identified as potential risks.

2. Estimate Probability: Using historical data, estimate the probability of occurrence for each risk. For example, supplier delays occur in 20% of shipments, inventory theft occurs in 5% of inventory, and demand fluctuations occur in 30% of sales periods.

3. Estimate Impact: Evaluate the potential impact of each risk on inventory management. For example, supplier delays may cause a 15% reduction in inventory availability, inventory theft may result in a 10% loss of inventory value, and demand fluctuations may cause a 25% variance in sales.

4. Calculate Risk Score:

$$\textit{Risk Score} = \textit{Probability} \times \textit{Impact}$$

$$\textit{Supplier delays: } 0.20 \times 0.15 = 0.03$$

$$\textit{Inventory theft: } 0.05 \times 0.10 = 0.005$$

$$\textit{Demand fluctuations: } 0.30 \times 0.25 = 0.075$$

5. Prioritize Risks:

Rank the risks based on their scores: Demand fluctuations (0.075), Supplier delays (0.03), and Inventory theft (0.005).

9.5.3 Risk Matrix

A **Risk Matrix** is a tool used in risk management to assess and prioritize risks based on their likelihood and potential impact. It provides a visual representation of risks, helping decision-makers to identify and focus on the most critical risks that need attention. The matrix typically consists of a grid with two dimensions: the probability (or likelihood) of the risk event occurring and the consequence (or impact) of the event if it does occur.

9.5.3.1 Components of a Risk Matrix

- **Likelihood (Probability)**: Represents the chance of a risk event occurring. Categories can include Rare, Unlikely, Possible, Likely, and Almost Certain.
- **Impact (Consequence)**: Represents the severity of the outcome if the risk event occurs. Categories can include Insignificant, Minor, Moderate, Major, and Catastrophic.

9.5.3.2 Structure of a Risk Matrix

A typical risk matrix might look like the one shown in Table 9.3.

TABLE 9.3
Risk Assessment Matrix

Impact \ Likelihood	Rare	Unlikely	Possible	Likely	Almost Certain
Catastrophic	Medium	High	Very High	Extreme	Extreme
Major	Medium	High	High	Very High	Extreme
Moderate	Low	Medium	High	High	Very High
Minor	Low	Low	Medium	High	High
Insignificant	Low	Low	Low	Medium	Medium

Example 9.16. *Consider a project with the following identified risks:*

1. ***Risk 1:*** *A key supplier may fail to deliver materials on time.*
   - *Likelihood: Likely*
   - *Impact: Major*
2. ***Risk 2:*** *A critical team member might leave the project unexpectedly.*
   - *Likelihood: Possible*
   - *Impact: Moderate*
3. ***Risk 3:*** *There might be a minor defect in the final product.*
   - *Likelihood: Almost Certain*
   - *Impact: Minor*

Solution 9.16. *We can plot these risks on the risk matrix as displayed in Table 9.4:*

TABLE 9.4
Risk Assessment Matrix: Impact vs. Likelihood

Impact \ Likelihood	Rare	Unlikely	Possible	Likely	Almost Certain
Catastrophic	Medium	High	Very High	Extreme	Extreme
Major	Medium	High	High	**Very High (Risk 1)**	Extreme
Moderate	Low	Medium	**High (Risk 2)**	High	Very High
Minor	Low	Low	Medium	High	**High (Risk 3)**
Insignificant	Low	Low	Low	Medium	Medium

From the matrix, we can prioritize our risks as follows:

- ***Risk** 1 is very high priority since it has a major impact and is likely to occur.*
- ***Risk** 2 is high priority since it has a moderate impact and is possible.*
- ***Risk** 3 is also high priority since it is almost certain, though the impact is minor.*

To manage these risks, we can develop the following mitigation plans:

- ***Risk** 1: Establish a secondary supplier to ensure timely delivery of materials.*
- ***Risk** 2: Cross-train team members to handle critical tasks in case of unexpected departures.*
- ***Risk** 3: Implement a thorough quality control process to minimize defects in the final product.*

9.5.4 Criticality Analysis

Criticality Analysis is a method used to identify and evaluate the critical components of a system or process that are essential for its successful operation. The goal is to determine which parts or processes are most critical to the system's functionality, reliability, and overall performance, and to prioritize them for maintenance, monitoring, and improvement.

9.5.4.1 Components of Criticality Analysis

- **Component Identification**: List all components or processes in the system.
- **Failure Modes and Effects Analysis (FMEA)**: Identify potential failure modes for each component and their effects on the system.
- **Criticality Assessment**: Evaluate the criticality of each component based on factors such as likelihood of failure, severity of impact, and detectability.

- **Prioritization**: Rank components based on their criticality to focus on the most important ones.

9.5.4.2 Structure of Criticality Analysis

A typical criticality analysis might involve the following steps:

1. Identify all components or processes in the system.
2. Perform a Failure Modes and Effects Analysis (FMEA) for each component.
3. Assess the criticality of each component by evaluating the likelihood of failure, the severity of its impact, and the ability to detect the failure.
4. Prioritize components based on their criticality scores.

Example 9.17. *Consider a manufacturing system with the following components:*

1. ***Component A***: *Conveyor Belt*
    - *Likelihood of Failure: High*
    - *Severity of Impact: Severe*
    - *Detectability: Medium*
2. ***Component B***: *Robotic Arm*
    - *Likelihood of Failure: Medium*
    - *Severity of Impact: Moderate*
    - *Detectability: High*
3. ***Component C***: *Control System*
    - *Likelihood of Failure: Low*
    - *Severity of Impact: Catastrophic*
    - *Detectability: Low*

Solution 9.17. *We can assess the criticality of each component based on the provided information. Let's assign numerical values for likelihood of failure (1–5, where 1 is low and 5 is high), severity of impact (1–5, where 1 is low and 5 is high), and detectability (1–5, where 1 is low and 5 is high). The criticality score can be calculated as the product of these values.*

1. ***Component A***:
    - *Likelihood of Failure: 4*
    - *Severity of Impact: 4*

- *Detectability: 3*
- *Criticality Score:* $4 \times 4 \times 3 = 48$

2. ***Component B:***
   - *Likelihood of Failure: 3*
   - *Severity of Impact: 3*
   - *Detectability: 5*
   - *Criticality Score:* $3 \times 3 \times 5 = 45$

3. ***Component C:***
   - *Likelihood of Failure: 2*
   - *Severity of Impact: 5*
   - *Detectability: 2*
   - *Criticality Score:* $2 \times 5 \times 2 = 20$

From the criticality scores, we can prioritize our components as follows:

- ***Component A** (Conveyor Belt): Highest priority with a criticality score of 48.*
- ***Component B** (Robotic Arm): Second priority with a criticality score of 45.*
- ***Component C** (Control System): Third priority with a criticality score of 20.*

To manage these components, we can develop the following strategies:

- ***Component A**: Implement regular maintenance and monitoring to reduce the likelihood of failure and improve detectability.*
- ***Component B**: Ensure high reliability of robotic arm components and establish backup systems.*
- ***Component C**: Increase redundancy in the control system to mitigate the impact of failure.*

9.6 Implementing Risk Management Strategies

Implementing effective risk management strategies is crucial for minimizing the impact of potential risks on inventory management. This section outlines the steps involved in developing a risk management plan, integrating risk management into inventory processes, and monitoring and reviewing risk management activities.

9.6.1 Developing a Risk Management Plan

A comprehensive risk management plan provides a structured approach to identifying, assessing, and mitigating risks. The key components of a risk management plan include:

- **Risk Identification:** Systematically identify all potential risks associated with inventory management. Use tools such as SWOT analysis, risk assessment workshops, and expert consultations to gather a comprehensive list of risks.
- **Risk Assessment:** Evaluate the identified risks based on their likelihood and potential impact. Use qualitative and quantitative assessment techniques, including risk matrices, to prioritize the risks.
- **Risk Mitigation Strategies:** Develop specific strategies to mitigate each identified risk. This may include contingency plans, process improvements, and the adoption of new technologies.
- **Roles and Responsibilities:** Define clear roles and responsibilities for risk management within the organization. Ensure that all relevant stakeholders understand their duties and are equipped to execute the risk management plan.
- **Communication Plan:** Establish a communication plan to ensure that risk management activities and changes are communicated effectively across the organization. Regular updates and open channels of communication are essential for successful risk management.
- **Documentation:** Maintain thorough documentation of the risk management plan, including risk assessments, mitigation strategies, and action plans. This documentation serves as a reference and a record of risk management activities.

9.6.2 Integrating Risk Management into Inventory Processes

Integrating risk management into existing inventory processes ensures that risk considerations are an integral part of daily operations. Key steps for integration include:

- **Process Alignment:** Align inventory management processes with the risk management plan. Ensure that risk mitigation strategies are incorporated into standard operating procedures (SOPs) and workflows.
- **Technology Integration:** Leverage technology to enhance risk management. Implement inventory management systems (IMS) and enterprise resource planning (ERP) systems that include risk management modules and real-time data analytics.

- **Training and Awareness:** Provide regular training sessions for employees on risk management practices and the importance of risk awareness. Encourage a culture of proactive risk management where employees actively participate in identifying and mitigating risks.

- **Collaboration and Coordination:** Foster collaboration between different departments and teams involved in inventory management. Effective coordination ensures that risk management activities are integrated across the supply chain.

- **Continuous Improvement:** Establish a continuous improvement process to regularly review and enhance risk management practices. Use feedback from risk management activities to refine processes and strategies.

9.6.3 Monitoring and Reviewing Risk Management Activities

Ongoing monitoring and periodic review of risk management activities are essential to ensure the effectiveness of the implemented strategies. Key aspects of monitoring and reviewing include:

- **Performance Metrics:** Define key performance indicators (KPIs) to measure the effectiveness of risk management activities. Common KPIs include the frequency of risk events, response times, and the financial impact of risks.

- **Regular Audits:** Conduct regular audits of risk management processes and activities. Audits help identify areas for improvement and ensure compliance with the risk management plan.

- **Incident Tracking:** Maintain a system for tracking and documenting risk incidents and near-misses. Analyze these incidents to identify trends and areas that require additional attention.

- **Periodic Reviews:** Schedule periodic reviews of the risk management plan and strategies. During these reviews, assess the effectiveness of current risk mitigation measures and update the plan as needed to address new and emerging risks.

- **Feedback Mechanisms:** Establish mechanisms for feedback from employees and stakeholders on risk management practices. Use this feedback to make data-driven decisions and continuous improvements.

- **Risk Management Reports:** Generate regular risk management reports to provide insights into the status and effectiveness of risk management activities. Share these reports with senior management and relevant stakeholders to ensure transparency and accountability.

In developing a comprehensive risk management plan, integrating it into inventory processes, and continuously monitoring and reviewing activities, organizations can effectively manage risks and enhance the resilience and efficiency of their inventory management systems.

9.7 Implications for Companies

Effective inventory risk management has far-reaching implications for companies, influencing various aspects of their operations and strategic positioning. This section explores the implications of inventory risk management for operational efficiency, financial stability, and competitive advantage.

9.7.1 Operational Efficiency

Implementing robust inventory risk management practices significantly enhances a company's operational efficiency. Key benefits include:

- **Streamlined Processes:** Risk management strategies often lead to the optimization of inventory management processes. Identifying and mitigating operational risks, companies can streamline workflows, reduce redundancies, and improve overall process efficiency.
- **Reduced Disruptions:** Effective risk management minimizes the impact of disruptions in the supply chain. With proactive measures in place, companies can quickly respond to issues such as supplier delays or inventory shortages, ensuring smooth operations.
- **Improved Inventory Accuracy:** Implementing risk mitigation strategies, such as enhanced quality control and inventory tracking technologies, improves inventory accuracy. This leads to better inventory visibility and reduces the likelihood of errors and discrepancies.
- **Increased Responsiveness:** Companies with strong risk management practices are more agile and responsive to changes in demand and market conditions. This flexibility enables them to adjust inventory levels and meet customer needs more effectively.

9.7.2 Financial Stability

Inventory risk management plays a crucial role in maintaining a company's financial stability. Key financial benefits include:

- **Cost Reduction:** Mitigating risks associated with inventory holding costs, stockouts, and overstocking, companies can significantly reduce unnecessary expenses. Effective risk management leads to optimal inventory levels, minimizing carrying costs and maximizing cost efficiency.

- **Improved Cash Flow:** Risk management practices help in managing cash flow more effectively by aligning inventory investment with actual demand. This ensures that capital is not unnecessarily tied up in excess inventory, leading to better liquidity and financial health.
- **Minimized Financial Losses:** Proactive risk management reduces the likelihood of financial losses due to inventory-related issues such as obsolescence, damage, or theft. Implementing insurance and hedging strategies, companies can protect themselves against significant financial risks.
- **Enhanced Profitability:** With efficient inventory management and reduced risks, companies can improve their profit margins. Better inventory control and cost management directly contribute to higher profitability.

9.7.3 Competitive Advantage

Effective inventory risk management provides companies with a competitive edge in the marketplace. Key advantages include:

- **Enhanced Customer Satisfaction:** Companies that manage inventory risks effectively can ensure timely delivery of products and maintain high service levels. This leads to increased customer satisfaction and loyalty, providing a competitive advantage.
- **Market Responsiveness:** With robust risk management practices, companies can quickly adapt to changes in market demand and conditions. This responsiveness enables them to seize market opportunities and stay ahead of competitors.
- **Innovation and Agility:** Effective risk management frees up resources and capital, allowing companies to invest in innovation and agile business practices. This fosters a culture of continuous improvement and helps companies stay competitive in a dynamic market environment.
- **Reputation and Trust:** Companies that demonstrate effective risk management build a reputation for reliability and trustworthiness. This enhances their brand image and strengthens relationships with customers, suppliers, and stakeholders.

The implications of inventory risk management for companies are profound. Enhancing operational efficiency, ensuring financial stability, and providing a competitive advantage, effective risk management practices contribute to the overall success and sustainability of an organization.

9.7.4 Future Directions in Inventory Risk Management

As companies continue to navigate an increasingly complex and dynamic business environment, the field of inventory risk management is poised for significant evolution. Future directions in inventory risk management are likely to be influenced by technological advancements, changing market conditions, and emerging trends in supply chain management. Key areas of focus for future developments include:

- **Advanced Analytics and Artificial Intelligence:** The integration of advanced analytics and artificial intelligence (AI) into inventory risk management systems will enable more precise risk prediction and mitigation. Machine learning algorithms can analyze vast amounts of data to identify patterns and trends, providing actionable insights for proactive risk management. Predictive analytics will allow companies to anticipate potential disruptions and implement preventive measures before issues arise.

- **Internet of Things (IoT) and Real-Time Monitoring:** The adoption of IoT technologies will revolutionize inventory management by providing real-time visibility into inventory levels, location, and condition. Sensors and connected devices will enable continuous monitoring of inventory assets, helping companies detect anomalies and respond swiftly to risks such as temperature fluctuations, theft, or damage. Real-time data will enhance decision-making and improve overall inventory control.

- **Blockchain for Supply Chain Transparency:** Blockchain technology offers the potential to enhance transparency and traceability across the supply chain. Creating a decentralized and immutable ledger of inventory transactions, blockchain can help mitigate risks related to counterfeiting, fraud, and supply chain disruptions. Companies can track the provenance of goods, verify authenticity, and ensure compliance with regulatory requirements.

- **Sustainability and Risk Management:** As sustainability becomes a strategic priority for many companies, integrating sustainability considerations into inventory risk management will be crucial. Companies will need to assess and mitigate risks associated with environmental impact, resource scarcity, and regulatory changes. Sustainable inventory practices, such as reducing waste and optimizing resource usage, will not only mitigate risks but also enhance corporate reputation and customer trust.

- **Collaborative Risk Management:** Future inventory risk management will emphasize collaboration and information sharing among supply chain partners. Companies will work more closely with suppliers, distributors, and customers to identify and address shared risks. Collaborative platforms and networks will facilitate data exchange, joint risk assessments, and coordinated response strategies, leading to more resilient supply chains.

- **Risk Management in a Global Context:** With globalization expanding the reach and complexity of supply chains, companies will need to adopt a more global perspective on risk management. This includes understanding geopolitical risks, trade policies, and cultural differences that can impact inventory management. Companies will need to develop flexible and adaptable risk management strategies to navigate the uncertainties of the global marketplace.

The future of inventory risk management will be shaped by technological innovation, collaboration, and a heightened focus on sustainability. Embracing these future directions, companies can enhance their ability to manage risks effectively, ensuring operational efficiency, financial stability, and competitive advantage in a rapidly changing world.

9.8 Conclusion

In conclusion, inventory risk management is a critical component of effective inventory management that significantly impacts a company's operational efficiency, financial stability, and competitive advantage. This chapter has explored the essential aspects of inventory risk management, including the identification and assessment of risks, the development of risk mitigation strategies, and the implementation and monitoring of these strategies. Systematically identifying risks such as demand fluctuations, supply chain disruptions, operational inefficiencies, and financial uncertainties, companies can develop comprehensive risk management plans that address these challenges. Implementing robust risk management practices helps streamline processes, reduce disruptions, improve inventory accuracy, and enhance responsiveness to market changes. From a financial perspective, effective risk management leads to cost reduction, improved cash flow, minimized financial losses, and enhanced profitability. Managing inventory-related risks, companies can optimize their inventory levels, reduce unnecessary expenses, and protect themselves against significant financial impacts. Furthermore, effective inventory risk management provides a competitive edge by enhancing customer satisfaction, increasing market responsiveness, fostering innovation and agility, and building a strong reputation for reliability and trustworthiness. These advantages contribute to the overall success and sustainability of an organization, enabling it to thrive in a dynamic and competitive marketplace. As we look to the future, advancements in technology, such as advanced analytics, IoT, blockchain, and sustainability initiatives, will continue to shape the landscape of inventory risk management. Embracing these developments and integrating them into risk management practices will be crucial for companies seeking to stay ahead of potential risks and capitalize on emerging opportunities. Overall, the integration of effective risk management practices into inventory management is essential for maintaining operational efficiency, ensuring financial stability, and achieving a competitive advantage. Companies that prioritize inventory risk management will be better positioned to navigate uncertainties, adapt to changes, and drive long-term success.

Exercises

Exercise 9.1. *A company uses the Analytic Hierarchy Process (AHP) to prioritize its strategic initiatives. The pairwise comparison matrix for the initiatives is given below:*

$$\mathbf{A} = \begin{pmatrix} 1 & 3 & 5 \\ \frac{1}{3} & 1 & 4 \\ \frac{1}{5} & \frac{1}{4} & 1 \end{pmatrix}$$

Calculate the Consistency Index (CI) and Consistency Ratio (CR) for this matrix. The Random Consistency Index (RI) for a matrix of size 3 is 0.58.

Exercise 9.2. *Consider the following pairwise comparison matrix for a project selection process:*

$$\mathbf{B} = \begin{pmatrix} 1 & 7 & 3 & 5 \\ \frac{1}{7} & 1 & \frac{1}{5} & \frac{1}{3} \\ \frac{1}{3} & 5 & 1 & 2 \\ \frac{1}{5} & 3 & \frac{1}{2} & 1 \end{pmatrix}$$

After performing the initial calculation, the Consistency Ratio (CR) was found to be greater than 0.1, indicating an inconsistent matrix. Adjust the comparison value between the first and third elements to achieve a CR less than 0.1. The Random Consistency Index (RI) for a matrix of size 4 is 0.90.

Exercise 9.3. *Consider a decision-making scenario where a company needs to select a supplier based on multiple criteria. The pairwise comparison matrix for the criteria is given as follows:*

$$\mathbf{C} = \begin{pmatrix} 1 & 2 & 5 & 7 \\ \frac{1}{2} & 1 & 3 & 5 \\ \frac{1}{5} & \frac{1}{3} & 1 & 2 \\ \frac{1}{7} & \frac{1}{5} & \frac{1}{2} & 1 \end{pmatrix}$$

Calculate the Consistency Index (CI) and Consistency Ratio (CR) for this matrix. Determine if the matrix is consistent according to Saaty's threshold of 0.10. The Random Consistency Index (RI) for a matrix of size 4 is 0.90.

Exercise 9.4. *In a research study, a group of experts has provided pairwise comparisons for different project options. The initial comparison matrix has a CR of 0.12, which is above the acceptable threshold of 0.10. Slightly adjusting* *the pairwise comparison values, the goal is to bring the CR below 0.10. The initial comparison matrix is:*

$$\mathbf{D} = \begin{pmatrix} 1 & 6 & 3 \\ \frac{1}{6} & 1 & \frac{1}{4} \\ \frac{1}{3} & 4 & 1 \end{pmatrix}$$

Adjust one of the comparison values and recalculate the CI and CR to achieve a CR less than 0.10. The Random Consistency Index (RI) for a matrix of size 3 is 0.58.

Exercise 9.5. *For a pairwise comparison matrix of size 5, the Random Consistency Index (RI) is 1.12. Given the following matrix:*

$$\mathbf{E} = \begin{pmatrix} 1 & 8 & 5 & 3 & 9 \\ \frac{1}{8} & 1 & 4 & 2 & 6 \\ \frac{1}{5} & \frac{1}{4} & 1 & 3 & 7 \\ \frac{1}{3} & \frac{1}{2} & \frac{1}{3} & 1 & 5 \\ \frac{1}{9} & \frac{1}{6} & \frac{1}{7} & \frac{1}{5} & 1 \end{pmatrix}$$

Calculate the Consistency Index (CI) and Consistency Ratio (CR) for this matrix. Discuss the implications of the CR value in the context of the decision-making process.

Exercise 9.6. *Thomas L. Saaty's empirical studies suggest that a CR below 0.10 is acceptable for decision-making. Given the pairwise comparison matrix:*

$$\mathbf{F} = \begin{pmatrix} 1 & 3 & 4 \\ \frac{1}{3} & 1 & 2 \\ \frac{1}{4} & \frac{1}{2} & 1 \end{pmatrix}$$

Calculate the Consistency Index (CI) and Consistency Ratio (CR). Verify if the matrix meets the empirical threshold of $CR < 0.10$. *The Random Consistency Index (RI) for a matrix of size 3 is 0.58.*

Exercise 9.7. *Explain the theoretical justification for the CR threshold of 0.10 in the Analytic Hierarchy Process (AHP). Provide a detailed analysis of how the threshold balances minor inconsistencies while ensuring reliable decision-making. Illustrate your explanation with a hypothetical example of a* 3×3 *pairwise comparison matrix and calculate its CI and CR to demonstrate the concept.*

Exercise 9.8. *A company uses historical data to assess the likelihood of machine failures in a production plant. The data indicates that the probability of a machine failure in a given month is 0.02. Calculate the probability that:*

1. *The machine will fail at least once in a year.*
2. *The machine will not fail in any given year.*

Exercise 9.9. *Using statistical modeling, a company assesses the potential impact of supply chain disruptions. The company models the time between disruptions using an exponential distribution with a mean of three months. Calculate:*

1. *The probability that the next disruption will occur within two months.*
2. *The probability that the next disruption will occur after five months.*

10

Inventory Sustainability

This chapter explores sustainable inventory management, focusing on reducing environmental and social impacts while maintaining operational efficiency. It highlights the integration of sustainable practices, emphasizing waste reduction, carbon footprint minimization, and ethical sourcing. The chapter examines the practical implementation of sustainable inventory management in supply chains, offering insights into current models and trends. It also outlines future research directions, stressing the importance of embedding sustainability into inventory practices. From an organizational perspective, the chapter presents a theoretical framework for sustainable inventory control. It serves as a valuable resource for practitioners, researchers, and students seeking to understand and apply sustainable inventory management in diverse contexts.

10.1 Introduction

Inventory sustainability refers to the strategic approach of managing inventory in a manner that minimizes environmental impact and promotes social responsibility while maintaining economic viability. This involves integrating sustainable practices throughout the inventory management process, from procurement and production to distribution and disposal. Key components of inventory sustainability include reducing waste, minimizing carbon emissions, ensuring ethical sourcing, and promoting the efficient use of resources.

Sustainable inventory management seeks to balance the traditional goals of inventory control, such as cost efficiency and customer satisfaction, with broader environmental and social objectives. Adopting sustainable practices, companies can reduce their ecological footprint, contribute to social welfare, and build a reputation for corporate responsibility. This holistic approach not only supports long-term business success but also aligns with global sustainability goals and regulatory requirements.

DOI: 10.1201/9781003534860-10

10.1.1 Importance of Sustainable Inventory Management

The importance of sustainable inventory management cannot be overstated, given the growing awareness of environmental issues and the increasing demand for socially responsible business practices. Here are some key reasons why sustainable inventory management is crucial:

- **Environmental Protection:** Sustainable inventory practices help reduce the negative impact of business operations on the environment. Minimizing waste, lowering carbon emissions, and promoting resource efficiency, companies can contribute to the preservation of natural ecosystems and the reduction of pollution.

- **Compliance and Risk Management:** Adopting sustainable practices ensures compliance with environmental regulations and standards, reducing the risk of legal penalties and reputational damage. It also helps companies anticipate and mitigate risks associated with resource scarcity and climate change.

- **Cost Savings:** Sustainable inventory management can lead to significant cost savings through improved efficiency and waste reduction. For instance, optimizing inventory levels to prevent overstocking and obsolescence reduces storage costs and the financial burden of excess inventory.

- **Market Competitiveness:** Consumers are increasingly favoring companies that demonstrate a commitment to sustainability. Adopting sustainable inventory practices, companies can enhance their brand image, attract environmentally conscious customers, and gain a competitive edge in the market.

- **Social Responsibility:** Sustainable inventory management reflects a company's commitment to ethical practices, including fair labor conditions and responsible sourcing. This contributes to social welfare and supports the development of sustainable communities.

- **Long-Term Viability:** Sustainable practices ensure the long-term viability of business operations by promoting the efficient use of resources and fostering innovation. Companies that invest in sustainability are better positioned to adapt to future challenges and opportunities in a rapidly changing global environment.

Sustainable inventory management is essential for addressing the environmental, social, and economic challenges of modern business. It enables companies to operate responsibly, meet regulatory requirements, achieve cost efficiencies, and build a sustainable competitive advantage. As such, it forms a critical component of a comprehensive strategy for corporate sustainability and long-term success.

10.1.2 Waste Reduction Strategies

Waste reduction is a fundamental aspect of sustainable inventory management. Implementing effective waste reduction strategies not only helps protect the environment but also enhances operational efficiency and cost savings. Key strategies include:

- **Inventory Optimization:** Using advanced inventory management techniques, such as just-in-time (JIT) and lean inventory systems, companies can reduce excess inventory and minimize waste. These methods ensure that inventory levels are closely aligned with actual demand, preventing overstocking and obsolescence.
- **Recycling and Reuse:** Implementing recycling programs and encouraging the reuse of materials can significantly reduce waste. Companies can establish systems for recycling packaging materials, reusing returned products, and repurposing by-products from the manufacturing process.
- **Efficient Production Processes:** Adopting efficient production processes, such as process optimization and waste minimization techniques, helps reduce waste at the source. Companies can implement practices like continuous improvement (Kaizen) and Six Sigma to identify and eliminate inefficiencies in their operations.
- **Sustainable Packaging:** Utilizing sustainable packaging solutions, such as biodegradable, recyclable, or reusable materials, can reduce packaging waste. Companies can also optimize packaging design to minimize material usage and improve transport efficiency.
- **Employee Training and Engagement:** Educating employees about the importance of waste reduction and involving them in sustainability initiatives can foster a culture of environmental responsibility. Encouraging employees to identify waste reduction opportunities and participate in sustainability programs can lead to significant improvements.

10.1.3 Minimizing Carbon Footprint

Minimizing the carbon footprint of inventory management is crucial for reducing greenhouse gas emissions and combating climate change. Key strategies include:

- **Energy-Efficient Warehousing:** Implementing energy-efficient practices in warehousing operations, such as using LED lighting, energy-efficient HVAC systems, and renewable energy sources, can reduce carbon emissions. Additionally, optimizing warehouse layout and material handling processes can improve energy efficiency.

- **Green Transportation:** Adopting green transportation practices, such as using fuel-efficient vehicles, optimizing delivery routes, and consolidating shipments, can significantly reduce the carbon footprint of logistics operations. Companies can also explore alternative transportation modes, such as rail or sea, which have lower emissions compared to road transport.
- **Sustainable Sourcing:** Partnering with suppliers who prioritize sustainability and have lower carbon footprints can help reduce the overall environmental impact of the supply chain. Companies can conduct supplier assessments and select partners based on their environmental performance.
- **Carbon Offsetting:** Investing in carbon offset programs, such as reforestation projects or renewable energy initiatives, can help compensate for unavoidable emissions. Companies can calculate their carbon footprint and purchase carbon credits to offset their emissions.
- **Supply Chain Collaboration:** Collaborating with supply chain partners to implement joint sustainability initiatives can amplify the impact of carbon reduction efforts. Shared goals and coordinated actions can lead to more significant environmental benefits.

10.1.4 Promoting Ethical Sourcing and Production

Promoting ethical sourcing and production practices ensures that inventory management supports social responsibility and sustainable development. Key strategies include:

- **Fair Labor Practices:** Ensuring that suppliers and manufacturers adhere to fair labor practices, including safe working conditions, fair wages, and respect for workers' rights, is essential for ethical sourcing. Companies can conduct regular audits and assessments to verify compliance with labor standards.
- **Sustainable Materials:** Sourcing materials that are environmentally sustainable, such as organic, recycled, or responsibly harvested resources, supports ethical production. Companies can establish criteria for sustainable materials and work with suppliers to meet these standards.
- **Transparent Supply Chains:** Increasing transparency in the supply chain allows companies to monitor and address social and environmental issues. Implementing traceability systems and requiring suppliers to provide detailed information about their practices can enhance accountability.
- **Local Sourcing:** Prioritizing local sourcing reduces the environmental impact of transportation and supports local economies. Companies can identify local suppliers who meet their sustainability criteria and establish long-term partnerships.

- **Supplier Development:** Engaging with suppliers to improve their sustainability practices through training, support, and capacity-building initiatives can promote ethical sourcing and production. Companies can work collaboratively with suppliers to implement best practices and achieve shared sustainability goals.

Integrating these sustainable practices into inventory management not only enhances environmental and social responsibility but also contributes to long-term business success. Adopting waste reduction strategies, minimizing carbon footprints, and promoting ethical sourcing and production, companies can build more resilient and sustainable supply chains.

10.1.5 Models of Sustainable Inventory Management

Sustainable inventory management models provide frameworks and strategies that integrate sustainability principles into supply chain operations. These models help companies balance economic, environmental, and social objectives, ensuring long-term sustainability. Key models include:

- **Closed-Loop Supply Chain:** This model focuses on creating a circular economy by reusing, refurbishing, and recycling products and materials. It aims to minimize waste and reduce resource consumption by extending the life cycle of products. Companies can implement reverse logistics systems to collect used products and reintroduce them into the supply chain.

- **Lean and Green Supply Chain:** Combining lean principles with sustainability, this model emphasizes waste reduction, efficiency, and continuous improvement. It involves identifying and eliminating non-value-added activities, optimizing resource use, and reducing environmental impact. Lean and green practices can be applied to various aspects of the supply chain, from production to distribution.

- **Sustainable Supply Chain Management (SSCM):** This comprehensive model integrates sustainability into all supply chain activities, from sourcing to end-of-life disposal. SSCM involves setting sustainability goals, developing sustainable procurement policies, and collaborating with suppliers and stakeholders to achieve these objectives. It also includes monitoring and reporting on sustainability performance.

- **Green Supply Chain Operations Reference (Green SCOR) Model:** An extension of the SCOR model, the Green SCOR model incorporates environmental and sustainability metrics into supply chain performance measurement. It provides a standardized framework for evaluating and improving sustainability practices across the supply chain, including processes like sourcing, production, delivery, and return.

- **Cradle-to-Cradle (C2C):** This model advocates for designing products with their entire life cycle in mind, ensuring that materials can be safely returned to the environment or reused in new products. It emphasizes the use of safe and sustainable materials, renewable energy, and water stewardship. The C2C model encourages companies to rethink product design and production processes to achieve closed-loop systems.

10.1.5.1 How to Implement a Closed-Loop Supply Chain

1. **Implement Reverse Logistics:** Establish a system for collecting used electronic gadgets from customers for refurbishment or recycling. This includes setting up collection points and logistics for retrieving used products.
2. **Refurbish and Resell:** Refurbish collected gadgets and resell them as certified pre-owned products. This process extends the lifecycle of the gadgets and reduces waste by giving them a second life.
3. **Recycle Components:** Disassemble unusable gadgets and recycle the components and materials. This reduces the need for new raw materials and minimizes overall resource consumption.
4. **Benefits:** Adopting a closed-loop supply chain model reduces waste, lowers raw material costs, and enhances the company's reputation for sustainability. It also contributes to environmental conservation by promoting recycling and efficient resource use.

10.1.5.2 How to Implement a Lean and Green Supply Chain

1. **Identify Waste:** Conduct a value stream mapping exercise to identify and eliminate non-value-added activities in the production process. This helps in streamlining operations and reducing unnecessary steps.
2. **Optimize Resource Use:** Implement energy-efficient technologies and practices to reduce energy consumption and greenhouse gas emissions. This includes upgrading machinery, improving insulation, and using renewable energy sources.
3. **Continuous Improvement:** Adopt a continuous improvement culture by encouraging employees to suggest and implement sustainability initiatives. Regularly review and refine processes to enhance efficiency and reduce environmental impact.
4. **Expected Outcomes:** The implementation of a lean and green supply chain model leads to improved operational efficiency, reduced environmental impact, and cost savings from waste reduction and optimized resource use.

Example 10.1. *A company is considering implementing a closed-loop supply chain system to recycle its products. The initial investment is $500,000, and the expected annual savings from reduced raw material costs and waste disposal are $100,000. Additionally, the system is expected to generate an annual revenue of $50,000 from recycled product sales. Calculate the payback period and the net present value (NPV) of the investment over a ten-year period, assuming a discount rate of 5.*

Solution 10.1. *To calculate the payback period and the net present value (NPV) of the investment:*

1. ***Payback Period:***

$$\textit{Annual Savings} = \$100{,}000 + \$50{,}000 = \$150{,}000$$

$$\textit{Payback Period} = \frac{\$500{,}000}{\$150{,}000} \approx 3.33 \textit{ years}$$

2. ***Net Present Value (NPV):*** *The NPV is calculated using the formula:*

$$NPV = \sum_{t=1}^{T} \frac{R_t - C_t}{(1+r)^t} - I_0$$

where R_t is the annual revenue ($150,000), C_t is the annual cost (0 in this case), r is the discount rate (5).

$$NPV = \sum_{t=1}^{10} \frac{150{,}000}{(1+0.05)^t} - 500{,}000$$

Using the annuity formula to simplify:

$$NPV = 150{,}000 \left(\frac{1-(1+0.05)^{-10}}{0.05} \right) - 500{,}000$$

$$NPV = 150{,}000 \left(\frac{1-0.6139}{0.05} \right) - 500{,}000$$

$$NPV = 150{,}000\,(7.7219) - 500{,}000$$

$$NPV = 1{,}158{,}285 - 500{,}000 = 658{,}285$$

Therefore, the NPV of the investment over a ten-year period is approximately $658,285.

Example 10.2. *Implementing a closed-loop supply chain system is projected to reduce the company's carbon footprint by 2,000 metric tons of CO_2 annually. If the social cost of carbon is valued at $50 per metric ton, calculate the total monetary value of the carbon reduction over a ten-year period. Additionally, determine the percentage reduction in carbon footprint if the company's current annual emissions are 10,000 metric tons of CO_2.*

Solution 10.2. *To calculate the total monetary value of the carbon reduction and the percentage reduction in carbon footprint:*

1. ***Total Monetary Value of Carbon Reduction:***

$$\textit{Annual Reduction} = 2{,}000 \textit{ metric tons of } CO_2$$

$$\textit{Social Cost of Carbon} = \$50 \textit{ per metric ton}$$

$$\textit{Total Value over 10 years} = 2{,}000 \times 50 \times 10 = 1{,}000{,}000$$

Therefore, the total monetary value of the carbon reduction over a ten-year period is $1,000,000.

2. ***Percentage Reduction in Carbon Footprint:***

$$\textit{Current Annual Emissions} = 10{,}000 \textit{ metric tons of } CO_2$$

$$\textit{Reduction} = 2{,}000 \textit{ metric tons of } CO_2$$

$$\textit{Percentage Reduction} = \left(\frac{2{,}000}{10{,}000}\right) \times 100\% = 20\%$$

Therefore, the percentage reduction in the company's carbon footprint is 20%.

Example 10.3. *A company is planning to implement a lean and green supply chain strategy. The initial implementation cost is $1,000,000. The strategy is expected to reduce operational costs by $200,000 annually and decrease carbon emissions by 1,500 metric tons annually. If the social cost of carbon is $50 per metric ton and the discount rate is 4%, calculate the net present value (NPV) of the strategy over a ten-year period. Also, determine the total environmental benefit in monetary terms over the same period.*

Solution 10.3. *To calculate the net present value (NPV) and the total environmental benefit:*

1. ***Net Present Value (NPV):*** *The NPV is calculated using the formula:*

$$NPV = \sum_{t=1}^{T} \frac{R_t - C_t}{(1+r)^t} - I_0$$

where R_t is the annual savings ($200,000), C_t is the annual cost (0 in this case), r is the discount rate (4%)

$$NPV = \sum_{t=1}^{10} \frac{200{,}000}{(1+0.04)^t} - 1{,}000{,}000$$

Using the annuity formula to simplify:

$$NPV = 200{,}000\left(\frac{1-(1+0.04)^{-10}}{0.04}\right) - 1{,}000{,}000$$

$$NPV = 200{,}000\left(\frac{1-0.6756}{0.04}\right) - 1{,}000{,}000$$

$$NPV = 200{,}000\,(8.109) - 1{,}000{,}000$$

$$NPV = 1{,}621{,}800 - 1{,}000{,}000 = 621{,}800$$

Therefore, the NPV of the strategy over a ten-year period is approximately $621,800.

2. *Total Environmental Benefit in Monetary Terms:* *The total reduction in carbon emissions over ten years:*

$$\textit{Total Reduction} = 1{,}500\ \textit{metric tons/year} \times 10 = 15{,}000\ \textit{metric tons}$$

The total monetary value of the carbon reduction:

$$\textit{Monetary Value} = 15{,}000\ \textit{metric tons} \times \$50/\textit{metric ton} = 750{,}000$$

Therefore, the total environmental benefit in monetary terms over the ten-year period is $750,000.

Example 10.4. *After implementing a lean and green supply chain, a company observed a reduction in average inventory levels from $500,000 to $300,000. The carrying cost of inventory is 10% per year. Calculate the annual savings in carrying costs due to the reduction in inventory. Additionally, assess the impact of this reduction on the company's cash flow over a five-year period, assuming a discount rate of 5%.*

Solution 10.4. *To calculate the annual savings in carrying costs and the impact on cash flow:*

1. *Annual Savings in Carrying Costs:* *The reduction in inventory levels:*

$$\textit{Inventory Reduction} = 500{,}000 - 300{,}000 = 200{,}000$$

The carrying cost savings:

$$\textit{Annual Savings} = 200{,}000 \times 0.10 = 20{,}000$$

Therefore, the annual savings in carrying costs due to the reduction in inventory is $20,000.

2. *Impact on Cash Flow Over Five Years:* *The present value of the savings over five years:*

$$PV = \sum_{t=1}^{5} \frac{20{,}000}{(1+0.05)^t}$$

Using the annuity formula to simplify:

$$PV = 20{,}000\left(\frac{1-(1+0.05)^{-5}}{0.05}\right)$$

$$PV = 20{,}000\left(\frac{1-0.7835}{0.05}\right)$$

$$PV = 20{,}000\,(4.33) = 86{,}600$$

Therefore, the impact of the inventory reduction on the company's cash flow over a five-year period, assuming a discount rate of 5%, is approximately $86,600.

10.1.6 Current Trends in Sustainable Supply Chains

Sustainable supply chains are evolving rapidly, driven by technological advancements, regulatory pressures, and changing consumer expectations. Current trends in sustainable supply chains include:

- **Digital Transformation:** The adoption of digital technologies, such as IoT, blockchain, and AI, is revolutionizing supply chain management. These technologies enable real-time tracking of inventory, improve transparency, and enhance decision-making. Digital tools also facilitate the implementation of sustainable practices, such as optimizing transportation routes and monitoring environmental impacts.
- **Circular Economy:** The shift towards a circular economy is gaining momentum, with companies increasingly focusing on product life cycle management and resource efficiency. Initiatives like product take-back programs, remanufacturing, and recycling are becoming more common, reducing waste and conserving resources.
- **Sustainable Sourcing:** There is a growing emphasis on ethical and sustainable sourcing practices. Companies are prioritizing suppliers who adhere to environmental and social standards, ensuring that raw materials are sourced responsibly. Certifications and standards, such as Fair Trade and FSC, are used to verify sustainable practices.
- **Carbon Footprint Reduction:** Reducing carbon emissions is a key priority for sustainable supply chains. Companies are investing in renewable

energy, optimizing logistics, and adopting energy-efficient technologies. Initiatives like the Science Based Targets initiative (SBTi) provide frameworks for setting and achieving carbon reduction goals.

- **Transparency and Traceability:** Consumers and regulators are demanding greater transparency in supply chains. Companies are implementing traceability systems to track the origin and journey of products and materials. This helps ensure compliance with sustainability standards and builds consumer trust.
- **Collaboration and Partnerships:** Collaboration among supply chain partners is essential for achieving sustainability goals. Companies are forming strategic partnerships with suppliers, NGOs, and industry groups to share best practices, develop sustainable innovations, and address common challenges.
- **Consumer Demand for Sustainability:** Increasing consumer awareness and demand for sustainable products are driving companies to adopt more sustainable practices. Brands are responding by offering eco-friendly products, reducing packaging waste, and communicating their sustainability efforts to consumers.

Incorporating these models and trends into supply chain operations helps companies achieve sustainability objectives, reduce environmental impact, and meet stakeholder expectations. Staying abreast of emerging trends and adopting innovative models, companies can build resilient and sustainable supply chains that contribute to long-term success.

10.1.7 Emerging Topics in Sustainable Inventory Management

The field of sustainable inventory management is continuously evolving, with new areas of research emerging as industries seek to enhance their sustainability practices. Some of the emerging topics include:

- **Circular Economy Implementation:** Further research is needed to develop and refine models for implementing circular economy principles in inventory management. This includes strategies for designing products for reuse, recycling, and remanufacturing, as well as systems for tracking and managing the life cycle of materials.
- **Sustainable Procurement:** Investigating methods to integrate sustainability criteria into procurement processes is a growing area of interest. Research can explore ways to assess supplier sustainability performance, develop sustainable procurement policies, and create incentives for suppliers to adopt sustainable practices.

- **Social Sustainability:** While much focus has been on environmental aspects, there is increasing recognition of the importance of social sustainability in inventory management. This includes fair labor practices, community impact, and human rights considerations. Research can explore how to measure and enhance social sustainability within supply chains.
- **Consumer Behavior and Sustainability:** Understanding how consumer preferences and behaviors influence sustainable inventory management is crucial. Research can examine the factors that drive consumer demand for sustainable products and how companies can effectively respond to these trends.
- **Policy and Regulation Impact:** Analyzing the impact of governmental policies and regulations on sustainable inventory management practices is essential. Research can explore how regulatory frameworks can support or hinder sustainability efforts and what policy changes could further promote sustainable practices.

10.1.8 Technological Innovations and Sustainability

Technological advancements play a critical role in advancing sustainable inventory management. Future research directions in this area include:

- **Blockchain Technology:** Exploring the use of blockchain for enhancing transparency and traceability in supply chains. Research can focus on how blockchain can ensure the authenticity of sustainability claims, track the life cycle of products, and reduce fraud.
- **Artificial Intelligence (AI) and Machine Learning:** Investigating how AI and machine learning can optimize inventory management for sustainability. This includes predictive analytics for demand forecasting, optimizing supply chain routes to reduce emissions, and identifying opportunities for waste reduction.
- **Internet of Things (IoT):** Examining the application of IoT devices for real-time monitoring and management of inventory. IoT can provide data on inventory levels, environmental conditions, and product movement, helping companies make more informed decisions to enhance sustainability.
- **Advanced Materials and Green Technologies:** Researching the development and use of advanced materials and green technologies in inventory management. This includes biodegradable materials, sustainable packaging solutions, and energy-efficient storage and transportation technologies.
- **Big Data Analytics:** Utilizing big data analytics to gather and analyze vast amounts of data from supply chains. Research can explore how big data can provide insights into sustainability performance, identify trends, and drive continuous improvement in sustainable practices.

10.1.9 Theoretical Framework for Sustainable Inventory Control

Developing a theoretical framework for sustainable inventory control involves integrating sustainability principles into traditional inventory management theories. This framework serves as a foundation for organizations to systematically incorporate sustainability into their inventory practices. Key components of the theoretical framework include:

- **Triple Bottom Line (TBL) Approach:** This approach emphasizes balancing economic, environmental, and social performance. Sustainable inventory control should consider not only financial outcomes but also the environmental impact and social implications of inventory management decisions.

- **Life Cycle Assessment (LCA):** LCA involves evaluating the environmental impacts of products throughout their entire life cycle, from raw material extraction to disposal. Incorporating LCA into inventory control helps organizations understand and minimize the environmental footprint of their products.

- **Resource Efficiency and Waste Minimization:** The framework should focus on optimizing resource use and reducing waste. This includes strategies such as just-in-time (JIT) inventory, which reduces excess inventory and waste, and lean manufacturing principles that streamline processes and eliminate inefficiencies.

- **Sustainable Procurement and Supplier Collaboration:** Sustainable inventory control requires collaboration with suppliers to ensure the sustainability of sourced materials. This involves setting sustainability criteria for suppliers, engaging in long-term partnerships, and promoting ethical sourcing practices.

- **Risk Management:** Integrating sustainability into inventory control involves identifying and mitigating sustainability-related risks. This includes risks associated with supply chain disruptions, regulatory compliance, and reputational damage due to unsustainable practices.

- **Performance Measurement and Reporting:** The framework should include metrics and KPIs to measure sustainability performance. This involves tracking indicators such as carbon footprint, waste generation, and resource consumption, and reporting these metrics to stakeholders.

10.1.10 Case Studies of Sustainable Inventory Practices

Examining real-world examples of sustainable inventory practices provides valuable insights into how organizations can implement sustainable inventory control. Case studies from various industries highlight successful strategies and lessons learned:

- **Case Study 1: Patagonia's Sustainable Supply Chain**
  Patagonia, an outdoor clothing and gear company, is renowned for its commitment to sustainability. The company has implemented a closed-loop supply chain, encouraging customers to return used products for recycling or repurposing. Patagonia also prioritizes the use of recycled materials and collaborates with suppliers to ensure ethical and sustainable practices [18–20].

- **Case Study 2: Unilever's Zero Waste to Landfill Initiative**
  Unilever, a global consumer goods company, has achieved significant progress in waste reduction through its "Zero Waste to Landfill" initiative. Optimizing inventory levels, improving packaging design, and enhancing recycling efforts, Unilever has minimized waste generation across its supply chain. The company also engages with suppliers to promote sustainable practices [21–23].

- **Case Study 3: IKEA's Sustainable Inventory Management**
  IKEA, a multinational furniture retailer, focuses on sustainable inventory management through its circular economy approach. The company designs products for durability, repairability, and recyclability. IKEA also offers a furniture take-back program, encouraging customers to return used items for resale or recycling. This initiative reduces waste and promotes resource efficiency [24–27].

- **Case Study 4: Walmart's Project Gigaton**
  Walmart, a global retail corporation, launched Project Gigaton to reduce greenhouse gas emissions in its supply chain. The initiative involves collaborating with suppliers to set and achieve emissions reduction targets. Walmart also invests in sustainable inventory management practices, such as optimizing transportation routes and increasing the use of renewable energy [28].

- **Case Study 5: Interface's Sustainable Inventory Practices**
  Interface, a global leader in modular flooring and carpet tile manufacturing, has been a pioneer in sustainability through its Mission Zero program. Launched with the ambitious goal of eliminating the company's negative environmental impact by 2020, the program focused on achieving a zero footprint worldwide, selling only products with a restorative impact on the planet, and embedding sustainability principles across its operations and supply chain, [**?**]. While Interface achieved many of its Mission Zero goals by 2020, the company has since shifted its focus to the Climate Take Back initiative, which aims to create a carbon-negative enterprise by 2040. This new phase builds on the lessons learned from Mission Zero and emphasizes restorative business practices, carbon neutrality, and circular economy principles.

These case studies illustrate how leading organizations successfully integrate sustainability into their inventory control practices. Adopting similar strategies and continuously improving their sustainability efforts, companies can achieve significant environmental, social, and economic benefits.

10.1.11 Implications for Practice and Research

The integration of sustainable inventory control practices has profound implications for both practitioners and researchers in the field of supply chain management.

Implications for Practice:

- **Enhanced Operational Efficiency:** Practitioners can leverage sustainable inventory control to streamline operations, reduce waste, and improve resource utilization. Implementing sustainable practices can lead to cost savings and increased operational efficiency.
- **Compliance and Risk Mitigation:** Companies can ensure compliance with environmental regulations and mitigate risks associated with unsustainable practices. This proactive approach helps avoid potential legal and reputational risks.
- **Competitive Advantage:** Adopting sustainable inventory practices, companies can differentiate themselves in the market. Sustainability initiatives can enhance brand reputation, attract eco-conscious consumers, and build stronger relationships with stakeholders.
- **Innovation and Collaboration:** Sustainable inventory control fosters innovation by encouraging the development of new materials, technologies, and processes. It also promotes collaboration with suppliers, customers, and other stakeholders to achieve common sustainability goals.
- **Long-term Viability:** Focusing on sustainability ensures the long-term viability of supply chains. Companies that invest in sustainable practices are better positioned to adapt to future challenges and opportunities, ensuring resilience and continuity.

Implications for Research:

- **New Research Avenues:** The evolving field of sustainable inventory management presents numerous research opportunities. Researchers can explore emerging topics such as the circular economy, social sustainability, and the impact of digital technologies on sustainability.
- **Interdisciplinary Approaches:** Sustainable inventory management requires an interdisciplinary approach, integrating insights from supply chain

management, environmental science, economics, and social sciences. Researchers can collaborate across disciplines to develop comprehensive sustainability frameworks.

- **Quantitative and Qualitative Studies:** Future research can employ both quantitative and qualitative methods to investigate the effectiveness of sustainable inventory practices. Quantitative studies can provide empirical data on the impact of these practices, while qualitative research can offer deeper insights into implementation challenges and best practices.
- **Policy and Regulation Impact:** Researchers can examine the influence of policies and regulations on sustainable inventory management. This includes analyzing the effectiveness of existing policies and proposing new regulatory frameworks to support sustainability initiatives.
- **Case Studies and Best Practices:** Conducting case studies and identifying best practices in sustainable inventory management can provide valuable guidance for practitioners. Research can document successful implementations, challenges faced, and lessons learned from various industries.

10.2 Conclusion

In conclusion, sustainable inventory management is essential for modern supply chains, providing significant benefits in terms of operational efficiency, compliance, risk mitigation, and competitive advantage. Integrating sustainability principles into inventory practices, companies can enhance their long-term viability and contribute positively to environmental and social goals. The chapter has explored key concepts and methodologies in sustainable inventory management, including theoretical frameworks, practical strategies, and case studies. The discussion on emerging topics and technological innovations highlights the dynamic nature of the field and the ongoing opportunities for improvement. Future research directions emphasize the need for interdisciplinary approaches, robust quantitative and qualitative studies, and the development of comprehensive sustainability frameworks. Continuing to explore and implement sustainable practices, both practitioners and researchers can contribute to the advancement of sustainable inventory management, ultimately leading to more resilient and responsible supply chains.

Exercises

Exercise 10.1. *A company has adopted digital technologies, such as IoT and AI, to enhance its supply chain management. The implementation costs are \$2,000,000, and the annual savings due to improved efficiency and reduced waste are estimated at \$400,000. If the discount rate is 6%, calculate the net present value (NPV) of the digital transformation over a ten-year period. Additionally, evaluate the break-even point in years.*

Exercise 10.2. *A manufacturing company has shifted towards a circular economy model, implementing a product take-back program and recycling initiatives. If the program costs \$500,000 annually but reduces waste disposal costs by \$150,000 and generates additional revenue of \$100,000 from recycled materials, calculate the annual net benefit. Using a discount rate of 5%, determine the present value of the net benefits over a five-year period.*

Exercise 10.3. *A company is considering switching to sustainable sourcing practices, which involves an upfront investment of \$1,500,000 for certification and compliance. The expected increase in annual operating costs is \$200,000, but the company anticipates gaining \$300,000 in additional annual revenue from environmentally conscious consumers. With a discount rate of 7%, determine the net present value (NPV) of this switch over a seven-year period. Also, identify the payback period.*

Exercise 10.4. *A logistics company aims to reduce its carbon footprint by investing \$3,000,000 in renewable energy and energy-efficient technologies. This investment is expected to save \$250,000 annually in energy costs and reduce carbon emissions by 2,000 metric tons annually. If the social cost of carbon is \$60 per metric ton and the discount rate is 4%, calculate the total economic benefit (including energy savings and carbon reduction) over a ten-year period. Also, determine the net present value (NPV) of the investment.*

Exercise 10.5. *A company plans to implement a traceability system to improve supply chain transparency. The system implementation cost is \$1,200,000, and it is expected to reduce regulatory compliance costs by \$100,000 annually and increase consumer trust, leading to an additional \$150,000 in annual revenue. Using a discount rate of 5%, calculate the net present value (NPV) of the traceability system over an eight-year period. Additionally, estimate the break-even point.*

11

Inventory Ethics and Social Responsibility

This chapter examines the ethical dimensions and social responsibilities of inventory management. It highlights the importance of adopting responsible practices to reduce environmental impact and contribute positively to society. Ethical considerations in sourcing, production, and distribution are explored, with a focus on transparency and fair treatment across the supply chain. The chapter also addresses broader societal implications, such as resource allocation, waste reduction, and social equity. Integrating theoretical frameworks, it provides valuable insights for fostering ethical decision-making and socially responsible inventory strategies. The role of technology in enhancing ethical practices is also discussed. This chapter serves as a valuable resource for practitioners, researchers, and students seeking to understand the ethical responsibilities embedded in inventory management and their impact on both business and society.

11.1 Introduction

Inventory ethics refers to the principles and standards governing the conduct of inventory management practices. These principles ensure that inventory-related activities are performed in a manner that is morally sound and aligns with broader ethical values. Key aspects of inventory ethics include:

- **Fair Sourcing:** Ensuring that the procurement of raw materials and products is conducted fairly, without exploitation or harm to individuals or communities.
- **Transparency:** Maintaining openness and honesty in inventory operations, including clear communication about sourcing, production processes, and product origins.
- **Environmental Responsibility:** Minimizing the environmental impact of inventory management practices through sustainable sourcing, efficient use of resources, and waste reduction.

DOI: 10.1201/9781003534860-11

- **Social Equity:** Promoting fairness and justice within the supply chain, ensuring that all stakeholders, from suppliers to consumers, are treated equitably.

- **Compliance with Laws and Regulations:** Adhering to all relevant laws, regulations, and standards that govern inventory management practices, including labor laws, environmental regulations, and trade policies.

Understanding and defining inventory ethics is crucial for establishing a foundation upon which responsible and sustainable inventory practices can be built.

11.1.1 Importance of Social Responsibility in Inventory Management

Social responsibility in inventory management involves the commitment of companies to conduct their operations in a way that positively impacts society and the environment. The importance of integrating social responsibility into inventory management is multifaceted:

- **Building Trust and Reputation:** Companies that prioritize social responsibility build trust with customers, suppliers, and other stakeholders. A reputation for ethical practices can enhance brand loyalty and attract socially conscious consumers.

- **Risk Mitigation:** Adopting socially responsible practices helps mitigate risks associated with unethical behavior, such as legal penalties, supply chain disruptions, and reputational damage.

- **Sustainable Growth:** Focusing on long-term sustainability rather than short-term profits, companies can ensure their growth is sustainable. This involves considering the environmental and social impact of their inventory management practices.

- **Employee Engagement:** Companies that demonstrate a commitment to social responsibility often experience higher levels of employee engagement and satisfaction. Employees are more likely to be motivated and loyal when they believe their company is making a positive impact.

- **Regulatory Compliance:** Socially responsible inventory management ensures compliance with increasingly stringent regulations related to environmental protection, labor rights, and ethical sourcing.

- **Community Impact:** Companies can contribute to the well-being of the communities in which they operate by ensuring that their inventory practices do not harm local environments or economies. This includes supporting local suppliers and engaging in community development initiatives.

Incorporating social responsibility into inventory management is not only ethically imperative but also beneficial for companies' long-term success and sustainability. It reflects a company's commitment to being a positive force in the world, balancing profit with purpose.

11.1.2 Ethical Sourcing Practices

Ethical sourcing practices ensure that materials and products are obtained in a responsible and sustainable manner. Key aspects of ethical sourcing include:

- **Supplier Audits:** Conducting regular audits of suppliers to ensure they adhere to ethical standards, including fair labor practices, environmental responsibility, and human rights protections.
- **Fair Trade Certification:** Sourcing products that are certified as fair trade, ensuring that producers in developing countries receive fair wages and work under safe conditions.
- **Sustainable Materials:** Prioritizing the use of sustainable and renewable materials to minimize environmental impact and promote resource conservation.
- **Local Sourcing:** Supporting local suppliers to reduce carbon footprints associated with transportation and to contribute to local economies.
- **Conflict-Free Sourcing:** Ensuring that sourced materials, particularly minerals, are not associated with conflict or exploitation in their regions of origin.

Implementing ethical sourcing practices, companies can ensure that their inventory management aligns with broader social and environmental goals, fostering a more sustainable and responsible supply chain.

11.1.3 Fair Treatment in Production

Ensuring fair treatment in production is crucial for maintaining ethical standards throughout the supply chain. This involves:

- **Labor Rights:** Upholding the rights of workers, including fair wages, reasonable working hours, and safe working conditions. Companies should ensure that their production partners comply with labor laws and international labor standards.
- **Non-Discrimination:** Promoting an inclusive and diverse workforce by prohibiting discrimination based on race, gender, age, religion, or other protected characteristics.

- **Health and Safety:** Implementing stringent health and safety standards to protect workers from hazards and ensure a safe working environment. Regular inspections and safety training are essential components of this practice.

- **Worker Empowerment:** Encouraging worker representation and participation in decision-making processes. This can include supporting the right to unionize and ensuring that workers have a voice in matters affecting their employment.

- **Ethical Treatment of Animals:** For companies involved in industries where animal welfare is a concern, ensuring humane treatment of animals and adherence to ethical standards in their use and care.

Fair treatment in production not only ensures compliance with legal and ethical standards but also contributes to a more motivated and productive workforce, ultimately enhancing the quality and sustainability of inventory management.

11.1.4 Transparency in Distribution

Transparency in distribution is essential for building trust and ensuring accountability throughout the supply chain. Key elements include:

- **Traceability:** Implementing systems that allow for the tracking of products from their origin to the final consumer. This includes maintaining detailed records of sourcing, production, and distribution processes.

- **Open Communication:** Providing clear and honest information to stakeholders about inventory practices, including sourcing policies, production methods, and distribution strategies. Transparency in communication helps build trust with customers, suppliers, and regulatory bodies.

- **Ethical Labeling:** Ensuring that product labels accurately reflect the ethical and sustainable attributes of the products. This can include certifications, such as organic, fair trade, or sustainably sourced labels.

- **Stakeholder Engagement:** Actively engaging with stakeholders, including customers, suppliers, and community members, to gather feedback and address concerns related to inventory management practices.

- **Accountability:** Establishing mechanisms for accountability, such as third-party audits and public reporting, to verify that ethical standards are being met throughout the supply chain.

Prioritizing transparency in distribution, companies can demonstrate their commitment to ethical practices, enhance their reputation, and foster trust with all stakeholders involved in the supply chain.

11.1.5 Resource Allocation Ethics

Resource allocation ethics deals with the fair and responsible distribution of resources within and across supply chains. This involves:

- **Equitable Distribution:** Ensuring that resources are distributed in a way that does not disproportionately benefit or disadvantage any particular group. This includes considering the needs of underserved or marginalized communities.
- **Prioritization of Needs:** Allocating resources based on the most pressing needs and potential impacts, rather than solely on profit or convenience. This can involve prioritizing essential goods for communities in need.
- **Efficient Utilization:** Using resources efficiently to minimize waste and reduce the environmental footprint. This includes optimizing inventory levels and reducing excess that could lead to unnecessary resource consumption.
- **Ethical Investment:** Investing in practices and technologies that support sustainable and responsible resource use, such as renewable energy sources and resource-efficient processes.
- **Transparency in Allocation:** Providing clear information about how resources are allocated and the criteria used for these decisions. Transparency helps build trust and ensures accountability.

Ethical resource allocation supports fairness and sustainability, contributing to more balanced and responsible inventory management practices.

11.1.6 Waste Reduction and Environmental Impact

Waste reduction and managing environmental impact are critical components of responsible inventory management. Key considerations include:

- **Minimizing Waste:** Implementing strategies to reduce waste throughout the inventory life cycle, from sourcing to end-of-life. This includes optimizing inventory levels to prevent overstocking and spoilage.
- **Sustainable Packaging:** Using eco-friendly packaging materials and reducing packaging waste. This can involve adopting recyclable or biodegradable materials and minimizing excessive packaging.
- **Energy and Resource Efficiency:** Enhancing the efficiency of operations to reduce energy and resource consumption. This includes adopting energy-saving technologies and practices.
- **Recycling and Reuse:** Encouraging recycling and the reuse of materials to reduce the need for new resources and lower environmental impact. This can include recycling inventory packaging and repurposing unsold goods.

- **Environmental Compliance:** Adhering to environmental regulations and standards to minimize adverse effects on the environment. This includes proper waste disposal and adherence to regulations related to emissions and pollutants.

Effective waste reduction and management of environmental impact are essential for reducing the ecological footprint of inventory management practices and supporting sustainability.

11.1.7 Social Equity and Fairness

Social equity and fairness in inventory management involve ensuring that all stakeholders are treated justly and equitably. This encompasses:

- **Fair Labor Practices:** Ensuring that all workers, regardless of their role or location, receive fair wages, safe working conditions, and respect for their rights. This is crucial for promoting social justice within the supply chain.
- **Community Impact:** Assessing and addressing the impact of inventory practices on local communities. This includes supporting local economies and minimizing negative social effects, such as displacement or exploitation.
- **Inclusive Decision-Making:** Involving diverse stakeholders in decision-making processes to ensure that various perspectives are considered and that policies reflect a broad range of interests and needs.
- **Accessibility:** Ensuring that products and services are accessible to all segments of society, including underserved and disadvantaged groups. This can involve pricing strategies, product design, and distribution practices.
- **Ethical Marketing:** Promoting products and services in a way that is honest and does not exploit vulnerable populations. Ethical marketing practices contribute to fairness and transparency.

Addressing social equity and fairness in inventory management helps to create a more just and inclusive supply chain, fostering positive social impact and contributing to broader societal well-being.

11.2 Fostering Ethical Decision-Making

Ethical decision-making is a critical aspect of inventory management that ensures practices align with the core values and principles of the organization. This involves developing ethical frameworks, providing training and education, and creating a culture of responsibility.

11.2.1 Developing Ethical Frameworks

Developing ethical frameworks involves creating structured guidelines and principles that guide decision-making within inventory management. Key aspects include:

- **Establishing Core Values:** Defining the fundamental ethical values and principles that the organization upholds. These core values should be integrated into all aspects of inventory management and decision-making processes.
- **Creating Ethical Policies:** Formulating specific policies that address key ethical issues related to inventory management, such as sourcing, production, and distribution practices. These policies should provide clear guidance on acceptable and unacceptable practices.
- **Implementing Decision-Making Frameworks:** Developing frameworks that help individuals evaluate ethical dilemmas and make informed decisions. These frameworks should include criteria for assessing the ethical implications of various options.
- **Reviewing and Updating Frameworks:** Regularly reviewing and updating ethical frameworks to ensure they remain relevant and effective in addressing emerging ethical challenges and industry developments.

Establishing robust ethical frameworks, organizations can provide clear guidance for making ethical decisions and ensure that inventory management practices align with their values.

11.2.2 Training and Education for Ethical Practices

Training and education are crucial for fostering ethical practices within inventory management. Key components include:

- **Ethics Training Programs:** Implementing training programs that educate employees about ethical standards, principles, and practices related to inventory management. These programs should include real-world scenarios and case studies to illustrate ethical decision-making.
- **Ongoing Education:** Providing continuous education opportunities to keep employees updated on new ethical issues, regulations, and best practices. This can include workshops, seminars, and online courses.
- **Creating Resources:** Developing and disseminating resources such as handbooks, guidelines, and toolkits that support ethical practices and provide practical advice for handling ethical dilemmas.

- **Encouraging Open Dialogue:** Promoting an environment where employees feel comfortable discussing ethical concerns and seeking guidance. This can be facilitated through regular meetings, forums, and feedback channels.

Training and education empower employees to make ethical decisions and contribute to a culture of responsibility within the organization.

11.2.3 Creating a Culture of Responsibility

Creating a culture of responsibility involves embedding ethical values and practices into the organizational culture. Key strategies include:

- **Leadership Commitment:** Demonstrating a strong commitment to ethical practices from top management. Leaders should model ethical behavior, communicate the importance of ethics, and actively support ethical initiatives.
- **Recognizing Ethical Behavior:** Acknowledging and rewarding employees who demonstrate ethical behavior and contribute to responsible inventory management practices. This can include formal recognition programs and incentives.
- **Integrating Ethics into Performance Metrics:** Incorporating ethical considerations into performance evaluations and decision-making processes. This ensures that ethical behavior is valued and aligned with organizational goals.
- **Encouraging Accountability:** Establishing mechanisms for holding individuals accountable for ethical lapses. This includes clear procedures for reporting and addressing unethical behavior and ensuring that consequences are consistently applied.
- **Promoting Transparency:** Fostering transparency in decision-making processes and operations. This includes openly sharing information about ethical practices, challenges, and outcomes with stakeholders.

Cultivating a culture of responsibility, organizations can embed ethical practices into their daily operations and ensure that inventory management is conducted with integrity and accountability.

11.3 Socially Responsible Inventory Management Strategies

Socially responsible inventory management strategies are essential for aligning business practices with environmental and social responsibility goals. This involves implementing sustainable sourcing and procurement methods, ensuring ethical labor practices, and supporting community development.

11.3.1 Sustainable Sourcing and Procurement

Sustainable sourcing and procurement involve selecting suppliers and materials that align with environmental and social responsibility goals. Key strategies include:

- **Green Procurement:** Prioritizing the purchase of products and materials that have minimal environmental impact. This includes selecting items with eco-friendly certifications and those made from recycled or sustainable materials.
- **Supplier Sustainability Assessments:** Evaluating suppliers based on their sustainability practices and performance. This can involve assessing their environmental impact, labor practices, and adherence to ethical standards.
- **Long-Term Supplier Relationships:** Building and maintaining long-term relationships with suppliers who demonstrate a commitment to sustainable practices. This fosters collaboration and encourages continuous improvement in sustainability.
- **Transparent Supply Chains:** Ensuring that the sourcing and procurement processes are transparent and traceable. This includes providing visibility into the supply chain and verifying the ethical standards of suppliers.
- **Support for Sustainable Innovations:** Investing in and supporting innovations that promote sustainability within the supply chain. This can include adopting new technologies or practices that reduce environmental impact.

Implementing these strategies, companies can ensure that their sourcing and procurement practices contribute to environmental sustainability and ethical responsibility.

11.3.2 Ethical Inventory Policies

Ethical inventory policies are designed to ensure that inventory management practices align with ethical and social responsibility standards. Key elements include:

- **Fair Trade Practices:** Implementing policies that support fair trade principles, including equitable treatment of workers and fair compensation throughout the supply chain.

- **Ethical Stock Management:** Managing inventory in a way that avoids practices such as deliberate obsolescence or planned shortages. Ensuring that inventory decisions are made based on ethical considerations and long-term sustainability.
- **Compliance with Regulations:** Adhering to all relevant regulations and standards related to inventory management, including those concerning product safety, labor rights, and environmental protection.
- **Ethical Disposal Methods:** Developing policies for the responsible disposal of surplus or obsolete inventory. This includes recycling, repurposing, or donating excess stock rather than sending it to landfills.
- **Inclusive Policies:** Ensuring that inventory policies support diversity and inclusion, providing equal opportunities for all stakeholders involved in the inventory process.

Ethical inventory policies help ensure that inventory management practices are conducted in a manner that is fair, responsible, and aligned with organizational values.

11.3.3 Community Engagement and Impact

Community engagement and impact involve actively working with and contributing to the well-being of local communities through inventory management practices. Key approaches include:

- **Local Sourcing Initiatives:** Supporting local businesses and suppliers by prioritizing local sourcing. This helps stimulate local economies and reduces the carbon footprint associated with transportation.
- **Community Investment:** Investing in community projects and programs that support social and environmental causes. This can include funding educational initiatives, supporting local charities, and participating in community development projects.
- **Stakeholder Involvement:** Engaging with community stakeholders to understand their needs and concerns. This can involve conducting consultations, surveys, and dialogues to gather feedback and incorporate it into inventory management practices.
- **Transparent Reporting:** Providing regular reports on the social and environmental impacts of inventory management practices. Transparency helps build trust with the community and demonstrates a commitment to social responsibility.

- **Volunteering and Philanthropy:** Encouraging employees to participate in volunteer activities and philanthropic efforts that benefit local communities. This fosters a culture of giving and social responsibility within the organization.

Adopting these strategies, companies can enhance their positive impact on communities and demonstrate a commitment to social responsibility in their inventory management practices.

11.4 Role of Technology in Enhancing Ethical Practices

Technology plays a pivotal role in enhancing ethical practices within inventory management and supply chains. Leveraging technological innovations, organizations can improve transparency, accountability, and efficiency in their operations.

11.4.1 Technological Innovations for Transparency

Technological innovations play a crucial role in enhancing transparency within inventory management and supply chains. Key advancements include:

- **Blockchain Technology:** Utilizing blockchain to create immutable records of transactions and movements within the supply chain. This technology enhances traceability and accountability, allowing stakeholders to verify the ethical origins and handling of products.

- **Smart Contracts:** Implementing smart contracts to automatically enforce and verify compliance with ethical standards and agreements. These contracts are self-executing and reduce the risk of fraud or misrepresentation.

- **IoT (Internet of Things):** Deploying IoT devices to monitor and track inventory in real time. This technology provides detailed visibility into the condition and location of products, ensuring adherence to ethical practices such as proper storage conditions.

- **Data Analytics:** Leveraging advanced data analytics to gain insights into supply chain operations and detect anomalies or unethical practices. Analytics tools can identify patterns and trends that indicate potential issues or areas for improvement.

- **Enhanced Reporting Systems:** Using technology to generate comprehensive and transparent reports on supply chain practices, including environmental and social impact assessments. These systems improve stakeholder access to relevant information and support informed decision-making.

These technological innovations help organizations enhance transparency, enabling more ethical and responsible inventory management practices.

11.4.2 Digital Tools for Ethical Supply Chain Management

Digital tools are essential for managing and improving ethical practices within the supply chain. Key tools include:

- **Ethical Sourcing Platforms:** Using platforms that connect businesses with suppliers who adhere to ethical and sustainable practices. These platforms often include certifications and ratings to guide procurement decisions.
- **Supply Chain Management Software:** Implementing software that integrates various aspects of supply chain management, including sourcing, logistics, and compliance. These tools help ensure that ethical standards are met across the entire supply chain.
- **Supplier Risk Management Tools:** Employing tools to assess and monitor supplier risks related to ethics and sustainability. These tools can evaluate supplier performance, track compliance with ethical standards, and identify potential issues.
- **Collaboration Platforms:** Utilizing platforms that facilitate communication and collaboration between stakeholders in the supply chain. These platforms support joint efforts to address ethical challenges and share best practices.
- **Training and Certification Systems:** Leveraging digital tools for employee training and certification programs focused on ethical practices. These systems ensure that staff are well-informed and equipped to uphold ethical standards.

Digital tools support the implementation and maintenance of ethical practices in supply chain management, promoting greater accountability and compliance.

11.4.3 Monitoring and Reporting Ethical Compliance

Monitoring and reporting systems are critical for ensuring ongoing ethical compliance in inventory management. Key approaches include:

- **Automated Monitoring Systems:** Using automated systems to continuously monitor supply chain activities and compliance with ethical standards. These systems can detect deviations and alert management to potential issues.

- **Compliance Dashboards:** Implementing dashboards that provide real-time visibility into ethical compliance metrics and performance indicators. Dashboards help track progress, identify trends, and address compliance issues promptly.
- **Regular Audits and Assessments:** Conducting regular audits and assessments of supply chain practices to ensure adherence to ethical standards. These audits can be facilitated by technology and provide a thorough evaluation of compliance.
- **Feedback Mechanisms:** Establishing mechanisms for stakeholders to provide feedback and report concerns related to ethical practices. Digital platforms can support anonymous reporting and facilitate prompt resolution of issues.
- **Transparent Reporting:** Generating and publishing reports on ethical compliance, including details on practices, incidents, and improvements. Transparent reporting enhances accountability and builds trust with stakeholders.

Effective monitoring and reporting systems ensure that ethical practices are upheld and provide a basis for continuous improvement in inventory management.

11.4.4 Theoretical Approaches to Inventory Ethics

Theoretical frameworks provide a foundation for understanding and analyzing ethical issues in inventory management. Key approaches include:

- **Ethical Theories:** Examining various ethical theories such as utilitarianism, deontology, and virtue ethics to guide decision-making in inventory management. Each theory offers a different perspective on what constitutes ethical behavior and how to address ethical dilemmas.
- **Corporate Social Responsibility (CSR) Framework:** Applying CSR principles to inventory management to ensure that business practices contribute positively to society and the environment. This framework emphasizes the integration of social and environmental considerations into business strategies.
- **Stakeholder Theory:** Analyzing inventory management practices through the lens of stakeholder theory, which focuses on balancing the interests of all stakeholders, including employees, customers, suppliers, and the community.
- **Sustainability Frameworks:** Utilizing sustainability frameworks, such as the Triple Bottom Line (TBL) or the UN Sustainable Development Goals (SDGs), to assess and improve the ethical dimensions of inventory management. These frameworks help align practices with broader sustainability objectives.

- **Ethical Supply Chain Management Models:** Exploring models that integrate ethical considerations into supply chain management, such as the Ethical Trading Initiative (ETI) or the Global Reporting Initiative (GRI). These models provide guidelines and standards for ethical practices.

These theoretical approaches offer valuable insights and guidelines for developing ethical inventory management practices.

11.4.5 Future Directions in Inventory Ethics and Social Responsibility

The field of inventory ethics and social responsibility is evolving, and several future directions are emerging. Key areas for exploration include:

- **Enhanced Regulatory Frameworks:** Development of more comprehensive and standardized regulations for ethical practices in inventory management. These frameworks could address gaps and inconsistencies in current standards and promote greater accountability.

- **Integration of Advanced Technologies:** Leveraging technologies such as artificial intelligence, blockchain, and big data analytics to further enhance transparency, traceability, and ethical decision-making in inventory management. These technologies can provide deeper insights and more robust solutions for managing ethical challenges.

- **Focus on Circular Economy:** Expanding the adoption of circular economy principles to minimize waste, enhance resource efficiency, and promote sustainability throughout the inventory life cycle. This approach emphasizes reducing, reusing, and recycling materials to achieve long-term environmental benefits.

- **Broader Stakeholder Engagement:** Increasing collaboration with a wider range of stakeholders, including consumers, non-governmental organizations (NGOs), and local communities, to address social responsibility issues and enhance ethical practices. Engaging diverse perspectives can lead to more inclusive and effective solutions.

- **Ethical Supply Chain Innovations:** Exploring new models and practices for ethical supply chain management, including innovative sourcing strategies, ethical certifications, and collaborative initiatives to improve social and environmental outcomes.

- **Education and Training Programs:** Developing comprehensive education and training programs for professionals in inventory management to raise awareness and promote ethical practices. These programs can equip individuals with the knowledge and skills needed to address ethical challenges effectively.

- **Impact Measurement and Reporting:** Advancing methods for measuring and reporting the impact of ethical and socially responsible practices. Improved metrics and reporting frameworks can help organizations track progress, demonstrate accountability, and communicate their commitment to stakeholders.

These future directions highlight the ongoing need for innovation, collaboration, and continuous improvement in inventory ethics and social responsibility.

11.5 Conclusion

The chapter on "Inventory Ethics and Social Responsibility" underscores the critical importance of incorporating ethical considerations into inventory management practices. As organizations navigate the complexities of global supply chains, adopting responsible and transparent practices is essential for mitigating environmental impact, promoting social equity, and enhancing overall business integrity. Exploring theoretical frameworks, real-world case studies, and future directions, this chapter provides a comprehensive examination of how ethical practices can be integrated into inventory management. Key takeaways include the need for robust ethical frameworks, the value of transparency and stakeholder engagement, and the role of technology in supporting responsible practices. The field continues to evolve, organizations are encouraged to embrace ethical principles and strive for continuous improvement in their inventory management practices.

Exercises

Exercise 11.1. *A company is faced with an ethical dilemma in inventory management: whether to overstate inventory levels to meet quarterly financial targets or to report accurate but lower levels, which may lead to a drop in stock prices. Analyze this situation using:*

1. ***Utilitarianism:*** *Consider the consequences of each action and determine which option maximizes overall happiness.*
2. ***Deontology:*** *Assess the actions based on moral rules or duties, regardless of the consequences.*

Exercise 11.2. *A company aims to enhance its inventory management practices by incorporating CSR principles. Describe how the company can:*

1. *Ensure that its inventory management practices positively impact society and the environment.*
2. *Integrate social and environmental considerations into its business strategies to align with CSR objectives.*

Exercise 11.3. *Analyze how a company can balance the interests of various stakeholders in its inventory management practices by:*

1. *Identifying key stakeholders and their respective interests.*
2. *Implementing inventory management practices that consider and address these interests.*

Exercise 11.4. *Evaluate a company's inventory management practices using sustainability frameworks by:*

1. *Applying the Triple Bottom Line (TBL) approach to assess the environmental, social, and economic impacts.*
2. *Aligning inventory practices with the UN Sustainable Development Goals (SDGs) to promote sustainable development.*

Exercise 11.5. *Discuss how a company can integrate ethical considerations into its supply chain management by:*

1. *Following guidelines from the Ethical Trading Initiative (ETI) to ensure fair labor practices and ethical sourcing.*
2. *Adopting standards from the Global Reporting Initiative (GRI) to enhance transparency and accountability in supply chain practices.*

12

Inventory Simulation and Gaming

This chapter explores the use of simulation and gaming techniques in inventory management, highlighting their effectiveness in improving decision-making. Through examples and scenarios, it demonstrates how these tools create dynamic, interactive environments that reflect real-world inventory challenges. The chapter examines practical applications of simulation models and gaming approaches, emphasizing their value in both educational and strategic contexts. It shows how these techniques enhance learning for practitioners, researchers, and students, fostering a deeper understanding of complex inventory dynamics and aiding in strategy development. This chapter is a valuable resource for those seeking to leverage simulation and gaming to improve inventory management decision-making.

12.1 Introduction

Simulation and gaming techniques offer innovative approaches to understanding and managing inventory systems. These techniques create dynamic and interactive environments that mimic real-world scenarios, allowing users to explore various inventory management strategies and their outcomes. Simulation models provide a virtual representation of inventory processes, enabling users to experiment with different variables and observe the effects in a controlled setting. Gaming techniques, on the other hand, introduce competitive and engaging elements to the learning process, fostering deeper engagement and practical experience.

Key aspects of simulation and gaming in inventory management include:

- **Dynamic Environment:** Both simulation and gaming create environments that evolve in response to user actions and decisions, reflecting the complexities of real-world inventory systems.
- **Scenario Exploration:** These techniques allow users to test various scenarios and strategies, providing insights into potential outcomes and helping to identify optimal inventory management practices.

DOI: 10.1201/9781003534860-12

- **Interactive Learning:** Actively participating in simulations or games, users gain hands-on experience and develop a better understanding of inventory dynamics and decision-making processes.

12.1.1 Benefits of Simulation and Gaming Techniques

Simulation and gaming techniques offer several benefits for inventory management, including:

- **Enhanced Understanding:** These techniques provide a deeper understanding of inventory systems and their complexities by allowing users to experiment with different scenarios and see the results in realtime.
- **Risk-Free Experimentation:** Simulations and games enable users to explore various strategies and decision-making approaches without the risks and costs associated with real-world implementation.
- **Improved Decision-Making Skills:** Engaging in interactive simulations or games, users can develop and refine their decision-making skills, learning to make informed choices based on data and outcomes.
- **Increased Engagement:** The interactive and often competitive nature of games can enhance engagement and motivation, making the learning process more enjoyable and effective.
- **Cost-Effective Training:** Simulation and gaming provide cost-effective training solutions by offering a scalable and repeatable method for developing inventory management skills and strategies.

12.1.2 Objectives of Using Simulation and Gaming

The primary objectives of using simulation and gaming techniques in inventory management are:

- **To Develop Practical Skills:** Simulation and gaming aim to enhance practical skills by providing users with hands-on experience in managing inventory systems and making real-time decisions.
- **To Test and Validate Strategies:** These techniques allow users to test and validate different inventory management strategies and approaches, helping to identify effective solutions and optimize practices.
- **To Improve Learning Outcomes:** Engaging users in interactive and dynamic environments, simulation and gaming aim to improve learning outcomes and facilitate a deeper understanding of inventory management concepts.

- **To Prepare for Real-World Challenges:** Simulation and gaming prepare users for real-world inventory management challenges by simulating complex scenarios and providing insights into effective problem-solving techniques.
- **To Foster Innovation and Creativity:** These techniques encourage innovative thinking and creative problem-solving by allowing users to experiment with new ideas and approaches in a risk-free environment.

12.1.3 Types of Simulation Models

Simulation models in inventory management come in various forms, each serving different purposes and offering unique insights. The main types of simulation models include:

- **Discrete Event Simulation (DES):** DES models inventory systems as a series of discrete events that occur at specific times. These models are useful for analyzing systems with complex interactions and dependencies, such as order fulfillment processes and inventory replenishment.
- **System Dynamics (SD):** SD models use stocks, flows, and feedback loops to represent the behavior of inventory systems over time. These models are effective for understanding the impact of changes in inventory policies and external factors on the overall system.
- **Agent-Based Simulation (ABS):** ABS models simulate the behavior of individual agents within an inventory system, such as suppliers, customers, and inventory managers. These models are useful for exploring how interactions between agents influence inventory performance.
- **Monte Carlo Simulation (MCS):** MCS models use random sampling and statistical methods to estimate the impact of uncertainty and variability on inventory systems. These models are useful for assessing the risks associated with inventory decisions and planning.
- **Queuing Models:** Queuing models analyze inventory systems as queues of items waiting to be processed or fulfilled. These models help to evaluate performance metrics such as wait times, service levels, and throughput.

12.1.4 Developing and Implementing Simulation Models

Developing and implementing simulation models in inventory management involves several key steps:

- **Defining Objectives:** Clearly define the objectives of the simulation model, including the specific inventory management issues to be addressed and the desired outcomes.

- **Model Design:** Design the simulation model based on the chosen type and objectives. This includes defining the system components, relationships, and parameters, as well as specifying the simulation logic and rules.
- **Data Collection:** Gather relevant data to populate the model, including historical inventory data, demand patterns, lead times, and other relevant variables.
- **Model Calibration:** Calibrate the model by adjusting parameters and validating its accuracy against real-world data. This ensures that the model accurately represents the inventory system and its behavior.
- **Simulation Execution:** Run the simulation model to generate results and analyze the behavior of the inventory system under different scenarios and conditions.
- **Result Analysis:** Analyze the simulation results to gain insights into inventory performance, identify potential issues, and evaluate the impact of different strategies.
- **Model Refinement:** Refine the model based on the results and feedback to improve its accuracy and effectiveness. This may involve adjusting parameters, incorporating additional data, or modifying the model design.

12.1.5 Case Studies of Simulation in Practice

Case studies provide valuable insights into the practical application of simulation models in inventory management. Examples include:

- **Retail Inventory Management:** A major retailer used a discrete event simulation model to optimize its inventory replenishment processes. The model helped to identify bottlenecks and improve stock levels, resulting in reduced stockouts and increased sales.
- **Manufacturing Supply Chain:** A manufacturing company implemented a system dynamics model to analyze the impact of changes in production schedules on inventory levels and lead times. The model enabled the company to better manage its production and inventory policies, reducing costs and improving service levels.
- **Healthcare Inventory Management:** A hospital used an agent-based simulation model to evaluate the effectiveness of different inventory management strategies for medical supplies. The model provided insights into the interactions between suppliers, medical staff, and inventory levels, leading to improved supply chain efficiency.
- **Logistics and Distribution:** A logistics company employed a Monte Carlo simulation model to assess the impact of variability in delivery times on

inventory levels and customer service. The model helped the company to optimize its distribution network and minimize service disruptions.

12.1.6 Types of Inventory Management Games

Inventory management games offer an interactive way to understand and improve inventory control practices. Different types of inventory management games include:

- **Simulation Games:** These games replicate real-world inventory management scenarios, allowing players to make decisions and see their effects in a controlled environment. Examples include managing stock levels, dealing with supply chain disruptions, and optimizing order quantities.
- **Role-Playing Games (RPGs):** RPGs place players in specific roles within an inventory management system, such as a warehouse manager or supply chain coordinator. Players make decisions based on their roles and responsibilities, providing insights into different perspectives within the inventory process.
- **Strategy Games:** Strategy games focus on long-term planning and decision-making. Players must develop and execute inventory strategies to achieve goals such as minimizing costs, maximizing service levels, or optimizing inventory turnover.
- **Educational Games:** These games are designed specifically for learning purposes. They often include tutorials, quizzes, and feedback mechanisms to help players understand inventory management concepts and best practices.
- **Competitive Games:** Competitive games introduce elements of competition among players or teams. They simulate competitive scenarios where players must manage inventory effectively to outperform their competitors and achieve superior results.

12.1.7 Designing Effective Inventory Management Games

Designing effective inventory management games involves several key considerations:

- **Define Objectives:** Clearly establish the learning objectives and goals of the game. Determine what inventory management concepts and skills the game is intended to teach or reinforce.
- **Create Realistic Scenarios:** Develop realistic and relevant scenarios that reflect common inventory management challenges and decisions. Ensure that scenarios are complex enough to provide meaningful learning experiences but manageable for players.

- **Incorporate Key Inventory Concepts:** Integrate important inventory management concepts such as demand forecasting, order fulfillment, inventory control, and supply chain management into the game mechanics.
- **Provide Feedback and Scoring:** Implement feedback mechanisms and scoring systems to help players understand the impact of their decisions and track their progress. Provide constructive feedback to enhance learning and improvement.
- **Ensure Engagement:** Design the game to be engaging and interactive. Use elements such as competition, rewards, and challenges to maintain player interest and motivation.
- **Test and Refine:** Pilot the game with a sample audience to identify any issues or areas for improvement. Gather feedback and refine the game to enhance its effectiveness and usability.

12.2 Educational and Strategic Dimensions

Simulation and gaming techniques play a significant role in education by offering interactive and immersive learning experiences. Their contributions include:

- **Interactive Learning:** Simulation and gaming provide hands-on, interactive experiences that enhance understanding of complex concepts. Students and practitioners can engage with realistic scenarios that reflect real-world inventory challenges, promoting active learning.
- **Practical Application:** These techniques allow learners to apply theoretical knowledge in a controlled environment. Experimenting with different strategies and observing outcomes, students gain practical insights into inventory management principles and practices.
- **Critical Thinking:** Simulation and gaming encourage critical thinking and problem-solving skills. Learners must analyze situations, make decisions, and adapt their strategies based on feedback, fostering deeper comprehension and analytical abilities.
- **Risk-Free Experimentation:** Participants can explore various scenarios and decisions without facing real-world consequences. This risk-free environment enables them to experiment with different approaches, learn from mistakes, and refine their skills.
- **Engagement and Motivation:** The interactive nature of simulations and games enhances engagement and motivation. Gamification elements, such

as scoring, rewards, and competition, make learning more enjoyable and stimulating.

12.2.1 Strategic Applications of Simulation and Gaming

Simulation and gaming techniques are valuable tools for strategic applications in inventory management, including:

- **Scenario Planning:** These techniques allow organizations to simulate various strategic scenarios and evaluate their potential impacts. This helps in understanding how different strategies and decisions will affect inventory performance and overall business outcomes.
- **Policy Testing:** Simulation and gaming can be used to test and refine inventory management policies and procedures. Modeling different policy options and their effects, organizations can identify the most effective strategies for achieving their goals.
- **Strategic Decision Support:** These techniques provide valuable insights and data that support strategic decision-making. Analyzing simulation results, organizations can make informed decisions about inventory investments, resource allocation, and operational improvements.
- **Change Management:** Simulation and gaming can aid in managing organizational change by modeling the effects of new strategies or processes on inventory management. This helps in preparing for and mitigating the impact of change on inventory systems and operations.
- **Competitive Analysis:** Organizations can use simulation and gaming to assess competitive strategies and market dynamics. Simulating competitive scenarios, they can develop strategies to enhance their competitive position and respond to market challenges.

12.2.2 Enhancing Decision-Making Skills Through Simulation and Gaming

Simulation and gaming techniques are effective for enhancing decision-making skills in inventory management by:

- **Providing Realistic Practice:** They offer a realistic environment for practicing decision-making skills. Participants make decisions based on simulated data and scenarios, gaining practical experience in managing inventory effectively.
- **Improving Analytical Abilities:** Participants develop strong analytical skills by analyzing simulation results, identifying patterns, and interpreting data. This improves their ability to make informed decisions based on quantitative and qualitative information.

- **Building Strategic Thinking:** Simulation and gaming promote strategic thinking by challenging participants to develop and implement long-term strategies. They learn to anticipate the consequences of their decisions and adjust their strategies accordingly.
- **Enhancing Problem-Solving Skills:** These techniques foster problem-solving skills by presenting complex and dynamic scenarios. Participants learn to identify problems, develop solutions, and evaluate the effectiveness of their strategies.
- **Facilitating Collaboration:** Many simulations and games involve collaborative elements, where participants work together to achieve common goals. This enhances teamwork and communication skills, which are crucial for effective decision-making in inventory management.

12.3 Challenges and Limitations

While simulation and gaming techniques offer significant benefits, they also face several common challenges:

- **Complexity and Realism:** Creating realistic and complex simulation models can be challenging. Ensuring that simulations accurately reflect real-world inventory management scenarios requires extensive development and validation, which can be resource-intensive.
- **Data Accuracy:** The effectiveness of simulations and games depends on the accuracy and relevance of the data used. Inaccurate or outdated data can lead to misleading results and reduce the reliability of the learning experience.
- **Participant Engagement:** Maintaining participant engagement can be difficult, especially if the simulation or game is overly complex or not sufficiently interactive. Ensuring that the experience remains engaging and motivating is crucial for effective learning.
- **Technical Issues:** Technical problems, such as software bugs or hardware malfunctions, can disrupt the simulation or gaming experience. These issues need to be addressed promptly to avoid impacting the learning outcomes.
- **Integration With Curriculum:** Aligning simulations and games with educational objectives and curriculum requirements can be challenging. Ensuring that the content and learning outcomes are well-integrated with academic or training programs is essential for maximizing their value.

12.3.1 Limitations and Considerations

Several limitations and considerations must be acknowledged when using simulation and gaming techniques:

- **Scope and Focus:** Simulations and games often focus on specific aspects of inventory management and may not cover all relevant areas. This limitation requires careful consideration of which aspects are addressed and how they align with the intended learning objectives.
- **Resource Requirements:** Developing and implementing simulations and games can be resource-intensive, requiring time, expertise, and financial investment. Organizations must weigh the benefits against the costs and resources required.
- **Learning Transfer:** There is no guarantee that skills and knowledge gained from simulations and games will directly transfer to real-world situations. Participants may need additional training and practice to apply what they have learned effectively.
- **Bias and Simplification:** Simulations and games may simplify complex inventory management scenarios, potentially leading to biased or incomplete understanding. It is important to recognize the limitations of these models and complement them with real-world experiences and additional learning resources.
- **Customization Needs:** Generic simulations and games may not fully address the unique needs of every organization or educational program. Customization may be required to ensure that the content and scenarios are relevant and applicable.

12.3.2 Overcoming Challenges in Practical Applications

To effectively overcome the challenges and limitations of simulation and gaming techniques, consider the following strategies:

- **Continuous Improvement:** Regularly update and refine simulation models and games based on feedback and new developments in inventory management. Continuous improvement ensures that the content remains relevant and effective.
- **Validation and Testing:** Conduct thorough validation and testing to ensure the accuracy and reliability of simulation models. Use real-world data and scenarios to enhance the realism and credibility of the simulations.
- **Engagement Strategies:** Incorporate engaging elements, such as interactive features, gamification, and competition, to maintain participant interest and motivation. Regularly assess and adjust engagement strategies to meet participants' needs.

- **Technical Support:** Provide robust technical support to address any issues promptly. Ensure that participants have access to assistance and troubleshooting resources to minimize disruptions.
- **Alignment With Objectives:** Ensure that simulations and games are well-aligned with educational or organizational objectives. Clearly define learning outcomes and integrate the simulations with relevant curriculum or training programs.
- **Complementary Learning:** Use simulations and games as a complementary learning tool, combined with other educational methods and real-world experiences. This approach helps bridge the gap between simulated scenarios and practical application.

12.4 Future Directions in Inventory Simulation and Gaming

The field of inventory simulation and gaming is evolving rapidly, with several emerging trends and technologies shaping its future:

- **Artificial Intelligence (AI) and Machine Learning (ML):** AI and ML technologies are being integrated into simulation models to enhance their predictive capabilities and adaptability. These technologies can provide more accurate simulations by analyzing large datasets and identifying patterns that might be missed by traditional methods.
- **Virtual Reality (VR) and Augmented Reality (AR):** VR and AR technologies offer immersive simulation experiences that enhance user engagement and provide more realistic scenarios. These technologies can simulate complex inventory environments and processes, allowing users to interact with them in a more intuitive way.
- **Big Data Analytics:** The use of big data analytics in simulations can improve the accuracy and depth of the models by incorporating vast amounts of real-time data. This enables more detailed and dynamic simulations, reflecting current market conditions and trends.
- **Blockchain Technology:** Blockchain can be utilized to enhance transparency and traceability in supply chains. Integrating blockchain into simulations, organizations can model and analyze the impact of blockchain on inventory management, including issues related to authenticity and security.
- **Cloud Computing:** Cloud-based simulation platforms offer scalability and accessibility, allowing users to run complex simulations without the need for

extensive local computing resources. Cloud computing also facilitates collaboration by enabling multiple users to access and interact with simulations from different locations.

12.4.1 Potential Areas for Research and Development

Several areas present opportunities for research and development in inventory simulation and gaming:

- **Enhanced Modeling Techniques:** Research into advanced modeling techniques, such as hybrid models that combine various simulation approaches, can lead to more accurate and comprehensive simulations. Exploring new methods for representing complex inventory scenarios will improve the effectiveness of simulations.
- **User Experience and Engagement:** Developing strategies to enhance user experience and engagement in simulations and games is crucial. Research into user interface design, gamification elements, and interactive features can lead to more effective and enjoyable learning experiences.
- **Integration with Enterprise Systems:** Investigating how simulation and gaming can be integrated with enterprise resource planning (ERP) and other business systems can improve the alignment between simulated scenarios and real-world operations. This integration can lead to more actionable insights and better decision-making.
- **Customization and Adaptability:** Developing techniques for customizing simulations to meet specific organizational needs or educational objectives is an important area of research. Flexible and adaptable simulation models can provide tailored learning experiences and address unique challenges.
- **Cross-Disciplinary Applications:** Exploring the application of simulation and gaming techniques across different disciplines and industries can lead to new insights and innovations. Research into how these techniques can be applied beyond traditional inventory management scenarios will broaden their applicability.

12.4.2 Integrating Simulation and Gaming With Other Tools

Integrating simulation and gaming with other tools can enhance their effectiveness and provide more comprehensive solutions:

- **Data Visualization Tools:** Combining simulation models with data visualization tools can help users better understand and interpret simulation results. Interactive dashboards and visualizations can provide real-time insights and facilitate more informed decision-making.

- **Decision Support Systems (DSS):** Integrating simulations with decision support systems can enhance their utility by providing additional analytical capabilities and supporting complex decision-making processes. DSS can complement simulations by offering additional data analysis and optimization features.

- **Collaboration Platforms:** Using collaboration platforms in conjunction with simulations can improve teamwork and communication among users. These platforms can facilitate the sharing of simulation results, feedback, and strategies, enhancing the overall learning and decision-making process.

- **Training and Learning Management Systems (LMS):** Integrating simulations with LMS can streamline the training process and track learning progress. This integration enables seamless access to simulations as part of broader educational programs and facilitates the assessment of learning outcomes.

- **Real-Time Data Integration:** Incorporating real-time data into simulations can enhance their relevance and accuracy. Integrating live data from inventory systems, simulations can reflect current conditions and provide more actionable insights.

12.4.3 Implications for Practitioners and Researchers

The integration of simulation and gaming techniques in inventory management presents several implications for practitioners and researchers:

- **Enhanced Decision-Making:** Practitioners can leverage simulation and gaming tools to make more informed decisions by exploring various scenarios and outcomes. These tools offer a risk-free environment to test different strategies and understand their potential impacts.

- **Skill Development:** For researchers and practitioners, engaging with simulations and games provides opportunities to develop and refine skills related to inventory management, strategic thinking, and problem-solving. These tools facilitate experiential learning and practical application of theoretical concepts.

- **Innovation and Creativity:** The use of simulation and gaming encourages innovation and creative problem-solving. Practitioners can experiment with new approaches and solutions in a controlled setting, fostering a culture of continuous improvement and innovation.

- **Research Opportunities:** Researchers can explore various aspects of inventory simulation and gaming, including the effectiveness of different models, the impact of emerging technologies, and the development of new methodologies. These research opportunities contribute to the advancement of knowledge in the field.

- **Collaboration and Networking:** Simulation and gaming environments can facilitate collaboration among practitioners, researchers, and students. Working together on simulation projects or participating in gaming scenarios, individuals can share insights, experiences, and best practices.

12.4.4 Final Thoughts and Recommendations

In conclusion, simulation and gaming techniques offer valuable tools for enhancing inventory management practices and decision-making. To maximize their effectiveness, consider the following recommendations:

- **Stay Updated with Emerging Technologies:** Continuously explore and adopt emerging technologies that can enhance simulation and gaming experiences. Incorporate advancements such as AI, VR, and big data analytics to improve the accuracy and relevance of simulations.
- **Invest in Training and Development:** Ensure that practitioners and researchers receive adequate training to effectively use simulation and gaming tools. Invest in professional development programs that focus on the application of these techniques in inventory management.
- **Focus on Real-World Relevance:** Design simulations and games that closely mirror real-world inventory scenarios and challenges. Ensure that the models and scenarios are relevant to the specific needs and contexts of users.
- **Encourage Collaboration and Feedback:** Foster collaboration among users and encourage feedback to continuously improve simulation and gaming tools. Create platforms for sharing experiences, insights, and best practices.
- **Evaluate and Refine:** Regularly evaluate the effectiveness of simulation and gaming tools and refine them based on user feedback and performance outcomes. Ensure that the tools remain aligned with evolving industry needs and research advancements.

12.5 Conclusion

Simulation and gaming techniques represent a powerful approach to enhancing inventory management by providing dynamic and interactive environments for exploring complex scenarios and decision-making processes. As the field continues to evolve with technological advancements and innovative approaches, the integration of these techniques will play an increasingly important role in refining inventory management practices. Staying informed about emerging trends, investing in training, and focusing on real-world relevance, practitioners and researchers can leverage simulation and gaming to achieve more effective and strategic inventory management outcomes.

Exercises

Exercise 12.1. ***Discrete Event Simulation (DES):*** *Consider a warehouse inventory system where incoming orders arrive according to a Poisson process with a rate of five orders per hour. Each order takes an exponentially distributed time with a mean of ten minutes to be fulfilled. Using a Discrete Event Simulation (DES) model, analyze the average number of orders waiting to be fulfilled and the average waiting time for an order. Additionally, determine the impact of doubling the arrival rate on the system's performance.*

Exercise 12.2. ***Monte Carlo Simulation (MCS):*** *A company is evaluating its safety stock levels for a product with normally distributed demand. The mean daily demand is 100 units with a standard deviation of 15 units. The lead time for replenishment is seven days. Using a Monte Carlo Simulation (MCS) model, estimate the probability that the company will experience a stockout if it maintains a safety stock of 200 units. Perform 10,000 simulation runs to ensure the accuracy of your estimate.*

Exercise 12.3. ***Ethical Theories:*** *Consider a company facing an ethical dilemma in inventory management where they must choose between two suppliers: Supplier A, who offers lower costs but has questionable labor practices, and Supplier B, who offers higher costs but adheres to ethical labor standards. Using utilitarianism, deontology, and virtue ethics, evaluate which supplier the company should choose.*

Exercise 12.4. ***Corporate Social Responsibility (CSR) Framework:*** *A retail company is evaluating its inventory management practices to align with CSR principles. The company needs to assess the social and environmental impacts of its current inventory policies. Develop a CSR-based inventory management strategy that addresses these impacts and propose key performance indicators (KPIs) to measure the effectiveness of the strategy.*

Exercise 12.5. ***Probability Analysis:*** *A company uses probability analysis to assess the likelihood of stockouts for a particular product. The historical data shows that the product has a 10% chance of a stockout each month. Using a binomial distribution, calculate the probability that the company will experience at least one stockout in the next six months.*

Exercise 12.6. ***Monte Carlo Simulation:*** *A retailer wants to estimate the expected profit from selling a seasonal product. The demand for the product* *follows a normal distribution with a mean of 500 units and a standard deviation of 100 units. The selling price per unit is \$50, and the cost per unit is \$30. Using Monte Carlo Simulation, estimate the expected profit over 10,000 simulation runs.*

Exercise 12.7. ***Closed-Loop Supply Chain Implementation:*** *A company is considering implementing a closed-loop supply chain to enhance its sustainability efforts. Describe the key steps involved in designing and implementing a closed-loop supply chain, including the roles of reverse logistics and remanufacturing.*

Exercise 12.8. ***Closed-Loop Supply Chain Implementation:*** *Evaluate the potential challenges and benefits of implementing a closed-loop supply chain for an electronics manufacturer. Consider factors such as cost, technology, customer acceptance, and regulatory compliance.*

Exercise 12.9. ***Digital Transformation:*** *Discuss how the adoption of digital technologies, such as IoT, blockchain, and AI, is revolutionizing supply chain management. Provide examples of how these technologies are being used to enhance sustainability in supply chains.*

Exercise 12.10. ***Circular Economy:*** *Analyze the shift towards a circular economy in supply chains. Discuss the initiatives such as product take-back programs, remanufacturing, and recycling, and their impact on reducing waste and conserving resources.*

13

Inventory Education and Training

This chapter explores the educational and training aspects of inventory management, emphasizing the importance of continuous learning in an evolving business environment. It highlights strategies and methodologies for teaching inventory control principles, providing insights into the effectiveness of training programs. The chapter also examines the practical impact of education, showing how well-trained individuals enhance organizational efficiency. Additionally, it discusses emerging trends in inventory education, offering valuable perspectives for educators, trainers, and learners. With its comprehensive focus, this chapter serves as a valuable resource for those involved in developing or participating in inventory training programs, underscoring the critical role of education in optimizing inventory processes.

13.1 Introduction

Inventory education and training encompass a wide range of activities and programs designed to impart knowledge and skills related to inventory management. These programs aim to enhance the understanding of inventory principles, techniques, and best practices, ensuring that individuals involved in inventory management can make informed decisions and effectively manage inventory systems. The scope of inventory education and training includes formal academic courses, professional certification programs, workshops, seminars, and on-the-job training. Providing a structured learning environment, these educational initiatives help individuals develop the competencies needed to optimize inventory processes and contribute to overall organizational success.

DOI: 10.1201/9781003534860-13

13.1.1 Importance of Continuous Learning in Inventory Management

In today's rapidly changing business environment, continuous learning in inventory management is crucial. The dynamics of global supply chains, advancements in technology, and evolving customer expectations necessitate that inventory professionals stay current with the latest trends and developments. Continuous learning ensures that individuals are equipped with the latest knowledge and skills to adapt to these changes, improving their ability to respond to new challenges and opportunities. It also fosters a culture of innovation and improvement within organizations, leading to better inventory management practices, reduced costs, and enhanced service levels.

13.1.2 Objectives of Inventory Education and Training

The primary objectives of inventory education and training programs are:

- **Knowledge Enhancement:** To provide a deep understanding of inventory management concepts, techniques, and best practices.
- **Skill Development:** To develop practical skills for effectively managing inventory, including forecasting, inventory control, and optimization.
- **Process Improvement:** To equip individuals with the tools and methodologies needed to analyze and improve inventory processes, reducing waste and increasing efficiency.
- **Adaptability:** To prepare inventory professionals to adapt to changes in the business environment, including technological advancements and shifts in market demand.
- **Professional Growth:** To support the career development of inventory professionals by providing opportunities for certification and advanced training.
- **Organizational Success:** To contribute to the overall success of the organization by ensuring that inventory is managed in a way that supports business objectives and customer satisfaction.

13.2 Strategies for Effective Inventory Education

Effective inventory education strategies are essential for equipping individuals with the knowledge and skills necessary for efficient inventory management. These strategies encompass traditional training methods, modern technological approaches, and blended learning techniques.

13.2.1 Traditional Training Methods

Traditional training methods in inventory education include classroom-based instruction, textbooks, and workshops. These methods provide structured learning environments where students can engage directly with instructors and peers. Key components include:

- **Classroom Lectures:** Instructors deliver lectures on various inventory management topics, providing foundational knowledge and theoretical frameworks. These lectures allow for real-time interaction, where students can ask questions and engage in discussions to deepen their understanding.
- **Textbooks and Study Materials:** Students use textbooks and other printed materials to supplement classroom learning and gain a deeper understanding of inventory concepts. These resources provide comprehensive coverage of topics, including case studies, examples, and exercises that reinforce learning.
- **Workshops and Seminars:** Interactive sessions where participants can apply their knowledge through exercises, discussions, and case studies, facilitating practical understanding and skill development. Workshops and seminars often include hands-on activities, allowing students to practice inventory management techniques in simulated environments.

Traditional training methods offer a structured and interactive approach to learning, fostering a thorough understanding of inventory management principles and practices.

13.2.2 Modern Educational Approaches

Modern educational approaches leverage technology and innovative pedagogical methods to enhance inventory education. These approaches offer flexibility, accessibility, and engagement, making learning more effective. Key elements include:

- **Online Courses:** Digital platforms offering courses on inventory management, allowing learners to study at their own pace and convenience. These courses often include video lectures, quizzes, and interactive content.
- **Webinars and Virtual Workshops:** Live or recorded sessions conducted over the internet, providing opportunities for real-time interaction with experts and peers.
- **Simulation and Gamification:** Use of simulation tools and game-based learning to create realistic inventory management scenarios. These methods help learners practice decision-making and problem-solving in a risk-free environment.

13.2.3 Blended Learning Techniques

Blended learning combines traditional and modern educational approaches to create a comprehensive and flexible learning experience. This method integrates face-to-face instruction with online resources and activities, providing the best of both worlds. Key strategies include:

- **Flipped Classroom:** Students access lecture materials online before class, using classroom time for discussions, hands-on activities, and collaborative projects.
- **Hybrid Courses:** Courses that mix in-person sessions with online components, allowing for a balanced approach to learning and greater flexibility for students.
- **Interactive Learning Platforms:** Platforms that offer a variety of learning tools, such as discussion forums, interactive assignments, and multimedia resources, to support and enhance the learning experience.

Implementing these strategies, educational programs can effectively equip learners with the knowledge and skills necessary for successful inventory management, adapting to diverse learning styles and evolving industry needs.

13.3 Methodologies for Teaching Inventory Management

13.3.1 Case-Based Learning

Case-based learning is a methodology that uses real-world scenarios and case studies to teach inventory management concepts. This approach helps students understand the practical application of theoretical knowledge and develop critical thinking skills. Key aspects include:

- **Real-World Scenarios:** Students analyze real-life inventory management challenges faced by companies, learning to apply theoretical concepts to practical problems.
- **Discussion and Analysis:** Instructors facilitate discussions around the cases, encouraging students to explore different perspectives and develop well-rounded solutions.
- **Problem-Solving Skills:** Case-based learning emphasizes problem-solving, helping students develop the ability to identify issues, evaluate options, and implement effective strategies.

13.3.2 Simulation and Gaming Techniques

Simulation and gaming techniques create interactive and immersive learning environments where students can practice inventory management in a risk-free setting. These methods enhance engagement and deepen understanding through experiential learning. Key components include:

- **Inventory Simulations:** Computer-based simulations that model real-world inventory scenarios, allowing students to make decisions and see the outcomes of their actions.
- **Game-Based Learning:** Educational games designed to teach inventory management principles in a fun and interactive way. These games often involve challenges and competitions that motivate learners.
- **Role-Playing Activities:** Students take on different roles within a simulated inventory management environment, experiencing various aspects of the process and learning through hands-on practice.

13.3.3 Hands-On Workshops and Practicals

Hands-on workshops and practicals provide students with opportunities to apply inventory management concepts in real or simulated environments. These activities enhance practical skills and reinforce theoretical knowledge. Key elements include:

- **Interactive Workshops:** Facilitated sessions where students participate in activities such as inventory audits, stocktaking exercises, and demand forecasting.
- **Lab Practicals:** Practical sessions conducted in a controlled environment, allowing students to work with inventory management software, tools, and technologies.
- **Project-Based Learning:** Students undertake projects that require them to apply inventory management principles to real-world problems, fostering practical skills and experiential learning.

Incorporating these methodologies, educators can provide a comprehensive and engaging learning experience, equipping students with the knowledge and skills necessary for effective inventory management.

13.3.4 Measuring Training Effectiveness

Measuring the effectiveness of inventory management training programs is crucial to ensure that the training objectives are being met and that participants are gaining the necessary knowledge and skills. Key methods for measuring training effectiveness include:

- **Pre- and Post-Training Assessments:** Evaluating participants' knowledge and skills before and after the training to measure learning gains.
- **Performance Metrics:** Monitoring key performance indicators (KPIs) related to inventory management to assess the impact of the training on job performance.
- **Participant Surveys:** Gathering feedback from participants about the training content, delivery, and overall experience.

13.3.5 Feedback and Continuous Improvement

Collecting and analyzing feedback from training participants is essential for continuous improvement. Effective strategies for gathering and utilizing feedback include:

- **Post-Training Surveys:** Distributing surveys to participants immediately after the training to collect their insights and suggestions.
- **Focus Groups:** Conducting focus group discussions with participants to gain in-depth feedback and identify areas for improvement.
- **Action Plans:** Developing action plans based on feedback to address identified gaps and enhance future training sessions.

13.3.6 Best Practices in Training Evaluation

Implementing best practices in the evaluation of training programs ensures that the assessment process is effective and yields actionable insights. Key best practices include:

- **Clear Objectives:** Defining clear and measurable training objectives to guide the evaluation process.
- **Comprehensive Evaluation Framework:** Using a multi-level evaluation framework that includes reaction, learning, behavior, and results.
- **Ongoing Monitoring:** Continuously monitoring and evaluating the training program to ensure its relevance and effectiveness over time.
- **Stakeholder Involvement:** Involving key stakeholders, such as trainers, managers, and participants, in the evaluation process to gain diverse perspectives.

Effectively evaluating training programs, organizations can ensure that their inventory management training is impactful and aligned with their overall business objectives.

13.3.7 Technological Advancements in Training Tools

Technological advancements are revolutionizing the way inventory management training is delivered and experienced. Key trends include:

- **Virtual Reality (VR) and Augmented Reality (AR):** Utilizing VR and AR to create immersive training environments where participants can engage in realistic inventory management scenarios.
- **Interactive Simulations:** Developing interactive simulation tools that allow learners to practice and hone their inventory management skills in a risk-free virtual environment.
- **Mobile Learning Applications:** Leveraging mobile apps to provide on-the-go training resources and facilitate continuous learning for inventory management professionals.

13.3.8 Online Learning Platforms and Resources

Online learning platforms and resources are making inventory management education more accessible and flexible. Key features of these platforms include:

- **E-Learning Modules:** Offering comprehensive e-learning modules that cover various aspects of inventory management, allowing learners to study at their own pace.
- **Webinars and Online Workshops:** Conducting live and recorded webinars and workshops that provide expert insights and practical knowledge on inventory management topics.
- **Collaborative Learning Communities:** Creating online forums and communities where learners can share experiences, ask questions, and collaborate on inventory management challenges.

13.3.9 The Role of Artificial Intelligence and Machine Learning

Artificial Intelligence (AI) and Machine Learning (ML) are transforming inventory management education by providing personalized and data-driven learning experiences. Key applications include:

- **Adaptive Learning Systems:** Implementing AI-powered adaptive learning systems that tailor training content to the individual needs and progress of each learner.
- **Predictive Analytics:** Using ML algorithms to analyze training data and predict future learning needs, enabling proactive and targeted training interventions.

- **Intelligent Tutoring Systems:** Developing AI-driven tutoring systems that provide real-time feedback, guidance, and support to learners as they navigate through inventory management training programs.

Embracing these emerging trends and innovations, organizations can enhance the effectiveness and accessibility of their inventory management education and training programs.

13.4 Practical Implications of Inventory Education

13.4.1 Impact on Organizational Efficiency

Effective inventory education and training have significant practical implications for organizational efficiency:

- **Optimized Inventory Processes:** Well-trained employees can streamline inventory processes, reduce errors, and ensure smoother operations, leading to higher efficiency.
- **Reduced Operational Costs:** Proper training helps in minimizing waste, reducing holding costs, and preventing stockouts, thereby lowering overall operational expenses.
- **Improved Inventory Accuracy:** Enhanced skills in inventory management result in more accurate tracking and recording of inventory levels, reducing discrepancies and improving reliability.

13.4.2 Enhancing Employee Competence and Skills

Investing in inventory education enhances the competence and skills of employees, contributing to the overall success of the organization:

- **Increased Expertise:** Continuous learning opportunities help employees develop specialized knowledge and expertise in inventory management.
- **Higher Job Satisfaction:** Providing training and development programs boosts employee morale and job satisfaction, leading to higher retention rates.
- **Adaptability and Innovation:** Educated employees are more adaptable to changes and are better equipped to implement innovative solutions to inventory challenges.

13.4.3 Aligning Training With Organizational Goals

Aligning inventory education and training programs with organizational goals ensures that training efforts contribute directly to the company's success:

- **Strategic Alignment:** Training programs should be designed to support the strategic objectives of the organization, ensuring that employees are equipped to achieve key business goals.
- **Performance Metrics:** Establishing clear performance metrics helps in measuring the impact of training on organizational goals, enabling continuous improvement.
- **Collaborative Planning:** Involving key stakeholders in the planning and execution of training programs ensures that the training is relevant and aligned with the needs of the organization.

Focusing on these practical implications, organizations can maximize the benefits of their inventory education and training programs, leading to improved efficiency, enhanced employee skills, and alignment with organizational objectives.

13.5 Challenges and Opportunities in Inventory Training

13.5.1 Common Challenges in Training Implementation

Implementing effective inventory training programs can be challenging due to various factors:

- **Resource Constraints:** Limited budget and resources can hinder the development and delivery of comprehensive training programs.
- **Resistance to Change:** Employees may resist new training initiatives, preferring established methods over new approaches.
- **Lack of Engagement:** Ensuring that employees are engaged and motivated during training sessions can be difficult.
- **Scalability Issues:** Scaling training programs to accommodate large or geographically dispersed teams presents logistical challenges.

13.5.2 Opportunities for Improvement and Innovation

Despite these challenges, there are numerous opportunities to enhance and innovate inventory training programs:

- **Leveraging Technology:** Utilizing advanced technologies such as virtual reality (VR), augmented reality (AR), and e-learning platforms can make training more interactive and effective.
- **Continuous Feedback:** Implementing mechanisms for continuous feedback can help improve training content and delivery methods.
- **Customizable Training:** Developing modular and customizable training programs allows for personalized learning experiences tailored to individual needs.
- **Collaborative Learning:** Encouraging collaborative learning through group activities and peer-to-peer training can enhance knowledge sharing and retention.

13.5.3 Addressing Diverse Learning Needs

To ensure that inventory training programs are inclusive and effective, it is important to address the diverse learning needs of employees:

- **Varied Learning Styles:** Recognizing and accommodating different learning styles—visual, auditory, kinesthetic—can make training more accessible and effective.
- **Multilingual Training:** Providing training materials in multiple languages ensures that all employees, regardless of their linguistic background, can benefit from the training.
- **Accessibility Considerations:** Ensuring that training programs are accessible to employees with disabilities is crucial for inclusivity.
- **Flexible Learning Options:** Offering flexible learning options, such as on-demand online courses, can accommodate employees' varying schedules and learning paces.

13.6 Future Directions in Inventory Education and Training

13.6.1 Emerging Areas of Focus

As the field of inventory management continues to evolve, several emerging areas of focus are gaining importance in education and training:

- **Sustainability and Green Inventory Practices:** Emphasizing sustainable practices in inventory management, including waste reduction, eco-friendly sourcing, and minimizing the carbon footprint.

- **Data Analytics and Big Data:** Training programs increasingly incorporate data analytics to enhance decision-making and optimize inventory levels.
- **Risk Management:** Developing expertise in identifying and mitigating risks associated with inventory, such as demand fluctuations, supply chain disruptions, and financial risks.
- **Global Supply Chain Dynamics:** Understanding the complexities of managing inventory in a global context, including cross-border logistics, tariffs, and cultural differences.

13.6.2 Integration of New Technologies

The integration of advanced technologies is transforming inventory education and training:

- **Artificial Intelligence (AI) and Machine Learning:** Utilizing AI and machine learning to predict demand, optimize stock levels, and automate routine inventory tasks.
- **Virtual Reality (VR) and Augmented Reality (AR):** Implementing VR and AR to create immersive training environments that simulate real-world inventory scenarios.
- **Blockchain Technology:** Educating trainees on the use of blockchain for enhancing transparency, traceability, and security in inventory management.
- **Internet of Things (IoT):** Incorporating IoT devices for real-time inventory tracking and management, providing hands-on experience with cutting-edge technology.

13.6.3 Developing Comprehensive Training Programs

To meet the evolving needs of the industry, training programs must be comprehensive and adaptable:

- **Holistic Curriculum Design:** Creating a curriculum that covers all aspects of inventory management, from basic principles to advanced strategies and technologies.
- **Continuous Professional Development:** Encouraging continuous learning through certifications, workshops, and online courses to keep pace with industry advancements.
- **Interdisciplinary Approach:** Integrating knowledge from related fields such as logistics, supply chain management, and business analytics to provide a well-rounded education.

- **Collaborative Learning Environments:** Promoting collaboration through group projects, peer learning, and industry partnerships to enhance practical understanding and skills application.

13.6.4 Recommendations for Practitioners and Educators

To enhance the effectiveness of inventory education and training, practitioners and educators should consider the following recommendations:

- **Embrace Technological Advancements:** Incorporate the latest technologies such as AI, machine learning, VR, AR, and blockchain into training programs to provide hands-on experience with cutting-edge tools and techniques.
- **Promote Sustainability:** Integrate sustainability principles into the curriculum, emphasizing eco-friendly practices and the importance of minimizing the environmental impact of inventory management.
- **Foster a Culture of Continuous Learning:** Encourage continuous professional development through certifications, workshops, online courses, and other learning opportunities to ensure that skills remain current and relevant.
- **Utilize Real-World Case Studies:** Incorporate case studies and practical examples into the curriculum to bridge the gap between theory and practice, providing students with insights into real-world challenges and solutions.
- **Adopt an Interdisciplinary Approach:** Design training programs that integrate knowledge from related fields such as logistics, supply chain management, business analytics, and risk management to provide a comprehensive understanding of inventory management.
- **Engage with Industry Partners:** Establish partnerships with industry leaders to provide students with access to internships, mentorships, and networking opportunities, enhancing their practical knowledge and career prospects.
- **Implement Feedback Mechanisms:** Regularly collect feedback from trainees and industry professionals to continuously improve training programs and ensure they meet the evolving needs of the industry.
- **Focus on Ethical Practices:** Highlight the importance of ethics and social responsibility in inventory management, encouraging students to adopt fair and transparent practices in their professional roles.

13.7 Conclusion

Inventory education and training are critical components for the development of effective and efficient inventory management practices. This chapter has explored various aspects of inventory education, from traditional and modern teaching strategies to the integration of emerging technologies and the importance of continuous learning. Focusing on these elements, educators and practitioners can ensure that individuals are well-prepared to meet the challenges of inventory management in a dynamic and evolving business environment. The future of inventory education and training lies in embracing technological advancements, promoting sustainability, fostering continuous learning, and integrating interdisciplinary approaches. Implementing the recommendations provided, educators and practitioners can create robust and comprehensive training programs that not only enhance individual competencies but also contribute to the overall efficiency and success of organizations. In conclusion, the ongoing evolution of inventory management necessitates a commitment to continuous education and training. Staying abreast of emerging trends and technologies and prioritizing practical, ethical, and sustainable practices, professionals in the field can drive innovation and excellence in inventory management.

Exercises

Exercise 13.1. *Explain how a deep understanding of inventory management concepts, techniques, and best practices can benefit an organization's overall efficiency and effectiveness. Provide specific examples of how enhanced knowledge in this area can lead to improved decision-making and process optimization.*

Exercise 13.2. *Identify and discuss the key practical skills required for effectively managing inventory. How can these skills be developed through targeted training programs, and what are the expected outcomes of such skill development?*

Exercise 13.3. *Discuss the tools and methodologies that can be used to analyze and improve inventory processes. How do these tools help in reducing waste and increasing efficiency? Provide examples of successful process improvement initiatives.*

Exercise 13.4. *How can inventory professionals prepare for changes in the business environment, including technological advancements and shifts in market demand? Discuss strategies for developing adaptability in inventory management.*

Exercise 13.5. *Outline the importance of certification and advanced training for the career development of inventory professionals. What opportunities are available for such certifications, and how do they contribute to professional growth?*

Exercise 13.6. *Explain how effective inventory management contributes to the overall success of an organization. How does it support business objectives and customer satisfaction? Provide examples to illustrate your points.*

Exercise 13.7. *What are the advantages of classroom lectures in teaching inventory management? Discuss how real-time interaction and discussion during lectures can enhance understanding of theoretical frameworks.*

Exercise 13.8. *Describe the role of textbooks and printed study materials in inventory education. How do these resources complement classroom learning, and what features make them effective for deepening understanding of inventory concepts?*

14

Inventory Research and Development

This chapter provides an in-depth exploration of the latest research and development trends in inventory management. It highlights the evolving landscape of the field by drawing from various studies and systematic reviews. The chapter emphasizes the importance of staying updated on emerging concepts, techniques, and innovations, underscoring the role of ongoing research efforts. It also discusses the practical implications of these advancements, offering insights for practitioners, researchers, and students aiming to expand their understanding of inventory management. With its broad coverage, the chapter serves as a valuable resource for those looking to stay informed on the latest developments and their impact on organizational operations.

14.1 Introduction

Inventory research and development (R&D) involves systematic investigation into the methods, tools, and techniques used to manage and optimize inventory systems. This field encompasses the exploration of new theories, models, and technologies aimed at enhancing inventory management practices. The primary goal of inventory R&D is to identify innovative solutions that can improve efficiency, reduce costs, and adapt to evolving market demands.

The landscape of inventory management is continually evolving, driven by technological advancements, changes in consumer behavior, and shifts in supply chain dynamics. Researchers in this field explore various dimensions of inventory management, including inventory control systems, forecasting methods, and supply chain integration. They employ a range of methodologies, from theoretical modeling to empirical analysis, to address contemporary challenges and opportunities in inventory management.

DOI: 10.1201/9781003534860-14

14.1.1 Importance of Continuous Research in Inventory Management

Continuous research in inventory management is crucial for several reasons:

- **Adapting to Technological Advances:** The rapid pace of technological innovation necessitates ongoing research to incorporate new tools and techniques into inventory management practices. Technologies such as artificial intelligence, machine learning, and big data analytics are transforming how inventory is managed, and staying updated with these developments is essential for maintaining competitive advantage.

- **Responding to Market Changes:** Consumer preferences, market conditions, and supply chain dynamics are constantly changing. Continuous research helps organizations stay responsive to these changes by providing insights into new inventory management strategies that can address emerging trends and challenges.

- **Improving Efficiency and Reducing Costs:** Research helps identify ways to optimize inventory levels, streamline processes, and reduce holding costs. Exploring new methods and technologies, organizations can enhance their inventory management practices and achieve significant cost savings.

- **Enhancing Decision-Making:** Ongoing research provides valuable data and insights that support better decision-making. It enables organizations to make informed choices about inventory strategies, supplier relationships, and resource allocation.

- **Fostering Innovation:** Continuous research encourages innovation in inventory management practices. Exploring novel approaches and testing new ideas, organizations can develop more effective and efficient inventory management systems.

14.1.2 Objectives of Research and Development in Inventory Management

The primary objectives of research and development in inventory management include:

- **Enhancing Inventory Optimization:** Developing new models and techniques to optimize inventory levels, reduce stockouts, and improve order fulfillment.

- **Advancing Forecasting Methods:** Improving accuracy and reliability of demand forecasting through innovative statistical methods, machine learning algorithms, and data integration.

- **Integrating Technology:** Exploring the application of emerging technologies, such as IoT, blockchain, and advanced analytics, to enhance inventory management processes.
- **Improving Supply Chain Coordination:** Investigating strategies to better align inventory management with supply chain operations, including supplier collaboration, logistics optimization, and inventory visibility.
- **Addressing Environmental and Social Impacts:** Researching sustainable practices and ethical considerations in inventory management to minimize environmental impact and promote social responsibility.
- **Developing Practical Solutions:** Translating research findings into practical solutions that can be implemented in real-world inventory management scenarios, including the design of user-friendly tools and systems.

14.1.3 Advancements in Inventory Optimization Techniques

Recent advancements in inventory optimization techniques are significantly enhancing how businesses manage their inventory. These advancements include:

- **Advanced Algorithms:** The development of sophisticated algorithms, such as genetic algorithms and particle swarm optimization, allows for more precise and efficient inventory optimization. These algorithms can handle complex constraints and multiple objectives, leading to better decision-making in inventory management.
- **Real-Time Optimization:** Innovations in real-time optimization technologies enable businesses to adjust inventory levels dynamically based on current market conditions and operational data. This capability improves responsiveness and reduces the risk of stockouts or excess inventory.
- **Automated Replenishment Systems:** Advances in automation technology have led to the creation of automated replenishment systems that use predictive analytics to determine optimal reorder points and quantities. This reduces manual intervention and enhances inventory control.
- **Multi-Echelon Inventory Optimization:** Researchers are focusing on multi-echelon inventory optimization to address challenges across various levels of the supply chain. This approach integrates inventory decisions from different stages, such as suppliers, warehouses, and retailers, to optimize overall inventory performance.
- **Integration With Supply Chain Networks:** New techniques are being developed to optimize inventory within broader supply chain networks, considering factors such as transportation costs, supplier lead times, and demand variability. This holistic approach helps to balance inventory levels across the entire supply chain.

14.1.4 Innovations in Inventory Forecasting and Planning

Innovations in forecasting and planning are transforming how businesses predict and plan their inventory needs. Key developments include:

- **Machine Learning and AI:** Machine learning algorithms and artificial intelligence (AI) are being used to enhance forecasting accuracy. These technologies analyze large datasets to identify patterns and trends, leading to more reliable demand forecasts.

- **Dynamic Forecasting Models:** Advanced dynamic forecasting models, including Bayesian networks and time-series analysis, adapt to changing conditions and improve the accuracy of inventory predictions. These models incorporate real-time data and adjust forecasts as new information becomes available.

- **Integration With External Data Sources:** Innovations in data integration allow businesses to incorporate external data sources, such as market trends, economic indicators, and social media sentiment, into their forecasting models. This comprehensive approach provides a more accurate picture of future demand.

- **Collaborative Forecasting:** Collaborative forecasting techniques, involving coordination between suppliers, retailers, and other stakeholders, are gaining traction. These techniques improve forecast accuracy by sharing information and aligning forecasts across the supply chain.

- **Scenario Planning:** Advanced scenario planning methods help businesses prepare for various demand scenarios by modeling different potential outcomes. This approach enhances flexibility and enables better decision-making in the face of uncertainty.

14.1.5 Integration of Big Data and Analytics

The integration of big data and analytics is revolutionizing inventory management by providing deeper insights and more effective decision-making tools. Key aspects include:

- **Big Data Analytics:** Big data analytics involves analyzing vast amounts of structured and unstructured data to gain insights into inventory trends, customer behavior, and supply chain performance. This information helps optimize inventory levels and improve overall efficiency.

- **Predictive Analytics:** Predictive analytics uses historical data and statistical models to forecast future inventory needs and identify potential issues before they arise. This proactive approach enhances inventory planning and reduces the risk of stockouts and overstocking.

- **Data Visualization Tools:** Advanced data visualization tools allow for the intuitive presentation of complex inventory data. These tools help stakeholders understand trends, patterns, and anomalies, facilitating better decision-making.
- **Real-Time Data Integration:** The ability to integrate and analyze real-time data from various sources, such as point-of-sale systems and supply chain sensors, provides up-to-date insights into inventory levels and demand fluctuations. This real-time capability enhances responsiveness and agility.
- **Advanced Analytics Platforms:** New analytics platforms offer powerful capabilities for analyzing inventory data, including machine learning algorithms, data mining, and optimization techniques. These platforms support advanced inventory management strategies and improve overall performance.

14.2 Research Methodologies in Inventory Management

Research methodologies in inventory management are crucial for developing effective strategies and improving overall efficiency. Employing various research methods, organizations can gain insights into inventory dynamics, optimize stock levels, reduce costs, and enhance service quality. This section explores the key research methodologies utilized in the field of inventory management, providing a comprehensive overview of both quantitative and qualitative approaches. These methodologies help in understanding the complexities of inventory systems, identifying patterns and trends, and formulating data-driven decisions that align with organizational goals. The integration of advanced research techniques enables businesses to stay competitive in an ever-evolving market landscape by continuously refining their inventory management practices.

14.2.1 Quantitative Research Methods

Quantitative research methods in inventory management involve the use of statistical, mathematical, and computational techniques to analyze numerical data and derive insights. These methods are crucial for developing models, testing hypotheses, and making data-driven decisions. Key quantitative research methods include:

- **Statistical Analysis:** This includes techniques such as regression analysis, time-series analysis, and hypothesis testing to understand relationships between inventory variables and to forecast future inventory needs. Statistical analysis helps in quantifying the impact of various factors on inventory performance.

- **Optimization Models:** Mathematical optimization techniques, such as linear programming and integer programming, are used to determine the most efficient inventory levels, reorder points, and order quantities. These models help in minimizing costs and improving inventory efficiency.
- **Simulation:** Simulation methods involve creating models that replicate real-world inventory systems to analyze their behavior under different scenarios. Techniques such as Monte Carlo simulation are used to assess the impact of variability and uncertainty on inventory performance.
- **Data Analytics:** Advanced data analytics techniques, including machine learning algorithms and predictive analytics, analyze large datasets to identify patterns, trends, and correlations. These techniques support inventory forecasting and decision-making by providing actionable insights from historical data.
- **Quantitative Surveys:** Surveys with structured questionnaires are used to collect numerical data from stakeholders, such as suppliers and customers. This data helps in understanding inventory management practices, challenges, and preferences across different industries.

14.2.2 Qualitative Research Approaches

Qualitative research approaches focus on understanding the underlying reasons, motivations, and experiences related to inventory management. These approaches provide in-depth insights into the qualitative aspects of inventory practices. Key qualitative research methods include:

- **Interviews:** Semi-structured or unstructured interviews with key stakeholders, such as inventory managers and suppliers, offer rich, detailed information about inventory practices, challenges, and decision-making processes. Interviews help in exploring complex issues and obtaining personal perspectives.
- **Focus Groups:** Focus group discussions with groups of individuals involved in inventory management provide diverse viewpoints and facilitate the exploration of common themes and issues. This method helps in understanding group dynamics and collective attitudes towards inventory practices.
- **Case Studies:** In-depth case studies of specific organizations or inventory systems provide a comprehensive understanding of how inventory management practices are implemented and how they impact organizational performance. Case studies highlight real-world examples and lessons learned.
- **Observations:** Direct observations of inventory processes and practices in organizations offer insights into how inventory management tasks are performed in practice. Observations help in identifying inefficiencies and understanding the context in which inventory decisions are made.

- **Document Analysis:** Analyzing documents such as inventory reports, policy manuals, and procedure guidelines provides insights into the formal aspects of inventory management. Document analysis helps in understanding how inventory practices are documented and communicated within organizations.

14.2.3 Case Studies and Empirical Research

Case studies and empirical research are valuable methods for exploring real-world applications of inventory management theories and practices. They provide practical insights and validate theoretical concepts. Key aspects include:

- **Case Studies:** Case studies involve detailed examinations of individual organizations or inventory systems to understand their inventory management practices, challenges, and outcomes. They offer practical examples of how theoretical concepts are applied in real-world settings.
- **Empirical Research:** Empirical research involves the collection and analysis of data from real-world inventory management practices. This research provides evidence-based insights into the effectiveness of different inventory management strategies and their impact on organizational performance.
- **Comparative Analysis:** Comparative analysis of different case studies or empirical research findings helps in identifying best practices and common challenges across various organizations. This method provides a broader understanding of inventory management trends and practices.
- **Action Research:** Action research involves a cyclical process of planning, acting, observing, and reflecting on inventory management practices. This participatory approach helps in solving practical problems and improving inventory practices through iterative experimentation and feedback.
- **Longitudinal Studies:** Longitudinal studies track changes in inventory management practices over time. These studies provide insights into the long-term effects of different strategies and practices on inventory performance and organizational outcomes.

14.2.4 Practical Implications for Inventory Control

Research in inventory management provides valuable insights that can significantly improve inventory control practices. The practical implications include:

- **Improved Inventory Accuracy:** Research findings on inventory tracking and data accuracy contribute to better inventory record-keeping. Techniques such as advanced barcoding and RFID technology, identified through research, help in reducing errors and discrepancies in inventory records.

- **Optimized Reorder Points and Order Quantities:** Studies on inventory optimization models enable organizations to determine optimal reorder points and order quantities. These insights help in minimizing stockouts and overstock situations, leading to more efficient inventory control.
- **Enhanced Inventory Visibility:** Research on inventory visibility technologies, such as real-time tracking and data integration, improves the ability to monitor inventory levels across different locations. This leads to better decision-making and responsiveness to changes in inventory status.
- **Reduction in Holding Costs:** Research on inventory holding costs and strategies for cost reduction helps organizations in managing inventory levels more effectively. Techniques such as Just-in-Time (JIT) inventory and Economic Order Quantity (EOQ) models help in minimizing carrying costs.
- **Increased Accuracy in Forecasting:** Advances in forecasting methods, derived from research, lead to more accurate predictions of future inventory needs. Improved forecasting helps in aligning inventory levels with actual demand, reducing the risk of excess or insufficient stock.

14.2.5 Impact on Supply Chain Efficiency

Research in inventory management has a profound impact on overall supply chain efficiency. Key impacts include:

- **Streamlined Supply Chain Operations:** Insights from inventory research help in streamlining supply chain operations by optimizing inventory flows, reducing lead times, and improving coordination between suppliers and retailers. This results in a more efficient and responsive supply chain.
- **Enhanced Collaboration:** Research on supply chain integration and collaborative planning leads to improved collaboration between different stakeholders. Enhanced communication and information sharing between suppliers, manufacturers, and distributors contribute to smoother supply chain processes.
- **Minimized Disruptions:** Research on risk management and contingency planning helps in identifying potential disruptions and developing strategies to mitigate them. This proactive approach reduces the impact of supply chain interruptions and enhances overall resilience.
- **Improved Demand-Supply Alignment:** Research on inventory forecasting and demand planning ensures better alignment between supply and demand. Accurate forecasts and inventory levels help in meeting customer demand more effectively and reducing the risk of stockouts or excess inventory.

- **Cost Efficiency:** Research findings on cost-saving measures, such as optimized transportation and warehousing strategies, lead to reduced operational costs. Efficient supply chain practices, driven by research, contribute to lower overall supply chain expenses.

14.2.6 Enhancements in Inventory Decision-Making

Research advancements in inventory management contribute to improved decision-making processes. Key enhancements include:

- **Data-Driven Decision Making:** Research on data analytics and decision support systems enables inventory managers to make informed decisions based on accurate and comprehensive data. Data-driven insights lead to better inventory planning and management.
- **Advanced Analytical Tools:** The development of advanced analytical tools and techniques, such as machine learning algorithms and predictive modeling, enhances the ability to analyze complex inventory data. These tools support better forecasting, optimization, and risk assessment.
- **Strategic Inventory Management:** Research on strategic inventory management practices provides insights into aligning inventory decisions with organizational goals and strategies. This strategic approach ensures that inventory management supports overall business objectives.
- **Enhanced Risk Management:** Research on inventory risk management helps in identifying potential risks and developing strategies to address them. Improved risk management practices lead to more resilient inventory systems and better decision-making in uncertain environments.
- **Innovation in Inventory Techniques:** Ongoing research leads to the development of innovative inventory management techniques and technologies. Adoption of these innovations improves decision-making capabilities and keeps inventory practices up-to-date with industry advancements.

14.2.7 Common Challenges in Inventory Research

Inventory research faces several common challenges that can impact the effectiveness and applicability of findings:

- **Data Quality and Availability:** One of the major challenges in inventory research is obtaining high-quality, accurate, and comprehensive data. Inconsistent or incomplete data can undermine the validity of research outcomes and limit the applicability of findings.
- **Complexity of Inventory Systems:** Inventory systems can be highly complex, involving numerous variables and interdependencies. Modeling and

analyzing such complex systems can be challenging, making it difficult to capture all relevant factors and dynamics.

- **Dynamic and Uncertain Environments:** The inventory management landscape is constantly evolving, with frequent changes in market conditions, consumer behavior, and technological advancements. Research must account for these dynamic and uncertain environments, which can complicate the development of robust and applicable models.
- **Integration of Emerging Technologies:** Incorporating new technologies such as artificial intelligence, machine learning, and big data analytics into inventory research can be challenging. Ensuring that research methodologies and tools are compatible with these technologies requires ongoing adaptation and expertise.
- **Scalability of Solutions:** Research findings and proposed solutions must be scalable and adaptable to different organizational sizes and types. Ensuring that recommendations are practical and applicable across various contexts can be challenging.

14.2.8 Opportunities for Future Research

There are several promising opportunities for advancing inventory research, which can contribute to more effective and innovative inventory management practices:

- **Advanced Analytics and Big Data:** Leveraging big data and advanced analytics presents opportunities to gain deeper insights into inventory patterns, demand forecasting, and supply chain optimization. Research in these areas can lead to more accurate predictions and improved decision-making.
- **Integration of Artificial Intelligence and Machine Learning:** Incorporating AI and machine learning techniques into inventory management research offers opportunities to develop more sophisticated models and algorithms for inventory optimization, risk management, and demand forecasting.
- **Sustainability and Environmental Impact:** Research focused on sustainability and the environmental impact of inventory practices presents opportunities to develop more eco-friendly inventory management strategies. This includes reducing waste, minimizing carbon footprint, and promoting ethical sourcing.
- **Customization and Personalization:** Exploring ways to tailor inventory management solutions to specific industries or organizational needs can lead to more effective and relevant practices. Customization and personalization can improve inventory accuracy and efficiency.

- **Cross-Disciplinary Approaches:** Integrating insights from related disciplines such as supply chain management, operations research, and behavioral science can enhance inventory research. Cross-disciplinary approaches can provide a more holistic understanding of inventory challenges and solutions.

14.2.9 Addressing Gaps in Current Research

Addressing existing gaps in inventory research is crucial for advancing the field and improving inventory management practices:

- **Empirical Validation of Theoretical Models:** Many theoretical models in inventory research require empirical validation to ensure their practical applicability. Conducting field studies and real-world experiments can help validate and refine theoretical models.
- **Inclusion of Emerging Trends:** Current research should incorporate emerging trends and technologies, such as blockchain, Internet of Things (IoT), and smart logistics. Addressing these trends can provide a more comprehensive view of modern inventory management challenges and solutions.
- **Enhanced Focus on Small and Medium Enterprises (SMEs):** Much of the existing research focuses on large organizations. More research is needed to address the unique inventory management challenges faced by small and medium enterprises (SMEs) and to develop solutions tailored to their needs.
- **Longitudinal Studies:** Longitudinal studies that track inventory management practices over time can provide valuable insights into the long-term effects of different strategies and technologies. Such studies can help in understanding the sustainability and effectiveness of various approaches.
- **Integration of Human Factors:** Research should also consider human factors, such as decision-making processes, organizational culture, and training, which can significantly impact inventory management. Addressing these factors can lead to more effective and practical inventory management solutions.

14.2.10 Technological Innovations in Inventory Management

Technological innovations have significantly transformed inventory management, leading to more efficient and effective practices. Key technological advancements include:

- **Automated Inventory Systems:** Automation technologies, such as barcode scanning and RFID (Radio Frequency Identification), have revolutionized inventory tracking and management. These systems enable real-time monitoring of inventory levels, reduce manual errors, and improve accuracy in stock management.

- **Advanced Forecasting Tools:** New forecasting tools leverage advanced algorithms and big data analytics to predict demand more accurately. These tools use historical data, market trends, and external factors to generate precise forecasts, helping organizations manage inventory levels more effectively.
- **Cloud-Based Inventory Solutions:** Cloud computing has facilitated the development of scalable and flexible inventory management systems. Cloud-based solutions allow for real-time data access, collaboration across multiple locations, and integration with other business systems, enhancing overall inventory control.
- **Internet of Things (IoT):** IoT devices and sensors provide continuous data collection and monitoring of inventory conditions. This technology enables the tracking of environmental factors such as temperature and humidity, which is crucial for managing perishable goods and maintaining optimal storage conditions.
- **Blockchain Technology:** Blockchain technology offers enhanced transparency and traceability in the supply chain. Providing a secure and immutable ledger of transactions, blockchain can improve the accuracy of inventory records and reduce fraud and discrepancies in inventory management.

14.2.11 The Impact of Artificial Intelligence and Machine Learning

Artificial Intelligence (AI) and Machine Learning (ML) are increasingly being integrated into inventory management to enhance decision-making and optimize operations. Their impact includes:

- **Demand Forecasting:** AI and ML algorithms analyze vast amounts of historical data and identify patterns that may not be apparent through traditional methods. This leads to more accurate demand forecasting and better alignment of inventory levels with actual market needs.
- **Inventory Optimization:** AI-driven tools can optimize inventory levels by predicting stockouts and overstock situations. These tools use real-time data to adjust inventory policies dynamically, ensuring optimal stock levels and reducing carrying costs.
- **Automated Replenishment:** Machine learning models can automate the replenishment process by predicting when and how much inventory should be reordered. This reduces manual intervention and minimizes the risk of human error in the ordering process.
- **Anomaly Detection:** AI systems can detect anomalies in inventory data, such as unusual fluctuations in stock levels or discrepancies between recorded

and actual inventory. Early detection of these issues allows for prompt corrective actions.

- **Enhanced Decision Support:** AI and ML provide advanced decision support by analyzing complex datasets and generating actionable insights. This enhances strategic decision-making, from optimizing supply chain strategies to improving warehouse operations.

14.2.12 Utilization of Simulation and Modeling Tools

Simulation and modeling tools are valuable in inventory research and development, offering insights into complex inventory systems and enabling the testing of various scenarios. Key aspects include:

- **Scenario Analysis:** Simulation tools allow for the creation and analysis of various inventory scenarios, such as changes in demand, supply disruptions, or new policies. This helps organizations understand potential impacts and make informed decisions based on simulated outcomes.
- **Optimization Modeling:** Modeling tools enable the development of optimization models to determine the best inventory policies and strategies. These models consider various constraints and objectives, such as minimizing costs or maximizing service levels, to find optimal solutions.
- **Risk Management:** Simulation techniques are used to assess risks and uncertainties in inventory management. Simulating different risk scenarios, organizations can develop strategies to mitigate potential issues and enhance resilience in their inventory systems.
- **Training and Education:** Simulation tools are also used for training and educational purposes, allowing practitioners to experience and manage inventory challenges in a controlled environment. This hands-on experience helps in understanding complex inventory dynamics and improving skills.
- **Process Improvement:** Modeling and simulation can identify inefficiencies and bottlenecks in inventory processes. Analyzing these models, organizations can implement process improvements and streamline operations for better performance.

14.2.13 Best Practices for Conducting Inventory Research

To ensure effective and impactful inventory research, practitioners should adhere to the following best practices:

- **Define Clear Objectives:** Clearly define the research objectives and scope to ensure that the study addresses relevant inventory management issues. This helps in focusing the research efforts and achieving actionable outcomes.

- **Utilize Comprehensive Data:** Use comprehensive and accurate data for research analysis. High-quality data is crucial for developing reliable models, forecasts, and recommendations.

- **Adopt Rigorous Methodologies:** Employ rigorous research methodologies, including quantitative and qualitative approaches, to ensure the validity and reliability of findings. Utilize statistical analysis, modeling techniques, and empirical research methods to support robust conclusions.

- **Engage Stakeholders:** Involve key stakeholders in the research process to gain valuable insights and ensure that the research addresses practical challenges. Collaboration with industry experts, practitioners, and other relevant parties enhances the relevance and applicability of the research.

- **Iterate and Validate:** Continuously iterate and validate research findings through testing and simulation. This iterative approach helps refine models and recommendations, ensuring that they are effective and applicable in real-world scenarios.

14.2.14 Emerging Areas of Focus

As inventory management continues to evolve, several emerging areas of focus are shaping the future of research in this field:

- **Sustainability and Environmental Impact:** Research is increasingly concentrating on integrating sustainability into inventory management practices. This includes optimizing inventory to minimize waste, reduce carbon footprints, and promote sustainable sourcing and production practices.

- **Advanced Analytics and AI Integration:** The application of advanced analytics, machine learning, and artificial intelligence in inventory management is a growing area of research. This includes developing predictive analytics models for demand forecasting, optimizing inventory levels, and enhancing decision-making processes.

- **Real-Time Inventory Management:** The focus is shifting towards real-time inventory management solutions that leverage Internet of Things (IoT) technologies and real-time data analytics. This approach aims to improve inventory visibility, accuracy, and responsiveness to market changes.

- **Blockchain Technology:** Blockchain is being explored for its potential to enhance transparency and traceability in inventory management. Research in this area focuses on how blockchain can improve supply chain integrity, reduce fraud, and streamline transactions.

- **Human-Centric Inventory Systems:** Emerging research is addressing the human factors in inventory management, including user experience, decision support systems, and the impact of automation on workforce dynamics.

14.2.15 Potential for Interdisciplinary Research

The future of inventory research is poised to benefit from interdisciplinary approaches, integrating insights and methodologies from various fields:

- **Operations Research and Management Science:** Combining inventory management research with operations research techniques can lead to more sophisticated models and solutions for optimizing inventory and supply chain processes.
- **Behavioral Economics:** Insights from behavioral economics can help understand and influence decision-making in inventory management, including how cognitive biases affect inventory decisions and how to design interventions to improve decision quality.
- **Computer Science and Data Engineering:** Collaboration with computer science and data engineering can enhance the development of advanced algorithms, big data analytics, and real-time inventory management systems.
- **Environmental Science:** Integrating environmental science with inventory management research can address sustainability challenges, including reducing environmental impact and implementing eco-friendly practices.
- **Human-Computer Interaction (HCI):** Research in HCI can improve user interfaces and decision support systems in inventory management, making them more intuitive and effective for users.

14.2.16 Developing Collaborative Research Opportunities

To advance inventory research, fostering collaborative opportunities is essential:

- **Industry-Academia Partnerships:** Strengthening partnerships between academic researchers and industry practitioners can bridge the gap between theoretical research and practical applications. Collaborative projects can lead to innovative solutions and real-world impact.
- **Cross-Disciplinary Research Groups:** Establishing research groups that bring together experts from different disciplines can generate new insights and approaches to inventory management challenges. These groups can work on complex problems that require diverse expertise.
- **International Research Networks:** Engaging in international research networks can provide access to a broader range of perspectives and expertise. Collaborative international projects can address global inventory management issues and facilitate knowledge exchange.

- **Public-Private Partnerships:** Collaborations between public sector organizations, private companies, and research institutions can drive innovation in inventory management. These partnerships can support research initiatives, pilot projects, and the implementation of new technologies.
- **Funding and Support Initiatives:** Securing funding and support for collaborative research projects can enhance the capacity for innovation and development. Funding agencies, grants, and industry sponsors can play a crucial role in supporting research endeavors and translating findings into practical solutions.

14.2.17 Implications for Practitioners and Researchers

The ongoing developments and future directions in inventory research have significant implications for both practitioners and researchers:

- **Enhanced Decision-Making:** For practitioners, the insights gained from emerging trends and interdisciplinary research can lead to more informed decision-making. The application of advanced analytics, AI, and real-time data can optimize inventory management practices, improve forecasting accuracy, and enhance supply chain efficiency.
- **Innovation and Competitive Advantage:** Researchers and practitioners who stay abreast of the latest innovations and technologies can gain a competitive advantage. Implementing cutting-edge solutions, such as blockchain for transparency and IoT for real-time monitoring, can set organizations apart and drive innovation in inventory management.
- **Collaborative Opportunities:** The emphasis on collaborative research presents opportunities for practitioners to engage with academic researchers, contributing practical insights that can inform theoretical developments. Researchers, in turn, benefit from real-world applications and validation of their theories and models.
- **Sustainability and Responsibility:** Both practitioners and researchers must consider the growing importance of sustainability in inventory management. Embracing eco-friendly practices and integrating environmental considerations into inventory strategies are crucial for meeting regulatory requirements and addressing societal expectations.
- **Continued Learning and Adaptation:** Practitioners should prioritize ongoing education and adaptation to new methodologies and technologies. Researchers should focus on addressing gaps in current knowledge and exploring new areas of interest to keep inventory management practices aligned with evolving industry needs.

14.3 Conclusion

This chapter on emphasizes the critical role of continuous innovation in inventory management. As the field evolves, staying current with new trends, technologies, and methodologies is essential for maintaining a competitive edge and improving operational efficiency. Key advancements include optimization techniques, forecasting methods, and data analytics.Interdisciplinary collaboration—bringing together expertise from operations research, computer science, environmental science, and other fields—greatly enhances inventory management practices. Effective research translates into practical solutions, benefiting both practitioners and researchers. This ongoing effort will be crucial in advancing and sustaining effective, sustainable inventory management practices. Remaining engaged with current advancements and leveraging technological innovations, stakeholders can drive the evolution of inventory management.

Exercises

Exercise 14.1. *Explain how big data analytics can be used to optimize inventory levels and improve overall efficiency in inventory management. Provide examples of the types of data analyzed and the insights gained.*

Exercise 14.2. *Discuss the role of predictive analytics in forecasting future inventory needs. How can historical data and statistical models be utilized to reduce the risk of stockouts and overstocking? Provide examples of predictive analytics applications in inventory management.*

Exercise 14.3. *Describe the importance of data visualization tools in inventory management. How do these tools help stakeholders understand complex inventory data and facilitate better decision-making? Provide examples of effective data visualization techniques.*

Exercise 14.4. *Explain the benefits of integrating and analyzing real-time data from various sources in inventory management. How does real-time data integration enhance responsiveness and agility in managing inventory levels and demand fluctuations?*

Exercise 14.5. *Discuss the capabilities of advanced analytics platforms in inventory management. How do machine learning algorithms, data mining, and optimization techniques contribute to advanced inventory management strategies? Provide examples of how these platforms have improved inventory performance.*

15

Inventory Career Opportunities

This chapter offers a comprehensive overview of the diverse and evolving career opportunities in inventory management. It examines the range of positions available, from entry-level roles to advanced careers, and highlights the skills and qualifications necessary for success in this dynamic field. The chapter emphasizes the strategic importance of inventory professionals in optimizing supply chain operations, noting the growing demand for expertise in inventory planning and control. Additionally, it discusses how technological advancements and global market trends are reshaping job roles and skill requirements. Aimed at aspiring professionals, current practitioners, and career advisors, the chapter provides valuable insights into various pathways and growth trajectories in inventory management. With a focus on practical considerations and industry trends, it serves as a useful resource for individuals looking to navigate and excel in their careers within this field.

15.1 Introduction

The field of inventory management offers a diverse range of career opportunities, spanning from entry-level positions to advanced, specialized roles. Inventory management is critical to the effective operation of supply chains, and professionals in this field are responsible for ensuring that goods are stored, managed, and distributed efficiently. Career opportunities include roles such as Inventory Clerk, Inventory Analyst, Inventory Manager, Supply Chain Coordinator, and more specialized positions like Inventory Control Specialist and Demand Planner. These roles vary significantly in terms of responsibilities, required skills, and career progression. Entry-level positions often involve data entry and basic inventory tracking, while mid-level roles might focus on optimizing inventory levels and managing stock turnover. Senior and specialized positions may involve strategic decision-making, advanced data analysis, and leadership responsibilities.

DOI: 10.1201/9781003534860-15

15.1.1 Importance of Inventory Professionals in Supply Chain Operations

Inventory professionals play a crucial role in supply chain operations. Their primary responsibilities include managing inventory levels, forecasting demand, ensuring accurate stock records, and coordinating with suppliers and distribution centers. Effective inventory management directly impacts a company's ability to meet customer demands, control costs, and maintain smooth operations. Inventory professionals ensure that the right products are available at the right time and place, which helps in minimizing stockouts and overstock situations. They also contribute to reducing carrying costs and improving cash flow by optimizing inventory levels. In essence, inventory professionals help in balancing supply and demand, thereby enhancing the overall efficiency of the supply chain.

15.1.2 Objectives of Exploring Inventory Careers

Exploring career opportunities in inventory management serves several objectives:

- **Understanding Career Pathways:** Identifying the various roles available within inventory management and understanding the skills and qualifications required for each position.
- **Recognizing the Value of Inventory Professionals:** Appreciating the strategic importance of inventory professionals in maintaining efficient supply chains and supporting organizational goals.
- **Guiding Career Development:** Providing insights and guidance for individuals pursuing careers in inventory management, including advice on skills development, education, and career advancement.
- **Addressing Industry Trends:** Understanding how emerging trends and technological advancements impact career opportunities and skill requirements within the field.

15.1.3 Entry-Level Positions

Entry-level positions in inventory management are foundational roles that often involve direct interaction with inventory processes and systems. These positions typically require a basic understanding of inventory principles and operational tasks. Common entry-level roles include:

- **Inventory Clerk:** Responsible for maintaining accurate inventory records, performing stock counts, and assisting with stock replenishment. Inventory Clerks ensure that inventory data is recorded correctly and manage day-to-day inventory operations.

- **Inventory Assistant:** Supports inventory management activities by assisting with data entry, tracking shipments, and handling stock adjustments. Inventory Assistants may also help in managing supplier relations and preparing inventory reports.
- **Stockroom Associate:** Handles the physical aspects of inventory management, including receiving, organizing, and storing inventory items. Stockroom Associates ensure that products are stored properly and are easily accessible for order fulfillment.

These positions provide a solid foundation for understanding inventory management processes and offer opportunities for growth into more advanced roles.

15.1.4 Mid-Level Roles

Mid-level roles involve more responsibility and strategic involvement in inventory management. Individuals in these positions are typically responsible for optimizing inventory levels, analyzing data, and implementing improvements. Key mid-level roles include:

- **Inventory Analyst:** Analyzes inventory data to identify trends, forecast demand, and optimize inventory levels. Inventory Analysts use data-driven insights to make recommendations for inventory improvements and support decision-making.
- **Inventory Control Specialist:** Focuses on maintaining accurate inventory records and resolving discrepancies. Inventory Control Specialists implement procedures to ensure inventory accuracy and manage inventory audits.
- **Supply Chain Coordinator:** Oversees the flow of goods from suppliers to customers, ensuring that inventory levels are aligned with demand. Supply Chain Coordinators work to optimize supply chain processes and manage supplier relationships.

Mid-level roles require a deeper understanding of inventory management principles and often involve cross-functional collaboration with other departments.

15.1.5 Senior and Specialized Positions

Senior and specialized positions are high-level roles that involve strategic oversight, leadership, and advanced expertise in inventory management. These roles typically require extensive experience and specialized knowledge. Prominent senior and specialized positions include:

- **Inventory Manager:** Oversees all aspects of inventory management, including strategy development, inventory control, and team management. Inventory Managers are responsible for ensuring optimal inventory levels and implementing best practices.
- **Demand Planner:** Focuses on forecasting future inventory needs based on market trends, historical data, and sales projections. Demand Planners work closely with other departments to align inventory levels with business goals.
- **Supply Chain Director:** Manages the entire supply chain function, including inventory management, procurement, and logistics. Supply Chain Directors develop strategic plans to optimize supply chain performance and achieve organizational objectives.

These positions require strong leadership skills, strategic thinking, and a comprehensive understanding of inventory management and supply chain operations.

15.1.6 Career Pathways and Advancement Opportunities

Career pathways in inventory management typically follow a progression from entry-level positions to more advanced roles. Advancement opportunities depend on factors such as experience, education, and performance. Common pathways include:

- **Specialization:** Professionals may choose to specialize in areas such as inventory control, demand planning, or supply chain management. Specialization can lead to expert roles and opportunities for advancement within specific domains.
- **Certification and Education:** Pursuing relevant certifications (e.g., APICS, CPIM, CSCP) and advanced degrees (e.g., MBA) can enhance career prospects and open doors to higher-level positions.
- **Leadership Roles:** Experienced professionals may transition into leadership roles, such as Inventory Manager or Supply Chain Director, where they oversee larger teams and strategic initiatives.

Career advancement in inventory management involves continuous learning, skill development, and the ability to adapt to evolving industry trends and technologies.

15.1.7 Skills and Qualifications for Inventory Careers

Success in inventory management relies on a diverse set of core skills that are essential for effectively handling inventory processes and optimizing operations. Key skills include:

- **Analytical Skills:** The ability to analyze inventory data, identify trends, and make data-driven decisions is crucial. Analytical skills help in forecasting demand, managing stock levels, and optimizing inventory performance.
- **Attention to Detail:** Precision in managing inventory records, conducting stock counts, and processing transactions is essential. Attention to detail ensures accuracy in inventory tracking and reduces the risk of errors.
- **Problem-Solving Skills:** Inventory management often involves addressing issues such as stock discrepancies, supply chain disruptions, and process inefficiencies. Strong problem-solving skills are necessary for resolving these challenges effectively.
- **Organizational Skills:** Efficiently organizing inventory, managing multiple tasks, and maintaining systematic records are vital for smooth inventory operations. Organizational skills ensure that inventory processes are well-structured and manageable.
- **Technical Proficiency:** Familiarity with inventory management software, enterprise resource planning (ERP) systems, and other relevant technologies is important. Technical proficiency enables effective use of tools for tracking, analyzing, and managing inventory.

15.1.8 Educational and Professional Qualifications

Educational and professional qualifications play a significant role in preparing individuals for careers in inventory management. Relevant qualifications include:

- **Bachelor's Degree:** A bachelor's degree in fields such as supply chain management, logistics, business administration, or operations management is often required for entry-level and mid-level positions in inventory management.
- **Master's Degree:** Advanced degrees, such as a master's in supply chain management or an MBA with a focus on operations, can enhance career prospects and provide a competitive edge for senior roles.
- **Relevant Coursework:** Courses in inventory control, procurement, logistics, data analysis, and operations research provide a solid foundation for understanding inventory management principles and practices.

15.1.9 Certifications and Training Programs

Certifications and training programs can significantly boost career prospects and demonstrate expertise in inventory management. Key certifications and programs include:

- **APICS Certified in Production and Inventory Management (CPIM):** This certification focuses on production and inventory management and is highly regarded in the industry.
- **APICS Certified Supply Chain Professional (CSCP):** This certification covers a broad range of supply chain topics, including inventory management, and is valuable for those seeking advanced roles.
- **Certified Professional in Supply Management (CPSM):** Offered by the Institute for Supply Management (ISM), this certification emphasizes supply chain management, including inventory practices.
- **Training Programs:** Various training programs, workshops, and seminars are available to enhance specific skills, such as inventory optimization, data analysis, and software proficiency.

15.1.10 Soft Skills and Competencies

In addition to technical skills and qualifications, soft skills and competencies are critical for success in inventory management careers. Key soft skills include:

- **Communication Skills:** Effective communication with team members, suppliers, and other stakeholders is essential for coordinating inventory processes and resolving issues.
- **Leadership Skills:** For managerial and senior roles, leadership skills are important for guiding teams, making strategic decisions, and driving improvements in inventory management.
- **Adaptability:** The ability to adapt to changing market conditions, technological advancements, and evolving business needs is crucial for staying relevant in the dynamic field of inventory management.
- **Interpersonal Skills:** Building strong relationships with colleagues, vendors, and customers requires strong interpersonal skills, which contribute to effective collaboration and negotiation.

Developing a balanced combination of technical, educational, and soft skills is essential for thriving in inventory management careers and achieving long-term success in the field.

15.1.11 Impact of Technological Advancements on Inventory Careers

Technological advancements have significantly transformed the landscape of inventory management, leading to new opportunities and changes in job roles. Key emerging technologies influencing inventory careers include:

- **Automation and Robotics:** The integration of automation and robotics in inventory management has streamlined warehouse operations, reducing manual labor and increasing efficiency. Roles related to managing and maintaining automated systems, as well as analyzing their performance, have become increasingly important.
- **Artificial Intelligence (AI) and Machine Learning:** AI and machine learning algorithms are being used for predictive analytics, demand forecasting, and inventory optimization. Professionals are now required to interpret AI-driven insights and make data-informed decisions to enhance inventory strategies.
- **Internet of Things (IoT):** IoT technologies enable real-time tracking and monitoring of inventory through connected sensors and devices. This has created roles focused on managing IoT infrastructure, analyzing sensor data, and ensuring seamless integration with inventory systems.
- **Blockchain Technology:** Blockchain provides a transparent and secure method for tracking inventory and supply chain transactions. Professionals must understand blockchain's application in inventory management to enhance traceability and reduce fraud.
- **Cloud Computing:** Cloud-based inventory management systems offer scalability and flexibility, enabling remote access and data sharing. Roles involving the implementation, management, and optimization of cloud solutions have become more prevalent.

15.1.12 Skills Required for Technology-Driven Roles

As technology continues to shape inventory management, professionals need to acquire specific skills to thrive in technology-driven roles:

- **Technical Proficiency:** Familiarity with emerging technologies such as automation systems, AI tools, IoT devices, and cloud platforms is essential. Professionals must be adept at using and managing these technologies to optimize inventory operations.
- **Data Analysis and Interpretation:** The ability to analyze large volumes of data generated by advanced technologies is crucial. Skills in data analysis, statistical methods, and interpreting predictive models are required to make informed decisions based on technological insights.
- **Cybersecurity Awareness:** With the increased use of digital technologies, understanding cybersecurity principles and practices is important to protect inventory data and systems from cyber threats.
- **Adaptability and Continuous Learning:** As technology evolves rapidly, professionals must be adaptable and committed to continuous learning.

Staying updated with the latest technological trends and acquiring new skills are vital for remaining relevant in the field.

- **Project Management:** Managing technology-driven projects, such as implementing new systems or upgrading existing ones, requires strong project management skills. This includes planning, executing, and overseeing technology initiatives effectively.

15.1.13 Adapting to Technological Changes in Inventory Management

Adapting to technological changes involves several strategies to ensure effective integration and utilization of new technologies:

- **Embracing Continuous Education:** Engaging in ongoing education and training programs focused on emerging technologies helps professionals stay current and improve their technical skills. This includes attending workshops, webinars, and pursuing relevant certifications.
- **Collaborating with Technology Providers:** Working closely with technology providers and vendors ensures that inventory professionals are well-informed about the latest developments and best practices. This collaboration can also facilitate smoother implementation and support.
- **Fostering a Culture of Innovation:** Encouraging a culture that embraces innovation and technological advancements within the organization helps in adapting to changes. Promoting open communication about new technologies and their benefits can drive acceptance and integration.
- **Implementing Change Management Practices:** Effective change management practices help in transitioning to new technologies smoothly. This includes planning for change, communicating with stakeholders, and providing support during the implementation phase.
- **Evaluating and Adapting Technology Solutions:** Regularly evaluating the performance and impact of technology solutions helps in identifying areas for improvement and adaptation. Continuous assessment ensures that the technology remains aligned with organizational goals and needs.

15.1.14 Global Market Trends and Their Effect on Inventory Careers

Globalization has had a profound impact on inventory management careers, reshaping job roles and requirements in several ways:

- **Expanded Supply Chains:** Globalization has led to the expansion of supply chains across international borders. Inventory professionals are increasingly required to manage complex, multinational supply chains, necessitating skills in international logistics, cross-border regulations, and global sourcing.

- **Increased Demand for Coordination:** The need for seamless coordination between global suppliers, manufacturers, and distributors has increased. Inventory roles now often involve managing relationships with international partners and ensuring the alignment of inventory strategies across diverse regions.
- **Enhanced Focus on Risk Management:** Global supply chains are exposed to a range of risks including geopolitical tensions, trade disputes, and currency fluctuations. Professionals must develop expertise in risk assessment and mitigation strategies to handle these challenges effectively.
- **Need for Multicultural Competence:** As businesses operate in diverse cultural environments, inventory professionals must be adept at navigating cultural differences and understanding regional market dynamics to optimize inventory strategies and operations.
- **Technological Integration:** Global operations often require the integration of advanced technologies to manage inventory across different time zones and regions. This includes the adoption of global inventory management systems and advanced analytics to track inventory in real time.

15.1.15 Regional Variations in Career Opportunities

Career opportunities in inventory management can vary significantly based on regional factors:

- **Economic Development:** Regions with rapidly growing economies, such as emerging markets, often present abundant opportunities in inventory management due to increased industrial activity and demand for efficient supply chain solutions.
- **Industry Concentration:** Different regions may have varying concentrations of industries that impact inventory careers. For instance, technology hubs may offer more opportunities in high-tech inventory solutions, while manufacturing centers may focus on traditional inventory management roles.
- **Regulatory Environment:** Regional differences in regulatory environments can influence career opportunities. Professionals must be aware of local regulations related to inventory control, warehousing, and transportation, which may vary across countries and regions.
- **Educational and Training Facilities:** The availability of educational programs and training facilities in different regions can affect the development of inventory management careers. Regions with robust educational infrastructure may offer more specialized training and certification programs.

- **Job Market Dynamics:** Regional job market dynamics, including employment rates, salary levels, and job demand, can impact career opportunities. Professionals should consider these factors when exploring career options in different regions.

15.1.16 Cultural and Market Trends Shaping Inventory Careers

Cultural and market trends significantly influence inventory careers, driving changes in job roles and skill requirements:

- **Consumer Behavior Trends:** Shifts in consumer behavior, such as the rise of e-commerce and demand for fast delivery, impact inventory management practices. Professionals must adapt to new trends in order fulfillment, inventory replenishment, and customer service.
- **Sustainability and Ethical Practices:** Increasing awareness of environmental sustainability and ethical practices affects inventory management. Professionals are expected to implement sustainable practices, manage eco-friendly inventory solutions, and adhere to ethical sourcing guidelines.
- **Digital Transformation:** The growing emphasis on digital transformation in business operations influences inventory careers. Professionals must be proficient in digital tools, data analytics, and cloud-based inventory systems to stay competitive.
- **Global Trade and Economic Trends:** Fluctuations in global trade policies, economic conditions, and currency exchange rates impact inventory management practices. Professionals need to stay informed about these trends and their implications for inventory operations.
- **Innovation and Technology Adoption:** Rapid advancements in technology and innovation drive changes in inventory management practices. Professionals must stay abreast of new technologies and trends to effectively integrate them into inventory strategies.

Global market trends have a profound effect on inventory careers, shaping job roles, opportunities, and required skills. Inventory professionals must navigate the complexities of globalization, regional variations, and cultural trends to excel in their careers and contribute to effective inventory management practices.

15.2 Career Development and Professional Growth

Advancing in an inventory management career involves several key strategies:

- **Setting Clear Career Goals:** Define short-term and long-term career objectives to guide your professional development. Identify specific roles or positions you aim to achieve and outline the skills and experiences needed to reach those goals.
- **Gaining Relevant Experience:** Seek opportunities to work on diverse projects and roles within inventory management. Experience in different areas, such as logistics, supply chain management, or inventory optimization, can enhance your skill set and increase your career prospects.
- **Pursuing Advanced Education:** Consider pursuing advanced degrees or certifications relevant to inventory management, such as a master's in supply chain management or certification from professional bodies like APICS or the Institute for Supply Management (ISM).
- **Developing Specialized Skills:** Focus on developing specialized skills that are in high demand, such as proficiency in advanced inventory management software, data analytics, or project management.
- **Seeking Mentorship:** Find mentors within the industry who can provide guidance, share their experiences, and offer valuable career advice. Mentorship can help you navigate your career path and make informed decisions.
- **Building a Strong Professional Reputation:** Demonstrate reliability, professionalism, and expertise in your current role. Your reputation as a competent and dependable professional can lead to new opportunities and career advancement.
- **Pursuing Leadership Roles:** Aim for leadership or management positions by developing skills in team management, strategic planning, and decision-making. Taking on leadership roles can open doors to higher-level career opportunities.

15.2.1 Networking and Professional Associations

Networking and involvement in professional associations play a crucial role in career development:

- **Building a Professional Network:** Attend industry conferences, seminars, and workshops to connect with other professionals in the field. Networking with peers, mentors, and industry leaders can provide insights, opportunities, and support for career growth.
- **Joining Professional Associations:** Become a member of relevant professional organizations such as APICS, ISM, or the Chartered Institute of Procurement & Supply (CIPS). These associations offer access to industry resources, professional development opportunities, and a community of like-minded professionals.

- **Participating in Industry Events:** Engage in industry events, webinars, and forums to stay updated on the latest trends and best practices. Participation in these events can enhance your knowledge and provide opportunities to showcase your expertise.
- **Contributing to Professional Communities:** Get involved in industry groups, committees, or volunteer opportunities within professional associations. Contributing to these communities can help build your reputation and expand your professional network.
- **Leveraging Social Media:** Use platforms like LinkedIn to connect with industry professionals, share your achievements, and engage in discussions related to inventory management. Social media can be a powerful tool for building and maintaining a professional network.

15.2.2 Continuing Education and Professional Development

Ongoing education and professional development are essential for career advancement:

- **Participating in Training Programs:** Enroll in training programs or workshops to enhance your skills and stay current with industry developments. Look for programs that focus on new technologies, methodologies, and best practices in inventory management.
- **Pursuing Certifications:** Obtain relevant certifications to validate your expertise and enhance your qualifications. Certifications from recognized organizations can improve your career prospects and demonstrate your commitment to professional growth.
- **Engaging in Lifelong Learning:** Stay informed about the latest trends, technologies, and research in inventory management. Regularly read industry publications, attend webinars, and engage in online courses to keep your knowledge up-to-date.
- **Setting Personal Development Goals:** Identify areas for improvement and set personal development goals to address these areas. Create a plan for acquiring new skills and knowledge to support your career objectives.
- **Seeking Feedback and Reflecting on Performance:** Regularly seek feedback from supervisors, peers, and mentors. Reflect on your performance and use the feedback to identify areas for improvement and development.

15.2.3 Navigating the Job Market

Navigating the job market effectively involves several key strategies:

- **Researching Potential Employers:** Investigate companies and organizations within the inventory management field. Understand their business models, company culture, and the specific roles they offer. This knowledge will help tailor your job search and applications to match the needs of potential employers.
- **Utilizing Job Search Platforms:** Leverage job search engines, company websites, and professional networks like LinkedIn to find job openings. Regularly check these platforms for updates and opportunities in the inventory management sector.
- **Networking with Industry Professionals:** Build and maintain a professional network through industry events, conferences, and online forums. Networking can provide valuable job leads, insights into companies, and potential referrals.
- **Exploring Internships and Entry-Level Positions:** For those new to the field, internships and entry-level positions can provide essential experience and a foot in the door. These opportunities can lead to full-time positions and offer hands-on learning in inventory management.
- **Tailoring Applications:** Customize your resume, cover letter, and application to each job role. Highlight relevant skills, experiences, and achievements that align with the job description and demonstrate your fit for the role.
- **Staying Updated on Industry Trends:** Keep abreast of the latest trends, technologies, and best practices in inventory management. Knowledge of current industry developments can make you a more attractive candidate and demonstrate your commitment to the field.

15.2.4 Preparing for Interviews and Assessments

Effective preparation for interviews and assessments is crucial for success:

- **Understanding the Role and Company:** Research the specific role and company thoroughly. Familiarize yourself with the company's inventory management practices, culture, and recent developments to demonstrate your interest and preparedness during the interview.
- **Practicing Common Interview Questions:** Prepare for common inventory management interview questions such as those related to problem-solving, inventory control strategies, and experiences with specific software or systems. Practice your responses to articulate your skills and experiences clearly.
- **Preparing for Technical Assessments:** If the interview includes technical assessments or case studies, review relevant concepts, tools, and methodologies. Practice solving inventory management problems and be ready to discuss your thought process and solutions.

- **Demonstrating Soft Skills:** Emphasize soft skills such as communication, teamwork, and adaptability. Inventory management roles often require collaboration with various stakeholders and the ability to navigate complex situations.

- **Preparing Questions for the Interviewer:** Prepare thoughtful questions to ask the interviewer about the role, team dynamics, and company goals. This demonstrates your interest and helps you assess if the position aligns with your career goals.

- **Following Up After Interviews:** Send a thank-you note or email to express your appreciation for the opportunity to interview. Reinforce your interest in the position and briefly reiterate why you are a strong fit for the role.

15.2.5 Building a Strong Professional Resume and Portfolio

Creating a compelling resume and portfolio is essential for showcasing your qualifications:

- **Crafting a Tailored Resume:** Design a resume that highlights relevant experiences, skills, and achievements related to inventory management. Use clear and concise language, and ensure that your resume is formatted professionally.

- **Including Relevant Experience:** Emphasize experiences that demonstrate your proficiency in inventory management, such as previous roles, internships, or projects. Highlight specific accomplishments and contributions that illustrate your impact.

- **Showcasing Technical Skills:** Include any relevant technical skills and certifications related to inventory management software, tools, and methodologies. Clearly list your proficiency in these areas to attract the attention of hiring managers.

- **Developing a Portfolio:** Create a portfolio that showcases your work, such as successful projects, case studies, or examples of problem-solving in inventory management. A portfolio can provide tangible evidence of your skills and accomplishments.

- **Obtaining Recommendations:** Seek recommendations or endorsements from previous employers, colleagues, or mentors. Positive references can validate your qualifications and enhance your credibility.

- **Updating Your Documents Regularly:** Keep your resume and portfolio up-to-date with your latest experiences, skills, and achievements. Regular updates ensure that you are prepared for new opportunities as they arise.

15.2.6 Emerging Trends and New Opportunities

As the field of inventory management continues to evolve, several emerging trends and new opportunities are shaping the future career landscape. This subsection explores these trends and highlights potential areas where new career opportunities may arise.

- **Increased Focus on Sustainability:** With growing environmental awareness, there is a rising demand for inventory professionals who can implement sustainable practices. Opportunities are emerging in areas such as green logistics, eco-friendly packaging, and waste reduction initiatives.
- **Advancements in Technology:** The integration of advanced technologies like artificial intelligence, machine learning, and blockchain is transforming inventory management. Professionals skilled in these technologies will find new roles related to data analytics, system automation, and blockchain integration.
- **Global Supply Chain Management:** As global trade continues to expand, there is an increasing need for experts in global supply chain management who can navigate complex international logistics, compliance issues, and cross-border inventory challenges.
- **Enhanced Data Analytics:** The growth of big data and analytics is creating opportunities for professionals who can leverage data to drive decision-making, forecast trends, and optimize inventory levels. Roles related to data science and predictive analytics are becoming more prevalent.

15.2.7 Potential Areas for Growth and Specialization

Inventory management careers offer several avenues for growth and specialization. This subsection identifies key areas where professionals can focus their efforts to advance their careers and develop expertise.

- **Supply Chain Analytics:** Specializing in supply chain analytics involves leveraging data to improve inventory management practices. Professionals in this field focus on optimizing inventory levels, reducing costs, and enhancing supply chain efficiency through data-driven insights.
- **Technology Integration:** Roles that involve integrating new technologies into inventory management systems are growing. Specializing in areas such as system implementation, technology upgrades, and digital transformation can lead to advanced career opportunities.
- **Risk Management and Resilience Planning:** Specializing in risk management and resilience planning involves developing strategies to mitigate supply chain disruptions and manage inventory risks. This area focuses on ensuring continuity and stability in inventory operations.

- **Sustainable Practices:** Professionals who focus on sustainable inventory practices, such as minimizing environmental impact and promoting ethical sourcing, are increasingly in demand. This specialization includes roles in sustainability consulting and corporate social responsibility.

15.2.8 Preparing for Future Career Challenges

To thrive in the evolving landscape of inventory management, professionals must prepare for future challenges. This subsection outlines strategies for staying ahead and addressing potential career obstacles.

- **Continuous Learning and Adaptation:** Staying updated with the latest trends, technologies, and best practices is essential. Pursuing ongoing education, attending industry conferences, and engaging in professional development activities can help professionals remain competitive.
- **Developing Technical Skills:** Gaining expertise in emerging technologies, such as data analytics, automation, and artificial intelligence, will be crucial. Acquiring technical skills and certifications related to these areas can enhance career prospects.
- **Building a Strong Professional Network:** Networking with industry peers, mentors, and thought leaders can provide valuable insights and opportunities. Participating in professional associations and industry groups can facilitate connections and career growth.
- **Embracing Change and Innovation:** Being open to change and innovation will be vital for adapting to new challenges. Professionals should embrace a proactive approach to change management and be willing to explore new ideas and methodologies.

15.2.9 Recommendations for Career Seekers and Advisors

For those pursuing careers in inventory management, as well as career advisors guiding them, the following recommendations can help navigate the dynamic landscape of the field:

- **Stay Informed about Industry Trends:** Career seekers should keep abreast of the latest trends and technological advancements in inventory management. Regularly reading industry publications, attending webinars, and participating in professional forums can provide valuable insights and keep skills relevant.
- **Pursue Relevant Certifications:** Obtaining certifications related to inventory management, such as Certified in Production and Inventory Management (CPIM) or Certified Supply Chain Professional (CSCP), can enhance credibility and open doors to advanced career opportunities.

- **Gain Practical Experience:** Hands-on experience through internships, part-time roles, or project work is essential for building practical skills and understanding real-world applications of inventory management concepts. Seek opportunities that offer exposure to various aspects of the field.
- **Develop Soft Skills:** In addition to technical knowledge, soft skills such as communication, problem-solving, and leadership are critical for career advancement. Engaging in activities that enhance these skills can provide a competitive edge.
- **Network with Professionals:** Building a strong professional network through industry events, online platforms, and professional associations can provide valuable connections, mentorship opportunities, and career guidance.
- **Seek Guidance from Career Advisors:** Career advisors should provide tailored advice based on individual career goals and the evolving job market. They should help career seekers identify their strengths, explore career pathways, and develop personalized action plans.
- **Embrace Lifelong Learning:** The field of inventory management is continuously evolving. Career seekers should adopt a mindset of lifelong learning and regularly update their skills and knowledge to stay competitive and relevant.

15.3 Conclusion

In conclusion, the field of inventory management offers a broad spectrum of career opportunities, each influenced by ongoing advancements and changing industry dynamics. As technology continues to reshape inventory management practices, professionals need to remain adaptable, continuously update their skills, and embrace emerging trends. Understanding the diverse career paths available, acquiring relevant skills and qualifications, and staying informed about industry developments, career seekers can navigate the evolving landscape of inventory management effectively. Career advisors play a crucial role in guiding individuals through this process, helping them identify suitable career paths and providing the necessary support for achieving their professional goals. Overall, the insights provided in this chapter aim to empower individuals to make informed career decisions, leverage emerging opportunities, and prepare for future challenges in the dynamic field of inventory management.

Exercises

Exercise 15.1. *Describe how analytical skills are utilized in inventory management. What specific tasks or decisions benefit from strong analytical abilities, and how do these skills contribute to optimizing inventory performance?*

Exercise 15.2. *Explain the importance of attention to detail in managing inventory records and conducting stock counts. How does precision in these activities help reduce the risk of errors and ensure accurate inventory tracking?*

Exercise 15.3. *Discuss the role of problem-solving skills in inventory management. Provide examples of common inventory-related challenges and how effective problem-solving can address these issues.*

Exercise 15.4. *What are the key organizational skills needed for efficient inventory operations? How do these skills help in managing multiple tasks and maintaining systematic records in inventory management?*

Exercise 15.5. *Why is technical proficiency important in inventory management? Describe how familiarity with inventory management software and ERP systems enhances the effectiveness of inventory tracking, analysis, and management.*

Solutions

Solution 1.1. *exe1*

To calculate the safety stock, we use the formula:

$$SS = Z \times \sigma_L$$

where SS is the safety stock, Z is the Z-score corresponding to the desired service level, and σ_L is the standard deviation of demand during lead time.

For a service level of 95%, the Z-score is 1.65.

Given: $\sigma_L = 60$
Thus: $SS = 1.65 \times 60 = 99$
The safety stock required is 99 units.

Solution 1.2. *exe2*

For a service level of 90%, the Z-score is 1.28.
Given: $\sigma_L = 30$
Thus: $SS = 1.28 \times 30 = 38.4$
The safety stock required is approximately 38 units.

Solution 1.3. *exe3*

For a service level of 99%, the Z-score is 2.33.
Given: $\sigma_L = 80$
Thus: $SS = 2.33 \times 80 = 186.4$
The safety stock required is approximately 186 units.

Solution 1.4. *exe4*

For a service level of 85%, the Z-score is 1.04.
Given: $\sigma_L = 50$
Thus: $SS = 1.04 \times 50 = 52$
The safety stock required is 52 units.

Solution 1.5. *exe5*

For a service level of 98%, the Z-score is 2.05.

Given: $\sigma_L = 25$
Thus: $SS = 2.05 \times 25 = 51.25$
The safety stock required is approximately 51 units.

Solution 1.6. *exe6*

To calculate the Economic Order Quantity (EOQ), we use the formula:

$$EOQ = \sqrt{\frac{2DS}{H}}$$

where D is the annual demand, S is the ordering cost per order, and H is the holding cost per unit per year.
Given: $D = 10{,}000, \quad S = 50, \quad H = 2$
Thus: $EOQ = \sqrt{\frac{2 \times 10{,}000 \times 50}{2}} = \sqrt{500{,}000} \approx 707 \ units$
The EOQ is approximately 707 units.

Solution 1.7. *exe7*

First, calculate the EOQ:
Given: $D = 15{,}000, \quad S = 75, \quad H = 1.50$

$$EOQ = \sqrt{\frac{2 \times 15{,}000 \times 75}{1.50}} = \sqrt{1{,}500{,}000} \approx 1{,}225 \ units$$

Next, calculate the total annual cost of inventory:

$$Total\ Annual\ Cost = Ordering\ Cost + Holding\ Cost + Purchase\ Cost$$

$$Ordering\ Cost = \frac{D}{EOQ} \times S = \frac{15{,}000}{1{,}225} \times 75 \approx 919$$

$$Holding\ Cost = \frac{EOQ}{2} \times H = \frac{1{,}225}{2} \times 1.50 = 918.75$$

$$Purchase\ Cost = D \times Cost\ per\ unit = 15{,}000 \times 10 = 150{,}000$$

$$Total\ Annual\ Cost = 919 + 918.75 + 150{,}000 = 151{,}837.75$$

The EOQ is approximately 1,225 units, and the total annual cost of inventory is \$151,837.75.

Solution 1.8. *exe8*

Given: $D = 25{,}000, \quad S = 100, \quad H = 3$
First, calculate the EOQ:

$$EOQ = \sqrt{\frac{2 \times 25{,}000 \times 100}{3}} = \sqrt{1{,}666{,}667} \approx 1{,}291 \ units$$

Check feasibility: The storage constraint allows for a maximum of 1,200 units at a time. Since the calculated EOQ (1,291 units) exceeds this limit, the EOQ is not feasible under the storage constraint. Therefore, the company must consider ordering the maximum feasible quantity of 1,200 units or adjust its storage capacity.

Solution 1.9. *exe9*

Given: $D = 8{,}000, \quad S = 60, \quad H = 2.50$
First, calculate the EOQ:

$$EOQ = \sqrt{\frac{2 \times 8{,}000 \times 60}{2.50}} = \sqrt{384{,}000} \approx 620 \ units$$

Next, calculate the reorder point:

$$Lead\ Time = 2\ weeks$$
$$Weekly\ Demand = \frac{D}{50} = \frac{8{,}000}{50} = 160\ units$$
$$Reorder\ Point = Lead\ Time \times Weekly\ Demand = 2 \times 160 = 320\ units$$

The EOQ is approximately 620 units, and the reorder point is 320 units.

Solution 1.10. *exe10*

Given: $D = 12{,}000, \quad S = 40$ *(or 30 for orders less than 1,000),* $H = 2$
First, calculate the EOQ with the original ordering cost:

$$EOQ_{40} = \sqrt{\frac{2 \times 12{,}000 \times 40}{2}} = \sqrt{480{,}000} \approx 692 \ units$$

Since 692 units are less than 1,000 units, recalculate the EOQ with the discounted ordering cost:

$$EOQ_{30} = \sqrt{\frac{2 \times 12{,}000 \times 30}{2}} = \sqrt{360{,}000} \approx 600 \ units$$

Check if ordering 1,000 units is more cost-effective:

Calculate total annual cost for ordering 1,000 units:

$$\textit{Ordering Cost} = \frac{D}{1{,}000} \times 30 = \frac{12{,}000}{1{,}000} \times 30 = 360$$

$$\textit{Holding Cost} = \frac{1{,}000}{2} \times 2 = 1{,}000$$

$$\textit{Total Cost}_{1{,}000} = 360 + 1{,}000 = 1{,}360$$

Compare with total cost using EOQ (either EOQ_{40} or EOQ_{30}):

$$\textit{Ordering Cost}_{EOQ_{40}} = \frac{D}{EOQ_{40}} \times S = \frac{12{,}000}{692} \times 40 \approx 692$$

$$\textit{Holding Cost}_{EOQ_{40}} = \frac{EOQ_{40}}{2} \times H = \frac{692}{2} \times 2 = 692$$

$$\textit{Total Cost}_{EOQ_{40}} = 692 + 692 = 1{,}384$$

Since ordering 1,000 units results in a lower total cost, the optimal order quantity is 1,000 units. The optimal order quantity is 1,000 units, and the total annual cost of inventory is $1,360.

Solution 1.11. *exe11*

To determine the optimal order quantity using the Newsvendor Model, we need to find the critical ratio (CR) and the corresponding z-value.
Given: $\mu = 500, \quad \sigma = 100, \quad c = 10, \quad p = 20, \quad s = 5$
First, calculate the critical ratio (CR):

$$CR = \frac{p - c}{p - s} = \frac{20 - 10}{20 - 5} = \frac{10}{15} = \frac{2}{3} \approx 0.67$$

Next, find the z-value corresponding to a cumulative probability of 0.67:

$$z \approx 0.44$$

Finally, calculate the optimal order quantity (Q^*):

$$Q^* = \mu + z\sigma = 500 + 0.44 \times 100 \approx 544$$

The optimal order quantity is approximately 544 units.

Solution 1.12. *exe12*

Given: $\mu = 200, \quad \sigma = 40, \quad c = 1, \quad p = 3, \quad s = 0.50$
First, calculate the critical ratio (CR):

$$CR = \frac{p - c}{p - s} = \frac{3 - 1}{3 - 0.50} = \frac{2}{2.5} = 0.8$$

Next, find the z-value corresponding to a cumulative probability of 0.8:

$$z \approx 0.84$$

Finally, calculate the optimal order quantity (Q^*):

$$Q^* = \mu + z\sigma = 200 + 0.84 \times 40 \approx 234$$

The optimal daily production quantity is approximately 234 loaves.

Solution 1.13. *exe13*

Given: $\mu = 1{,}000, \quad \sigma = 200, \quad c = 150, \quad p = 300, \quad s = 50$
First, calculate the critical ratio (CR):

$$CR = \frac{p - c}{p - s} = \frac{300 - 150}{300 - 50} = \frac{150}{250} = 0.6$$

Next, find the z-value corresponding to a cumulative probability of 0.6:

$$z \approx 0.25$$

Finally, calculate the optimal order quantity (Q^*):

$$Q^* = \mu + z\sigma = 1,000 + 0.25 \times 200 \approx 1,050$$

The optimal order quantity is approximately 1,050 units.

Solution 1.14. *exe14*

Given: $\mu = 600, \quad \sigma = 150, \quad c = 80, \quad p = 200, \quad s = 40$
First, calculate the critical ratio (CR):

$$CR = \frac{p - c}{p - s} = \frac{200 - 80}{200 - 40} = \frac{120}{160} = 0.75$$

Next, find the z-value corresponding to a cumulative probability of 0.75:

$$z \approx 0.67$$

Finally, calculate the optimal order quantity (Q^*):

$$Q^* = \mu + z\sigma = 600 + 0.67 \times 150 \approx 700$$

The optimal order quantity is approximately 700 units.

Solution 1.15. *exe15*

Given: $\mu = 300, \quad \sigma = 50, \quad c = 0.50, \quad p = 1, \quad s = 0.10$
First, calculate the critical ratio (CR):

$$CR = \frac{p-c}{p-s} = \frac{1-0.50}{1-0.10} = \frac{0.50}{0.90} \approx 0.56$$

Next, find the z-value corresponding to a cumulative probability of 0.56:

$$z \approx 0.15$$

Finally, calculate the optimal order quantity (Q^*):

$$Q^* = \mu + z\sigma = 300 + 0.15 \times 50 \approx 308$$

The optimal daily order quantity is approximately 308 newspapers.

Solution 1.16. *exe16*

To determine the optimal production lot size using the EPQ model, we use the following formula:

$$Q^* = \sqrt{\frac{2DS}{H}\left(\frac{P}{P-D}\right)}$$

Given: $D = 2{,}000$ *units/month,* $P = 10{,}000$ *units/month,* $S = \$500, \quad H = \$2$ *per unit per month*
First, calculate the optimal production lot size:

$$Q^* = \sqrt{\frac{2 \times 2{,}000 \times 500}{2}\left(\frac{10{,}000}{10{,}000 - 2{,}000}\right)} = \sqrt{\frac{2 \times 2{,}000 \times 500}{2}\left(\frac{10{,}000}{8{,}000}\right)}$$

$$Q^* = \sqrt{1{,}000{,}000 \times 1.25} = \sqrt{1{,}250{,}000} \approx 1{,}118$$

The optimal production lot size is approximately 1,118 units.

Solution 1.17. *exe17*

Given: $D = 12{,}000$ *units/year,* $P = 24{,}000$ *units/year,* $S = \$800, \quad H = \$1.50$ *per unit per year*
First, calculate the optimal production lot size:

$$Q^* = \sqrt{\frac{2 \times 12{,}000 \times 800}{1.50}\left(\frac{24{,}000}{24{,}000 - 12{,}000}\right)} = \sqrt{\frac{19{,}200{,}000}{1.50}\left(\frac{24{,}000}{12{,}000}\right)}$$

$$Q^* = \sqrt{12{,}800{,}000 \times 2} = \sqrt{25600{,}000} \approx 5{,}059$$

The optimal production quantity is approximately 5,059 units.

Solution 1.18. *exe18*

Given: $D = 2{,}000\,units/quarter$, $P = 8{,}000\,units/quarter$, $S = \$600$, $H = \$3\,per\ unit\ per\ quarter$
First, calculate the optimal production lot size:

$$Q^* = \sqrt{\frac{2 \times 2{,}000 \times 600}{3}\left(\frac{8{,}000}{8{,}000 - 2{,}000}\right)} = \sqrt{\frac{2400{,}000}{3}\left(\frac{8{,}000}{6{,}000}\right)}$$

$$Q^* = \sqrt{800{,}000 \times 1.333} = \sqrt{1{,}066{,}667} \approx 1{,}033$$

The optimal production lot size is approximately 1,033 units.

Solution 1.19. *exe19*

Given: $D = 5{,}000\,units/year$, $P = 15{,}000\,units/year$, $S = \$1{,}000$, $H = \$2.50\,per\ unit\ per\ year$
First, calculate the optimal production lot size:

$$Q^* = \sqrt{\frac{2 \times 5{,}000 \times 1{,}000}{2.50}\left(\frac{15{,}000}{15{,}000 - 5{,}000}\right)} = \sqrt{\frac{10000{,}000}{2.50}\left(\frac{15{,}000}{10{,}000}\right)}$$

$$Q^* = \sqrt{4{,}000{,}000 \times 1.5} = \sqrt{6{,}000{,}000} \approx 2{,}449$$

The optimal production quantity is approximately 2,449 units.

Solution 1.20. *exe20*

Given: $D = 5{,}000\,units/month$, $P = 20{,}000\,units/month$, $S = \$700$, $H = \$1\,per\ unit\ per\ month$
First, calculate the optimal production lot size:

$$Q^* = \sqrt{\frac{2 \times 5{,}000 \times 700}{1}\left(\frac{20{,}000}{20{,}000 - 5{,}000}\right)} = \sqrt{7{,}000{,}000\left(\frac{20{,}000}{15{,}000}\right)}$$

$$Q^* = \sqrt{7{,}000{,}000 \times 1.333} = \sqrt{9{,}333{,}333} \approx 3{,}055$$

The optimal production lot size is approximately 3,055 units.

Solution 1.21. *exe21*

To perform an ABC analysis, follow these steps:

1. *Calculate the annual inventory value for each item:*

$$Value = Annual\ Demand \times Unit\ Price$$

Item	*Annual Inventory Value (\$)*
A	$5{,}000 \times 10 = 50{,}000$
B	$2{,}000 \times 15 = 30{,}000$
C	$1{,}500 \times 20 = 30{,}000$
D	$1{,}200 \times 25 = 30{,}000$
E	$800 \times 30 = 24{,}000$

2. Sort items by annual inventory value in descending order:

Item	*Annual Inventory Value (\$)*
A	*50,000*
B	*30,000*
C	*30,000*
D	*30,000*
E	*24,000*

3. Calculate the cumulative percentage of the total inventory value:

$$\textit{Total Inventory Value} = 50{,}000+30{,}000+30{,}000+30{,}000+24{,}000 = 164{,}000$$

$$\textit{Cumulative Percentage} = \frac{\textit{Cumulative Value}}{\textit{Total Inventory Value}} \times 100$$

4. Classify items:

- *A: Items contributing to the top 70% of total inventory value.*
- *B: Items contributing to the next 20% of total inventory value.*
- *C: Items contributing to the remaining 10% of total inventory value.*

Solution 1.22. *exe22*

To perform the ABC analysis, follow these steps:

1. Calculate the total annual cost for each product:

$$\textit{Total Cost} = \textit{Annual Demand} \times \textit{Unit Cost}$$

Product	*Total Annual Cost (\$)*
P1	$4{,}000 \times 12 = 48{,}000$
P2	$3{,}000 \times 10 = 30{,}000$
P3	$2{,}000 \times 8 = 16{,}000$
P4	$1{,}000 \times 15 = 15{,}000$

2. Sort products by total annual cost in descending order:

Product	*Total Annual Cost* ($)
P1	*48,000*
P2	*30,000*
P3	*16,000*
P4	*15,000*

3. Calculate cumulative percentages of total cost:

$$\textit{Total Annual Cost} = 48{,}000 + 30{,}000 + 16{,}000 + 15{,}000 = 109{,}000$$

$$\textit{Cumulative Percentage} = \frac{\textit{Cumulative Cost}}{\textit{Total Annual Cost}} \times 100$$

4. Classify products:

- *A: Products contributing to the top 70% of total cost.*
- *B: Products contributing to the next 20% of total cost.*
- *C: Products contributing to the remaining 10% of total cost.*

Solution 1.23. *exe23*

To perform the ABC analysis, follow these steps:

1. Calculate the annual inventory value for each item:

$$\textit{Value} = \textit{Annual Demand} \times \textit{Unit Price}$$

Item	*Annual Inventory Value* ($)
I1	$6{,}000 \times 5 = 30{,}000$
I2	$3{,}500 \times 7 = 24{,}500$
I3	$2{,}500 \times 10 = 25{,}000$
I4	$2{,}000 \times 12 = 24{,}000$
I5	$1{,}500 \times 8 = 12{,}000$
I6	$1{,}000 \times 6 = 6{,}000$

2. Sort items by annual inventory value in descending order:

Item	*Annual Inventory Value* ($)
I1	*30,000*
I3	*25,000*
I2	*24,500*
I4	*24,000*
I5	*12,000*
I6	*6,000*

3. Calculate cumulative percentages of total inventory value:

$$\text{Total Inventory Value} = 30{,}000 + 25{,}000 + 24{,}500 + 24{,}000 + 12{,}000 + 6{,}000 = 121{,}500$$

$$\text{Cumulative Percentage} = \frac{\text{Cumulative Value}}{\text{Total Inventory Value}} \times 100$$

4. Classify items:

- *A: Items contributing to the top 70% of total inventory value.*
- *B: Items contributing to the next 20% of total inventory value.*
- *C: Items contributing to the remaining 10% of total inventory value.*

Solution 1.24. *exe24*

To perform the ABC analysis, follow these steps:

1. Calculate the total inventory cost for each SKU:

$$\text{Total Cost} = \text{Annual Demand} \times \text{Unit Cost}$$

SKU	*Total Annual Cost* ($)
X1	5,000 × 20 = 100,000
X2	4,000 × 25 = 100,000
X3	3,000 × 15 = 45,000
X4	2,000 × 18 = 36,000
X5	1,000 × 22 = 22,000

2. Sort SKUs by total annual cost in descending order:

SKU	*Total Annual Cost* ($)
X1	*100,000*
X2	*100,000*
X3	*45,000*
X4	*36,000*
X5	*22,000*

3. Calculate cumulative percentages of total cost:

$$\text{Total Annual Cost} = 100{,}000 + 100{,}000 + 45{,}000 + 36{,}000 + 22{,}000 = 303{,}000$$

$$\text{Cumulative Percentage} = \frac{\text{Cumulative Cost}}{\text{Total Annual Cost}} \times 100$$

4. Classify SKUs:

- *A: SKUs contributing to the top 70% of total cost.*
- *B: SKUs contributing to the next 20% of total cost.*
- *C: SKUs contributing to the remaining 10% of total cost.*

Solution 1.25. *exe25*

To perform the ABC analysis, follow these steps:

1. Calculate the total annual cost for each item:

$$\textit{Total Cost} = \textit{Annual Usage} \times \textit{Unit Cost}$$

Item	*Total Annual Cost* ($)
A1	8,000 × 7 = 56,000
A2	6,000 × 6 = 36,000
A3	4,000 × 5 = 20,000
A4	3,000 × 8 = 24,000
A5	2,000 × 10 = 20,000

2. Sort items by total annual cost in descending order:

Item	*Total Annual Cost* ($)
A1	*56,000*
A2	*36,000*
A4	*24,000*
A3	*20,000*
A5	*20,000*

3. Calculate cumulative percentages of total cost:

$$\textit{Total Annual Cost} = 56{,}000+36{,}000+24{,}000+20{,}000+20{,}000 = 156{,}000$$

$$\textit{Cumulative Percentage} = \frac{\textit{Cumulative Cost}}{\textit{Total Annual Cost}} \times 100$$

4. Classify items:

- *A: Items contributing to the top 70% of total cost.*
- *B: Items contributing to the next 20% of total cost.*
- *C: Items contributing to the remaining 10% of total cost.*

Solution 1.26. *exe26*

To perform the VED analysis, follow these steps:

1. Classify items based on their usage frequency and criticality:

- ***Vital (V)**: Items that are essential for operations and have high usage frequency.*
- ***Essential (E)**: Items that are important but not critical, with moderate usage frequency.*
- ***Desirable (D)**: Items with low criticality and usage frequency.*

2. Categorize the items:

Item	*VED Category*
M1	*V*
M2	*E*
M3	*D*
M4	*V*
M5	*E*

3. Explanation:

- *Items M1 and M4 are classified as **Vital** due to their high usage frequency.*
- *Items M2 and M5 are classified as **Essential** due to their moderate usage frequency.*
- *Item M3 is classified as **Desirable** due to its lower usage frequency.*

4. Summary:

- ***V**: M1, M4*
- ***E**: M2, M5*
- ***D**: M3*

Solution 1.27. *exe27*

To perform the VED analysis, follow these steps:

1. Classify items based on their annual consumption and importance:

- ***Vital (V)**: Items with high annual consumption and high importance.*
- ***Essential (E)**: Items with moderate annual consumption and importance.*

- ***Desirable (D)**: Items with low annual consumption and importance.*

2. Categorize the items:

Item	*VED Category*
H1	*V*
H2	*E*
H3	*D*
H4	*V*
H5	*E*

3. Explanation:

- *Items H1 and H4 are classified as **Vital** due to their high annual consumption.*
- *Items H2 and H5 are classified as **Essential** due to their moderate annual consumption.*
- *Item H3 is classified as **Desirable** due to its lower annual consumption.*

4. Summary:

- ***V**: H1, H4*
- ***E**: H2, H5*
- ***D**: H3*

Solution 1.28. *exe28*

To perform the VED analysis, follow these steps:

1. Classify products based on their annual demand and criticality:

- ***Vital (V)**: Products with high annual demand and criticality.*
- ***Essential (E)**: Products with moderate annual demand and criticality.*
- ***Desirable (D)**: Products with lower annual demand and criticality.*

2. *Categorize the products:*

Product	*VED Category*
W1	*E*
W2	*D*
W3	*V*
W4	*E*
W5	*D*

3. *Explanation:*

- *Product W3 is classified as* ***Vital*** *due to its high annual demand.*
- *Products W1 and W4 are classified as* ***Essential*** *due to their moderate annual demand.*
- *Products W2 and W5 are classified as* ***Desirable*** *due to their lower annual demand.*

4. *Summary:*

- ***V****: W3*
- ***E****: W1, W4*
- ***D****: W2, W5*

Solution 1.29. *exe29*

To perform the VED analysis, follow these steps:

1. *Calculate annual usage for each item:*

$$\text{Annual Usage} = \text{Monthly Usage} \times 12$$

2. *Classify items based on their annual usage and importance:*

- ***Vital (V)****: Items with high annual usage and importance.*
- ***Essential (E)****: Items with moderate annual usage and importance.*
- ***Desirable (D)****: Items with lower annual usage and importance.*

3. *Categorize the items:*

Item	*VED Category*
D1	*V*
D2	*E*
D3	*D*
D4	*V*
D5	*E*

4. *Explanation:*

- *Items D1 and D4 are classified as* ***Vital*** *due to their high annual usage.*
- *Items D2 and D5 are classified as* ***Essential*** *due to their moderate annual usage.*
- *Item D3 is classified as* ***Desirable*** *due to its lower annual usage.*

5. *Summary:*

- ***V****: D1, D4*
- ***E****: D2, D5*
- ***D****: D3*

Solution 1.30. *exe30*

To perform the VED analysis, follow these steps:

Classify items based on their annual usage and criticality:

- ***Vital (V)****: Items with high annual usage and high criticality.*
- ***Essential (E)****: Items with moderate annual usage and criticality.*
- ***Desirable (D)****: Items with lower annual usage and criticality.*

Categorize the items:

Item	*VED Category*
R1	*D*
R2	*V*
R3	*E*
R4	*D*
R5	*V*

Explanation:

- *Items R2 and R5 are classified as* ***Vital*** *due to their high annual usage.*
- *Item R3 is classified as* ***Essential*** *due to its moderate annual usage.*
- *Items R1 and R4 are classified as* ***Desirable*** *due to their lower annual usage.*

Summary:

- ***V****: R2, R5*
- ***E****: R3*
- ***D****: R1, R4*

Solution 1.31. *exe31*

To calculate the reorder point for each product using the JIT approach:

The reorder point (ROP) is calculated as:

$$ROP = \text{Daily Demand} \times \text{Lead Time}$$

Convert annual demand to daily demand:

$$\text{Daily Demand} = \frac{\text{Annual Demand}}{365}$$

Apply the formula:

$$ROP = \frac{\text{Annual Demand}}{365} \times \text{Lead Time}$$

Product	ROP
P1	$\frac{2{,}400}{365} \times 10 = 65.8$
P2	$\frac{1{,}800}{365} \times 15 = 73.7$
P3	$\frac{1{,}200}{365} \times 5 = 16.4$
P4	$\frac{1{,}500}{365} \times 8 = 32.8$
P5	$\frac{3{,}000}{365} \times 12 = 98.6$

Explanation:

- *Calculate daily demand by dividing annual demand by 365.*
- *Multiply daily demand by the lead time to determine the reorder point.*

Solution 1.32. *exe32*

To determine the order quantity for each item using JIT principles:

For JIT, order quantities are typically minimal to reduce holding costs. We use the reorder point to determine the necessary order quantity.

Given reorder points directly inform us of the order quantity:

$$Order\ Quantity = Reorder\ Point$$

Item	*Order Quantity*
I1	*1,000*
I2	*500*
I3	*400*
I4	*1,400*
I5	*800*

Explanation:

- *The order quantity in JIT systems is set to the reorder point to ensure stock is replenished just in time.*

Solution 1.33. *exe33*

To calculate the safety stock level for each product:

Safety stock level is calculated as:

$$Safety\ Stock = Daily\ Sales \times Supplier\ Lead\ Time$$

Apply the formula:

$$Safety\ Stock = Daily\ Sales \times Supplier\ Lead\ Time$$

Product	*Safety Stock*
E1	$50 \times 5 = 250$
E2	$30 \times 7 = 210$
E3	$20 \times 3 = 60$
E4	$40 \times 4 = 160$
E5	$60 \times 6 = 360$

Explanation:

- *Safety stock is calculated by multiplying daily sales by the supplier lead time.*

Solution 1.34. *exe34*

To calculate the EOQ for each product:

Use the EOQ formula:

$$EOQ = \sqrt{\frac{2 \times Demand \times Ordering\ Cost}{Holding\ Cost}}$$

Apply the formula for each product:

Product	*EOQ*
R1	$\sqrt{\frac{2\times 10{,}000\times 200}{50}} = 400$
R2	$\sqrt{\frac{2\times 8{,}000\times 150}{40}} = 300$
R3	$\sqrt{\frac{2\times 6{,}000\times 100}{30}} = 300$
R4	$\sqrt{\frac{2\times 12{,}000\times 250}{60}} = 500$
R5	$\sqrt{\frac{2\times 9{,}000\times 180}{45}} = 400$

Explanation:

- *EOQ is calculated to minimize the total cost of ordering and holding inventory.*

Solution 1.35. *exe35*

To calculate the production cycle time for each item:

Production cycle time is calculated as:

$$Cycle\ Time = \frac{Daily\ Demand}{Daily\ Production\ Rate}$$

Apply the formula:

$$Cycle\ Time = \frac{Daily\ Demand}{Daily\ Production\ Rate}$$

Item	*Cycle Time (days)*
C1	$\frac{80}{100} = 0.8$
C2	$\frac{120}{150} = 0.8$
C3	$\frac{150}{200} = 0.75$
C4	$\frac{200}{250} = 0.8$
C5	$\frac{250}{300} = 0.83$

Explanation:

- *Production cycle time indicates how long it takes to produce the quantity needed to meet daily demand.*

Solution 1.36. *exe39*

To determine the Total Cost of Inventory (TCI) for each echelon:

Total Cost of Inventory includes ordering and holding costs. Calculate the total cost using:

$$TCI = \frac{Demand}{EOQ} \times Ordering\ Cost + \frac{EOQ}{2} \times Holding\ Cost$$

Calculate EOQ for each echelon using:

$$EOQ = \sqrt{\frac{2 \times Demand \times Ordering\ Cost}{Holding\ Cost}}$$

Apply the formula:

Echelon	*TCI*
Supplier	$\frac{20{,}000}{500} \times 400 + \frac{500}{2} \times 30 = 16{,}000 + 7{,}500 = 23{,}500$
DC	$\frac{12{,}000}{500} \times 300 + \frac{500}{2} \times 25 = 7{,}200 + 6{,}250 = 13{,}450$
Retail Store	$\frac{6{,}000}{400} \times 150 + \frac{400}{2} \times 20 = 9{,}000 + 4{,}000 = 13{,}000$

Explanation:

- *Calculate the total cost of inventory by summing ordering and holding costs.*

To calculate the Optimal Order Quantity (OOQ) for each level:

Use the EOQ formula:

$$EOQ = \sqrt{\frac{2 \times Demand \times Ordering\ Cost}{Holding\ Cost}}$$

Apply the formula for each level:

Level	*EOQ*
DC	$\sqrt{\frac{2 \times 5{,}000 \times 400}{25}} = 800$
Retail Store	$\sqrt{\frac{2 \times 2{,}000 \times 150}{15}} = 400$

Explanation:

- *The Optimal Order Quantity is computed to minimize the combined ordering and holding costs.*

Solution 1.37. *exe41*

To find the optimal order quantities, use the EOQ formula for both echelons.

For the central warehouse:

$$\begin{aligned}
Q_c^* &= \sqrt{\frac{2 \cdot K_c \cdot D}{h_c}} \\
&= \sqrt{\frac{2 \cdot 300 \cdot 800 \cdot 10}{1}} \\
&= \sqrt{4800{,}000} \\
&= 2{,}191.16 \ \textit{units}
\end{aligned}$$

For each retail store:

$$\begin{aligned}
Q_r^* &= \sqrt{\frac{2 \cdot K_r \cdot D}{h_r}} \\
&= \sqrt{\frac{2 \cdot 50 \cdot 800}{2}} \\
&= \sqrt{40{,}000} \\
&= 200 \ \textit{units}
\end{aligned}$$

Hence, the optimal order quantity for the central warehouse is approximately 2,191 units, and for each retail store is 200 units.

Solution 1.38. *exe42*

The objective function to minimize is:

$$\begin{aligned}
&\textit{Minimize Total Cost} \\
&\quad = K_s \cdot \frac{Q_s}{2} + h_s \cdot Q_s + K_d \cdot \frac{Q_d}{2} + h_d \cdot Q_d \\
&\quad + K_r \cdot \frac{Q_r}{2} + h_r \cdot Q_r + T_{s,d} \cdot \frac{Q_s}{2} + T_{d,r} \cdot \frac{Q_d}{2} \\
&\quad = 400 \cdot \frac{Q_s}{2} + 2 \cdot Q_s + 250 \cdot \frac{Q_d}{2} + 3 \cdot Q_d \\
&\quad + 100 \cdot \frac{Q_r}{2} + 4 \cdot Q_r + 50 \cdot \frac{Q_s}{2} + 50 \cdot \frac{Q_d}{2} \\
&\quad = 200 \cdot Q_s + 2 \cdot Q_s + 125 \cdot Q_d + 3 \cdot Q_d + 50 \cdot Q_r + 4 \cdot Q_r \\
&\quad = 202 \cdot Q_s + 128 \cdot Q_d + 54 \cdot Q_r
\end{aligned}$$

Constraints include ensuring supply meets demand with the appropriate safety stock levels and balancing inventory across echelons. The exact solution requires solving this linear programming model using software tools.

Solution 1.39. *exe43*

For the bullwhip effect simulation:

- *Implement the (s, S) policy at each echelon and simulate for 100 periods.*
- *For retail stores: Variance of Orders* = *Var*$(Orders)$, *Variance of Inventory* = *Var*$(Inventory)$
- *For distribution centers: Variance of Orders* = *Var*$(Orders)$, *Variance ofInventory* = *Var*$(Inventory)$

Example results might show higher variance in orders and inventory levels at distribution centers compared to retail stores, indicating a bullwhip effect. For instance, the variance at the retail stores might be 1,000 units^2 *and at distribution centers 2,500 units*^{2}.

Solution 1.40. *exe44*

In the Monte Carlo simulation:

- *Generate demand scenarios using* $N(150, 30^2)$ *for each of the 15 retail stores.*
- *Apply the inventory policy, calculate total costs and service levels.*

Example results:

- *Total cost: $250,000*
- *Average service level: 95%*

Key steps include generating random demand, calculating inventory levels, and aggregating costs and service levels over multiple simulation runs.

Solution 1.41. *exe45*

For sensitivity analysis:

- *Use lead time variability between 2 to 5 periods for suppliers, 3 to 6 periods for distribution centers, and 1 to 4 periods for retail stores.*
- *Calculate total costs and service levels for each scenario.*

Example results:

- *Lead time variability increase from 2 to 5 periods increases total cost by $20,000 and reduces service level by 5%.*

Mitigation strategies include improving forecasting accuracy and increasing safety stock.

Solution 1.42. *exe46*

1. ***Initialization:*** *Relax the integer constraints and solve the LP:*

$$\text{Minimize} \quad 10x_1 + 20x_2 + 15x_3 + 25x_4$$

$$\text{Subject to} \quad x_1 + x_2 + x_3 + x_4 \geq 1 \quad \text{and} \quad 0 \leq x_i \leq 1$$

Solution: $x_1 = 1, x_2 = 1, x_3 = 1, x_4 = 0.$

2. ***Branching:*** *Create subproblems by setting* $x_4 = 1$ *and* $x_4 = 0$.

3. ***Bounding:*** *Solve the relaxed problems:*

$$\text{If} \quad x_4 = 1, \quad \text{Objective} = 70$$

$$\text{If} \quad x_4 = 0, \quad \text{Objective} = 45$$

4. ***Pruning:*** *Prune the subproblem with* $x_4 = 1$ *as it doesn't improve the solution.*

5. ***Optimal Solution:*** $x_1 = 1, x_2 = 1, x_3 = 1, x_4 = 0.$

Solution 1.43. *exe47*

$$\text{Lead Time Demand} = d \times L = 30 \times 5 = 150 \ \text{units}$$

Solution 1.44. *exe48*

$$\text{Inventory Turnover Ratio} = \frac{\text{COGS}}{\text{Average Inventory}} = \frac{1{,}200{,}000}{300{,}000} = 4$$

Solution 1.45. *exe49*

$$D_t = T_t + S_t + R_t = 100 + 20 + 5 = 125$$

Solution 1.46. *exe50*

$$F_t = \alpha D_{t-1} + (1-\alpha)F_{t-1} = 0.4 \times 200 + (1-0.4) \times 190 = 80 + 114 = 194 \ \text{units}$$

Solution 1.47. *exe51*

$$\text{Inventory Accuracy} = \frac{\text{Accurate Inventory Records}}{\text{Total Inventory Records}} \times 100 = \frac{950}{1{,}000} \times 100 = 95\%$$

Solution 1.48. *exe52*

$$\text{Stockout Rate} = \frac{\text{Number of Stockouts}}{\text{Total Orders}} \times 100 = \frac{3}{150} \times 100 = 2\%$$

Solution 1.49. *exe53*

$$\text{Quality Control Cost} = Q_i \times C_i = 1{,}000 \times 5 = \$5{,}000$$

Solution 1.50. *exe54*

$$\text{Order Fulfillment Rate} = \frac{\text{Orders Fulfilled from Inventory}}{\text{Total Orders}} \times 100 = \frac{450}{500} \times 100 = 90\%$$

Solution 1.51. *exe55*
Given the loss function:

$$L(\theta) = \frac{1}{n}\sum_{i=1}^{n}(y_i - \theta_0 - \theta_1 x_i)^2$$

The gradients are:

$$\frac{\partial L}{\partial \theta_0} = -\frac{2}{n}\sum_{i=1}^{n}(y_i - \theta_0 - \theta_1 x_i)$$

$$\frac{\partial L}{\partial \theta_1} = -\frac{2}{n}\sum_{i=1}^{n}(y_i - \theta_0 - \theta_1 x_i)x_i$$

With the initial parameters $\theta_0 = 0$ *and* $\theta_1 = 0$, *the gradients are:*

$$\frac{\partial L}{\partial \theta_0} = -\frac{2}{4}[(2-0-0)+(3-0-0)+(4-0-0)+(5-0-0)] = -2\cdot 14 = -28$$

$$\frac{\partial L}{\partial \theta_1} = -\frac{2}{4}[(2-0-0)\cdot 1+(3-0-0)\cdot 2+(4-0-0)\cdot 3+(5-0-0)\cdot 4] = -2\cdot 30 = -60$$

Update the parameters:

$$\theta_0 = 0 - 0.01 \cdot (-28) = 0.28$$

$$\theta_1 = 0 - 0.01 \cdot (-60) = 0.6$$

Therefore, the updated parameters are $\theta_0 = 0.28$ *and* $\theta_1 = 0.6$.

Solution 1.52. *exe56*

Given the loss function:

$$L(\theta) = (y^{(1)} - \theta_0 - \theta_1 x^{(1)})^2$$

The gradients for the first data point $(x^{(1)}, y^{(1)}) = (1, 2)$ *are:*

$$\frac{\partial L}{\partial \theta_0} = -2(y^{(1)} - \theta_0 - \theta_1 x^{(1)}) = -2(2 - 0 - 0) = -4$$

$$\frac{\partial L}{\partial \theta_1} = -2(y^{(1)} - \theta_0 - \theta_1 x^{(1)})x^{(1)} = -2(2 - 0 - 0) \cdot 1 = -4$$

Update the parameters:

$$\theta_0 = 0 - 0.01 \cdot (-4) = 0.04$$

$$\theta_1 = 0 - 0.01 \cdot (-4) = 0.04$$

Therefore, the updated parameters after one iteration are $\theta_0 = 0.04$ *and* $\theta_1 = 0.04$.

Solution 1.53. *exe57*

Given the gradients $\nabla_\theta L(\theta) = (1, -1)$, *initial values* $m_0 = 0$, $v_0 = 0$, $\beta_1 = 0.9$, $\beta_2 = 0.999$, *and* $\eta = 0.001$*:*

$$m_1 = \beta_1 m_0 + (1 - \beta_1)\nabla_\theta L(\theta) = 0.9 \cdot 0 + (1 - 0.9) \cdot (1, -1) = (0.1, -0.1)$$

$$v_1 = \beta_2 v_0 + (1 - \beta_2)(\nabla_\theta L(\theta))^2 = 0.999 \cdot 0 + (1 - 0.999) \cdot (1, 1) = (0.001, 0.001)$$

Bias-corrected estimates:

$$\hat{m}_1 = \frac{m_1}{1 - \beta_1^1} = \frac{(0.1, -0.1)}{1 - 0.9} = (1, -1)$$

$$\hat{v}_1 = \frac{v_1}{1 - \beta_2^1} = \frac{(0.001, 0.001)}{1 - 0.999} = (1, 1)$$

Update the parameters:

$$\theta = \theta - \eta \frac{\hat{m}_1}{\sqrt{\hat{v}_1} + \epsilon} = \theta - 0.001 \cdot \frac{(1, -1)}{\sqrt{(1, 1)} + 10^{-8}}$$

$$= \theta - 0.001 \cdot (1, -1) = (\theta_1 - 0.001, \theta_2 + 0.001)$$

Therefore, the updated parameters are $\theta = (\theta_1 - 0.001, \theta_2 + 0.001)$.

Solution 1.54. *exe58*

Given the gradients:

$$\frac{\partial L}{\partial w} = 2w$$

$$\frac{\partial L}{\partial b} = 2b$$

Update the parameters:

$$w = w - \eta\frac{\partial L}{\partial w} = 0.5 - 0.01 \cdot 2 \cdot 0.5 = 0.5 - 0.01 = 0.49$$

$$b = b - \eta\frac{\partial L}{\partial b} = 0.5 - 0.01 \cdot 2 \cdot 0.5 = 0.5 - 0.01 = 0.49$$

Therefore, the updated parameters are $w = 0.49$ *and* $b = 0.49$.

Solution 1.55. *exe59*

Given the gradients $\nabla_\theta L(\theta) = (2w, 2b)$, *initial values* $m_0 = 0$, $v_0 = 0$, $\beta_1 = 0.9$, $\beta_2 = 0.999$, *and* $\eta = 0.001$:

$$m_1 = \beta_1 m_0 + (1 - \beta_1)\nabla_\theta L(\theta) = 0.9 \cdot 0 + (1 - 0.9) \cdot (2, 2) = (0.2, 0.2)$$

$$v_1 = \beta_2 v_0 + (1 - \beta_2)(\nabla_\theta L(\theta))^2 = 0.999 \cdot 0 + (1 - 0.999) \cdot (4, 4) = (0.004, 0.004)$$

Bias-corrected estimates:

$$\hat{m}_1 = \frac{m_1}{1 - \beta_1^1} = \frac{(0.2, 0.2)}{1 - 0.9} = (2, 2)$$

$$\hat{v}_1 = \frac{v_1}{1 - \beta_2^1} = \frac{(0.004, 0.004)}{1 - 0.999} = (4, 4)$$

Update the parameters:

$$\theta = \theta - \eta\frac{\hat{m}_1}{\sqrt{\hat{v}_1} + \epsilon} = \theta - 0.001 \cdot \frac{(2, 2)}{\sqrt{(4, 4)} + 10^{-8}} = \theta - 0.001 \cdot (1, 1) = (w - 0.001, b - 0.001)$$

Therefore, the updated parameters are $w = 1 - 0.001 = 0.999$ *and* $b = 1 - 0.001 = 0.999$.

Solution 1.56. *exe60*

Given the initial state is low stock (state 1), we need to calculate the probability distribution after two time steps.
Let the initial state vector be $\mathbf{v}_0 = [0, 1, 0]$.

The state vector after one time step is:

$$\mathbf{v}_1 = \mathbf{v}_0 P = [0,1,0] \begin{bmatrix} 0.1 & 0.6 & 0.3 \\ 0.2 & 0.5 & 0.3 \\ 0.3 & 0.4 & 0.3 \end{bmatrix} = [0.2, 0.5, 0.3]$$

The state vector after two time steps is:

$$\mathbf{v}_2 = \mathbf{v}_1 P = [0.2, 0.5, 0.3] \begin{bmatrix} 0.1 & 0.6 & 0.3 \\ 0.2 & 0.5 & 0.3 \\ 0.3 & 0.4 & 0.3 \end{bmatrix} = [0.22, 0.49, 0.29]$$

Therefore, the probability distribution of the inventory levels after two time steps is $[0.22, 0.49, 0.29]$.

Solution 1.57. *exe61*

To find the steady-state distribution π*, we need to solve* $\pi P = \pi$ *with the condition that* $\sum_i \pi_i = 1$.
Let $\pi = [\pi_0, \pi_1, \pi_2]$.
The system of equations is:

$$\begin{cases} 0.1\pi_0 + 0.2\pi_1 + 0.3\pi_2 = \pi_0 \\ 0.6\pi_0 + 0.5\pi_1 + 0.4\pi_2 = \pi_1 \\ 0.3\pi_0 + 0.3\pi_1 + 0.3\pi_2 = \pi_2 \\ \pi_0 + \pi_1 + \pi_2 = 1 \end{cases}$$

Solving this system, we get:

$$\pi_0 = 0.2857, \quad \pi_1 = 0.4286, \quad \pi_2 = 0.2857$$

Therefore, the steady-state distribution of the inventory levels is $[0.2857, 0.4286, 0.2857]$.

Solution 1.58. *exe62*

Adding the penalty term $\lambda \sum_{j=1}^{p} \beta_j^2$*, ridge regression reduces the impact of multicollinearity, leading to more stable demand forecasts.*

Solution 1.59. *exe63*

The ridge regression model:

$$D = \beta_0 + \beta_1 P + \beta_2 M + \epsilon$$

addresses multicollinearity by penalizing large coefficients, thus providing more reliable estimates and improving the forecasting model's reliability for inventory management.

Solution 1.60. *exe64*

Applying the penalty term $\lambda \sum_{j=1}^{p} |\beta_j|$, *LASSO might shrink some* β *coefficients to zero, effectively identifying the significant factors affecting demand.*

Solution 1.61. *exe65*

The LASSO regression model:

$$D = \beta_0 + \beta_1 P + \beta_2 M + \beta_3 T + \epsilon$$

can shrink some coefficients to zero, indicating that only significant predictors (e.g., P and T) are included in the model, simplifying the forecasting process and improving precision.

Solution 1.62. *exe66*

Combining both ridge ($\lambda_1 \sum_{j=1}^{p} \beta_j^2$) *and LASSO* ($\lambda_2 \sum_{j=1}^{p} |\beta_j|$) *penalties, Elastic Net provides a balance between regularization and variable selection, resulting in robust demand predictions.*

Solution 1.63. *exe67*

The Elastic Net model:

$$D = \beta_0 + \beta_1 P + \beta_2 M + \epsilon$$

can manage multicollinearity between predictors P and M by combining ridge and LASSO penalties, improving the reliability and accuracy of the demand forecasting model for inventory management.

Solution 1.64. *exe68*

The inventory discrepancy for each item is calculated as the difference between the recorded quantity and the physical quantity.
For Item A:

$$\textit{Discrepancy} = 150 - 140 = 10$$

For Item B:

$$\textit{Discrepancy} = 200 - 210 = -10$$

For Item C:

$$\textit{Discrepancy} = 250 - 245 = 5$$

The total discrepancy for all items combined is:

$$10 + (-10) + 5 = 5$$

Therefore, the discrepancies are 10 for Item A, -10 for Item B, 5 for Item C, and the total discrepancy is 5.

Solution 1.65. *exe69*

Let the physical quantity be x.
Since the recorded quantity is 10% higher than the physical count:

$$550 = 1.10x$$

Solving for x:

$$x = \frac{550}{1.10} = 500$$

The discrepancy is:

$$550 - 500 = 50$$

Therefore, the physical quantity is 500 units and the discrepancy is 50 units.

Solution 1.66. *exe70*

Calculate the discrepancy for each item and then the financial impact.
For Item W:

$$\textit{Discrepancy} = 300 - 295 = 5 \quad \textit{units}$$

For Item X:

$$\textit{Discrepancy} = 400 - 390 = 10 \quad \textit{units}$$

For Item Y:

$$\textit{Discrepancy} = 500 - 505 = -5 \quad \textit{units}$$

For Item Z:

$$\textit{Discrepancy} = 600 - 590 = 10 \quad \textit{units}$$

Financial impact for each item at \$20 per unit:
For Item W:

$$5 \times 20 = 100 \quad \textit{dollars}$$

For Item X:

$$10 \times 20 = 200 \quad \textit{dollars}$$

For Item Y:

$$-5 \times 20 = -100 \quad \textit{dollars}$$

For Item Z:

$$10 \times 20 = 200 \quad \textit{dollars}$$

Total financial impact:

$$100 + 200 - 100 + 200 = 400 \quad \textit{dollars}$$

Therefore, the total financial impact of the discrepancies is \$400.

Solution 1.67. *exe71*

If the actual inventory levels are lower by 5%:

$$\textit{Discrepancy Percentage} = 5\%$$

The actual inventory value is:

$$\textit{Actual Inventory Value} = (1-0.05)\times 100{,}000 = 0.95\times 100{,}000 = 95{,}000 \quad \textit{dollars}$$

The total discrepancy in value is:

$$100{,}000 - 95{,}000 = 5{,}000 \quad \textit{dollars}$$

Therefore, the actual inventory value is \$95,000 and the total discrepancy in value is \$5,000.

Solution 1.68. *exe72*

The shrinkage rate is 2% of the recorded inventory.

$$\textit{Shrinkage Value} = 0.02 \times 50{,}000 = 1{,}000 \quad \textit{dollars}$$

The adjusted inventory value after accounting for the shrinkage is:

$$\textit{Adjusted Inventory Value} = 50{,}000 - 1{,}000 = 49{,}000 \quad \textit{dollars}$$

Therefore, the value of the lost inventory is \$1,000 and the adjusted inventory value is \$49,000.

Solution 1.69. *exe73*

The discrepancy D_t for each item is calculated as:
For Item P:

$$D_P = 200 - 180 = 20$$

For Item Q:

$$D_Q = 150 - 155 = -5$$

For Item R:

$$D_R = 100 - 95 = 5$$

The total discrepancy TD is:

$$TD = |20| + |-5| + |5| = 20 + 5 + 5 = 30$$

The Inventory Accuracy Rate A is:

$$A = \left(1 - \frac{TD}{\sum_{t=1}^{n} I_t}\right) \times 100\%$$

$$\sum_{t=1}^{n} I_t = 180 + 155 + 95 = 430$$

$$A = \left(1 - \frac{30}{430}\right) \times 100\% = (1 - 0.0698) \times 100\% \approx 93\%$$

Therefore, the discrepancies are 20 for Item P, -5 for Item Q, 5 for Item R, the total discrepancy is 30, and the Inventory Accuracy Rate is approximately 93%.

Solution 1.70. *exe74*

The Root Mean Square Error (RMSE) is calculated as follows:

$$RMSE = \sqrt{\frac{1}{n}\sum_{t=1}^{n}(D_t)^2}$$

$$\sum_{t=1}^{4}(D_t)^2 = (-5)^2 + 10^2 + (-2)^2 + 3^2 = 25 + 100 + 4 + 9 = 138$$

$$RMSE = \sqrt{\frac{138}{4}} = \sqrt{34.5} \approx 5.87$$

Therefore, the RMSE for the discrepancies is approximately 5.87.

Solution 1.71. *exe75*

Calculate the Inventory Accuracy Rate A before and after implementation:
Before implementation:

$$A_{before} = \left(1 - \frac{TD_{before}}{\sum_{t=1}^{n} I_{t,before}}\right) \times 100\%$$

$$A_{before} = \left(1 - \frac{2{,}500}{50{,}000}\right) \times 100\% = (1 - 0.05) \times 100\% = 95\%$$

After implementation:

$$A_{after} = \left(1 - \frac{TD_{after}}{\sum_{t=1}^{n} I_{t,after}}\right) \times 100\%$$

$$A_{after} = \left(1 - \frac{1{,}000}{55{,}000}\right) \times 100\% = (1 - 0.0182) \times 100\% \approx 98.18\%$$

The Error Reduction Rate is:

$$Error\ Reduction\ Rate = \left(\frac{A_{after} - A_{before}}{A_{before}}\right) \times 100\%$$

$$Error\ Reduction\ Rate = \left(\frac{98.18 - 95}{95}\right) \times 100\% = \left(\frac{3.18}{95}\right) \times 100\% \approx 3.35\%$$

Therefore, the Inventory Accuracy Rate before implementation is 95%, after implementation is approximately 98.18%, and the Error Reduction Rate is approximately 3.35%.

Solution 1.72. *exe76*

Classify the discrepancies into categories and calculate the frequency of each category.
Total number of discrepancies = 6.
Theft:

$$Frequency_{Theft} = \frac{2}{6} = \frac{1}{3} \approx 33.33\%$$

Data Entry Errors:

$$Frequency_{Data\ Entry\ Errors} = \frac{2}{6} = \frac{1}{3} \approx 33.33\%$$

Spoilage:

$$Frequency_{Spoilage} = \frac{2}{6} = \frac{1}{3} \approx 33.33\%$$

Therefore, the frequencies of each category are approximately 33.33% for Theft, 33.33% for Data Entry Errors, and 33.33% for Spoilage.

Solution 1.73. *exe77*

The Error Rate Reduction is calculated as follows:

$$Error\ Rate\ Reduction = E_{before} - E_{after}$$

$$Error\ Rate\ Reduction = 8\% - 5\% = 3\%$$

Calculate the Improvement in Accuracy:

$$A_{before} = 92\%$$

$$A_{after} = 100\% - 5\% = 95\%$$

$$Improvement\ in\ Accuracy = \left(\frac{A_{after} - A_{before}}{A_{before}}\right) \times 100\%$$

$$Improvement\ in\ Accuracy = \left(\frac{95 - 92}{92}\right) \times 100\% = \left(\frac{3}{92}\right) \times 100\% \approx 3.26\%$$

Therefore, the Error Rate Reduction is 3% and the Improvement in Accuracy is approximately 3.26%.

Solution 1.74. *exe78*

The GMROI is calculated using the formula:

$$GMROI = \frac{Gross\ Margin}{Average\ Inventory}$$

For the Electronics category:

$$Gross\ Margin = Gross\ Sales - COGS = \$500{,}000 - \$350{,}000 = \$150{,}000$$

$$GMROI = \frac{\$150{,}000}{\$50{,}000} = 3$$

For the Clothing category:

$$Gross\ Margin = Gross\ Sales - COGS = \$300{,}000 - \$150{,}000 = \$150{,}000$$

$$GMROI = \frac{\$150{,}000}{\$40{,}000} = 3.75$$

Analysis: The Clothing category has a higher GMROI (3.75) compared to the Electronics category (3.0). This indicates that the Clothing category is generating more gross margin return on its inventory investment.

Solution 1.75. *exe79*

The GMROI is calculated using the formula:

$$GMROI = \frac{Gross\ Margin}{Average\ Inventory}$$

First, calculate the annual gross margin:

$$Gross\ Margin = Gross\ Sales \times Gross\ Margin\ Percentage$$

$$Gross\ Sales = Average\ Inventory \times Inventory\ Turnover\ Rate$$

$$Gross\ Sales = \$200{,}000 \times 5 = \$1{,}000{,}000$$

$$Gross\ Margin = \$1{,}000{,}000 \times 0.30 = \$300{,}000$$

Now, calculate the GMROI:

$$GMROI = \frac{\$300{,}000}{\$200{,}000} = 1.5$$

Discussion: Increasing the inventory turnover rate will increase the gross sales and therefore the gross margin, which in turn increases the GMROI. Conversely, a lower turnover rate decreases the GMROI.

Solution 1.76. *exe80*

First, calculate the initial GMROI for each product line:
For Product Line A:

$$Gross\ Margin = \$1{,}200{,}000 - \$800{,}000 = \$400{,}000$$

$$GMROI = \frac{\$400{,}000}{\$100{,}000} = 4$$

For Product Line B:

$$Gross\ Margin = \$800{,}000 - \$400{,}000 = \$400{,}000$$

$$GMROI = \frac{\$400{,}000}{\$90{,}000} \approx 4.44$$

For Product Line C:

$$Gross\ Margin = \$600{,}000 - \$300{,}000 = \$300{,}000$$

$$GMROI = \frac{\$300{,}000}{\$80{,}000} = 3.75$$

Now, increase the COGS by 10% and recalculate the GMROI:
For Product Line A:

$$New\ COGS = \$800{,}000 \times 1.10 = \$880{,}000$$

$$New\ Gross\ Margin = \$1{,}200{,}000 - \$880{,}000 = \$320{,}000$$

$$New\ GMROI = \frac{\$320{,}000}{\$100{,}000} = 3.2$$

For Product Line B:

$$New\ COGS = \$400{,}000 \times 1.10 = \$440{,}000$$

$$New\ Gross\ Margin = \$800{,}000 - \$440{,}000 = \$360{,}000$$

$$New\ GMROI = \frac{\$360{,}000}{\$90{,}000} = 4$$

For Product Line C:

$$New\ COGS = \$300{,}000 \times 1.10 = \$330{,}000$$

$$New\ Gross\ Margin = \$600{,}000 - \$330{,}000 = \$270{,}000$$

$$New\ GMROI = \frac{\$270{,}000}{\$80{,}000} = 3.375$$

Analysis: The increase in COGS leads to a decrease in GMROI for all product lines. This highlights the importance of managing COGS to maintain high GMROI. For example, Product Line A's GMROI decreased from 4 to 3.2, showing a significant impact.

Solution 1.77. *exe81*

Days Sales of Inventory (DSI) is calculated using the formula:

$$DSI = \frac{Average\ Inventory}{COGS\ per\ day}$$

First, calculate the annual COGS:

$$Annual\ COGS = 50{,}000 + 60{,}000 + 55{,}000 + 65{,}000 + 70{,}000 + 75{,}000$$
$$+ 80{,}000 + 85{,}000 + 90{,}000 + 95{,}000 + 100{,}000 + 105,000 = 930{,}000$$

Next, calculate the average inventory:

$$\begin{aligned} &Average\ Inventory \\ &= \frac{120{,}000 + 130{,}000 + 125{,}000 + 140{,}000 + 135{,}000 + 145{,}000}{12} \\ &\quad + \frac{150{,}000 + 155{,}000 + 160{,}000 + 165{,}000 + 170{,}000 + 175{,}000}{12} \\ &= 146{,}250 \end{aligned}$$

Calculate the daily COGS:

$$COGS\ per\ day = \frac{930{,}000}{365} \approx 2{,}548$$

Finally, calculate the DSI:

$$DSI = \frac{146{,}250}{2{,}548} \approx 57.4\ days$$

Solution 1.78. *exe82*

Days Sales of Inventory (DSI) is given by:

$$DSI = \frac{Average\ Inventory}{COGS\ per\ day}$$

Given:

$$Current\ DSI = 45\ days$$

$$Average\ Inventory = 500{,}000$$

$$Annual\ COGS = 6{,}000{,}000$$

First, calculate the daily COGS:

$$COGS\ per\ day = \frac{6{,}000{,}000}{365} \approx 16{,}438$$

If the company wants to reduce its DSI to 30 days, the new average inventory I_{new} can be calculated as:

$$DSI_{new} = \frac{I_{new}}{COGS\ per\ day}$$

$$I_{new} = DSI_{new} \times COGS\ per\ day = 30 \times 16{,}438 \approx 493{,}140$$

The reduction in average inventory:

$$Reduction\ in\ Inventory = 500{,}000 - 493{,}140 = 6{,}860$$

This reduction directly improves the cash flow by \$6,860.

Solution 1.79. *exe83*

Days Sales of Inventory (DSI) is calculated using the formula:

$$DSI = \frac{Average\ Inventory}{COGS\ per\ day}$$

First, calculate the DSI for each product line:
For Product Line A:

$$COGS\ per\ day = \frac{1{,}200{,}000}{365} \approx 3{,}288$$

$$DSI = \frac{200{,}000}{3{,}288} \approx 60.8\ days$$

For Product Line B:

$$COGS\ per\ day = \frac{1{,}000{,}000}{365} \approx 2{,}740$$

$$DSI = \frac{150{,}000}{2{,}740} \approx 54.7\ days$$

For Product Line C:

$$COGS\ per\ day = \frac{800{,}000}{365} \approx 2{,}192$$

$$DSI = \frac{100{,}000}{2{,}192} \approx 45.6 \ days$$

Now, increase the inventory by 10% for each product line and recalculate the DSI:
For Product Line A:

$$New\ Average\ Inventory = 200{,}000 \times 1.10 = 220{,}000$$

$$New\ DSI = \frac{220{,}000}{3{,}288} \approx 66.9 \ days$$

For Product Line B:

$$New\ Average\ Inventory = 150{,}000 \times 1.10 = 165{,}000$$

$$New\ DSI = \frac{165{,}000}{2{,}740} \approx 60.2 \ days$$

For Product Line C:

$$New\ Average\ Inventory = 100{,}000 \times 1.10 = 110{,}000$$

$$New\ DSI = \frac{110{,}000}{2{,}192} \approx 50.2 \ days$$

Analysis: Increasing the inventory by 10% results in a corresponding increase in DSI for each product line. This means the company is holding onto its inventory for a longer period before it is sold, which could lead to higher holding costs and potentially lower cash flow.

Solution 1.80. *exe84*

To calculate the Consistency Index (CI) and Consistency Ratio (CR), follow these steps:

1. *Calculate the principal eigenvector and eigenvalue:*

$$\mathbf{A} = \begin{pmatrix} 1 & 3 & 5 \\ \frac{1}{3} & 1 & 4 \\ \frac{1}{5} & \frac{1}{4} & 1 \end{pmatrix}$$

Using a numerical method or software (e.g., MATLAB, Python), we find the principal eigenvalue ($\lambda_{\max}$):

$$\lambda_{\max} \approx 3.093$$

2. Calculate the Consistency Index (CI):

$$CI = \frac{\lambda_{\text{max}} - n}{n - 1}$$

For a 3x3 matrix:

$$CI = \frac{3.093 - 3}{3 - 1} = \frac{0.093}{2} = 0.0465$$

3. Calculate the Consistency Ratio (CR):

$$CR = \frac{CI}{RI}$$

Given $RI = 0.58$:

$$CR = \frac{0.0465}{0.58} \approx 0.0802$$

The Consistency Ratio (CR) is approximately 0.0802, which is less than 0.1, indicating a consistent matrix.

Solution 1.81. *exe85*

To achieve a Consistency Ratio (CR) less than 0.1, follow these steps:

1. Calculate the initial Consistency Index (CI) and Consistency Ratio (CR) for matrix **B***:*

$$\mathbf{B} = \begin{pmatrix} 1 & 7 & 3 & 5 \\ \frac{1}{7} & 1 & \frac{1}{5} & \frac{1}{3} \\ \frac{1}{3} & 5 & 1 & 2 \\ \frac{1}{5} & 3 & \frac{1}{2} & 1 \end{pmatrix}$$

Using numerical methods, we find the principal eigenvalue (λ_{max}):

$$\lambda_{\text{max}} \approx 4.2$$

2. Calculate the initial Consistency Index (CI):

$$CI = \frac{\lambda_{\text{max}} - n}{n - 1}$$

For a 4x4 matrix:

$$CI = \frac{4.2 - 4}{4 - 1} = \frac{0.2}{3} \approx 0.0667$$

3. Calculate the initial Consistency Ratio (CR):

$$CR = \frac{CI}{RI}$$

Given $RI = 0.90$:

$$CR = \frac{0.0667}{0.90} \approx 0.0741$$

The initial CR is 0.0741, which is less than 0.1. Therefore, we must check if the CR was initially greater than 0.1 and improve it by adjusting the comparison value.

4. Adjust the comparison value between the first and third elements: Let the new comparison value be x. Update the matrix $\mathbf{B}$ with the adjusted value and recalculate $\lambda_{\max}$:

$$\mathbf{B}_{new} = \begin{pmatrix} 1 & 7 & x & 5 \\ \frac{1}{7} & 1 & \frac{1}{5} & \frac{1}{3} \\ \frac{1}{x} & 5 & 1 & 2 \\ \frac{1}{5} & 3 & \frac{1}{2} & 1 \end{pmatrix}$$

Suppose $x = 4$:

$$\mathbf{B}_{new} = \begin{pmatrix} 1 & 7 & 4 & 5 \\ \frac{1}{7} & 1 & \frac{1}{5} & \frac{1}{3} \\ \frac{1}{4} & 5 & 1 & 2 \\ \frac{1}{5} & 3 & \frac{1}{2} & 1 \end{pmatrix}$$

Using numerical methods, recalculate $\lambda_{\max}$:

$$\lambda_{\max} \approx 4.1$$

5. Calculate the new Consistency Index (CI):

$$CI = \frac{\lambda_{\max} - n}{n - 1}$$

$$CI = \frac{4.1 - 4}{4 - 1} = \frac{0.1}{3} \approx 0.0333$$

6. Calculate the new Consistency Ratio (CR):

$$CR = \frac{CI}{RI}$$

$$CR = \frac{0.0333}{0.90} \approx 0.037$$

The adjusted Consistency Ratio (CR) is approximately 0.037, which is less than 0.1, indicating an acceptable level of consistency in the matrix.

Solution 1.82. *exe86*

To calculate the Consistency Index (CI) and Consistency Ratio (CR) for the matrix $\mathbf{C}$*:*

1. *Calculate the principal eigenvalue* ($\lambda_{\max}$)*:*

$$\mathbf{C} = \begin{pmatrix} 1 & 2 & 5 & 7 \\ \frac{1}{2} & 1 & 3 & 5 \\ \frac{1}{5} & \frac{1}{3} & 1 & 2 \\ \frac{1}{7} & \frac{1}{5} & \frac{1}{2} & 1 \end{pmatrix}$$

Using numerical methods:

$$\lambda_{\max} \approx 4.1$$

2. *Calculate the Consistency Index (CI):*

$$CI = \frac{\lambda_{\max} - n}{n-1}$$

$$CI = \frac{4.1-4}{4-1} = \frac{0.1}{3} \approx 0.0333$$

3. *Calculate the Consistency Ratio (CR):*

$$CR = \frac{CI}{RI}$$

Given $RI = 0.90$*:*

$$CR = \frac{0.0333}{0.90} \approx 0.037$$

The Consistency Ratio (CR) is approximately 0.037, which is less than 0.10, indicating a consistent matrix.

Solution 1.83. *exe87*

To adjust the matrix $\mathbf{D}$ *and achieve a CR less than 0.10:*

1. *Calculate the initial CI and CR:*

$$\mathbf{D} = \begin{pmatrix} 1 & 6 & 3 \\ \frac{1}{6} & 1 & \frac{1}{4} \\ \frac{1}{3} & 4 & 1 \end{pmatrix}$$

Using numerical methods:

$$\lambda_{\max} \approx 3.2$$

2. Calculate the initial CI:

$$CI = \frac{\lambda_{\max} - n}{n - 1}$$

$$CI = \frac{3.2 - 3}{3 - 1} = \frac{0.2}{2} = 0.1$$

3. Calculate the initial CR:

$$CR = \frac{CI}{RI}$$

Given $RI = 0.58$:

$$CR = \frac{0.1}{0.58} \approx 0.172$$

The initial CR is 0.172, which is greater than 0.10.

4. Adjust the comparison value between the first and second elements: Let the new comparison value be x. Update the matrix $\mathbf{D}$ with the adjusted value:

$$\mathbf{D}_{new} = \begin{pmatrix} 1 & x & 3 \\ \frac{1}{x} & 1 & \frac{1}{4} \\ \frac{1}{3} & 4 & 1 \end{pmatrix}$$

Suppose $x = 4$:

$$\mathbf{D}_{new} = \begin{pmatrix} 1 & 4 & 3 \\ \frac{1}{4} & 1 & \frac{1}{4} \\ \frac{1}{3} & 4 & 1 \end{pmatrix}$$

Using numerical methods, recalculate $\lambda_{\max}$:

$$\lambda_{\max} \approx 3.1$$

5. Calculate the new CI:

$$CI = \frac{\lambda_{\max} - n}{n - 1}$$

$$CI = \frac{3.1 - 3}{3 - 1} = \frac{0.1}{2} = 0.05$$

6. Calculate the new CR:

$$CR = \frac{CI}{RI}$$

$$CR = \frac{0.05}{0.58} \approx 0.086$$

The adjusted Consistency Ratio (CR) is approximately 0.086, which is less than 0.10, indicating an acceptable level of consistency in the matrix.

Solution 1.84. *exe88*

To calculate the Consistency Index (CI) and Consistency Ratio (CR) for the matrix **E***:*

1. *Calculate the principal eigenvalue ($\lambda_{\max}$):*

$$\mathbf{E} = \begin{pmatrix} 1 & 8 & 5 & 3 & 9 \\ \frac{1}{8} & 1 & 4 & 2 & 6 \\ \frac{1}{5} & \frac{1}{4} & 1 & 3 & 7 \\ \frac{1}{3} & \frac{1}{2} & \frac{1}{3} & 1 & 5 \\ \frac{1}{9} & \frac{1}{6} & \frac{1}{7} & \frac{1}{5} & 1 \end{pmatrix}$$

Using numerical methods:

$$\lambda_{\max} \approx 5.3$$

2. *Calculate the Consistency Index (CI):*

$$CI = \frac{\lambda_{\max} - n}{n-1}$$

$$CI = \frac{5.3-5}{5-1} = \frac{0.3}{4} = 0.075$$

3. *Calculate the Consistency Ratio (CR):*

$$CR = \frac{CI}{RI}$$

Given $RI = 1.12$:

$$CR = \frac{0.075}{1.12} \approx 0.067$$

The Consistency Ratio (CR) is approximately 0.067, which is less than 0.10. This indicates that the matrix is sufficiently consistent for decision-making. The low CR value ensures that the pairwise comparisons are reliable and that minor inconsistencies do not significantly impact the decision-making process.

Solution 1.85. *exe89*

To calculate the Consistency Index (CI) and Consistency Ratio (CR) for the matrix **F***:*

1. *Calculate the principal eigenvalue ($\lambda_{\max}$):*

$$\mathbf{F} = \begin{pmatrix} 1 & 3 & 4 \\ \frac{1}{3} & 1 & 2 \\ \frac{1}{4} & \frac{1}{2} & 1 \end{pmatrix}$$

Using numerical methods:

$$\lambda_{\max} \approx 3.05$$

2. Calculate the Consistency Index (CI):

$$CI = \frac{\lambda_{\max} - n}{n-1}$$

$$CI = \frac{3.05 - 3}{3 - 1} = \frac{0.05}{2} = 0.025$$

3. Calculate the Consistency Ratio (CR):

$$CR = \frac{CI}{RI}$$

Given $RI = 0.58$*:*

$$CR = \frac{0.025}{0.58} \approx 0.043$$

The Consistency Ratio (CR) is approximately 0.043, which is less than 0.10, indicating that the matrix is consistent according to Saaty's empirical threshold.

Solution 1.86. *exe90*

The theoretical justification for the CR threshold of 0.10 in the Analytic Hierarchy Process (AHP) balances minor inconsistencies while ensuring reliable decision-making.

1. For $n = 1$ *or* $n = 2$*, any matrix is always consistent, thus* $CI = 0$ *and* $CR = 0$*.*

2. For larger matrices, the likelihood of inconsistency increases. The threshold of 0.10 represents a balance between allowing minor inconsistencies and ensuring that these do not significantly affect the outcome. This is based on empirical studies conducted by Thomas L. Saaty.

Let's illustrate this with a hypothetical example of a 3x3 pairwise comparison matrix:

$$\mathbf{G} = \begin{pmatrix} 1 & 2 & 5 \\ \frac{1}{2} & 1 & 3 \\ \frac{1}{5} & \frac{1}{3} & 1 \end{pmatrix}$$

3. Calculate the principal eigenvalue ($\lambda_{\max}$)*: Using numerical methods:*

$$\lambda_{\max} \approx 3.08$$

4. *Calculate the Consistency Index (CI):*

$$CI = \frac{\lambda_{\max} - n}{n-1}$$

$$CI = \frac{3.08 - 3}{3 - 1} = \frac{0.08}{2} = 0.04$$

5. *Calculate the Consistency Ratio (CR):*

$$CR = \frac{CI}{RI}$$

Given $RI = 0.58$*:*

$$CR = \frac{0.04}{0.58} \approx 0.069$$

The Consistency Ratio (CR) is approximately 0.069, which is less than 0.10, indicating that the matrix is sufficiently consistent for decision-making. The CR threshold of 0.10 ensures that minor inconsistencies are tolerated without significantly impacting the reliability of the decision-making process.

Solution 1.87. *exe91*

To calculate the probability of machine failures:

1. *Probability that the machine will fail at least once in a year:*

Let $P(failure) = 0.02$ *be the probability of failure in a given month. The probability of no failure in a month is* $1 - 0.02 = 0.98$.

The probability of no failure in a year (12 months) is:

$$P(\textit{no failure in a year}) = 0.98^{12}$$

$$P(\textit{no failure in a year}) \approx 0.78$$

Therefore, the probability of at least one failure in a year is:

$$P(\textit{at least one failure in a year}) = 1 - P(\textit{no failure in a year})$$

$$P(\textit{at least one failure in a year}) = 1 - 0.78 = 0.22$$

2. *Probability that the machine will not fail in any given year:*

$$P(\textit{no failure in a year}) \approx 0.78$$

Thus, the probability that the machine will not fail in any given year is approximately 0.78.

Solution 1.88. *exe92*

To calculate the probability of supply chain disruptions:
Let the time between disruptions T be exponentially distributed with a mean of 3 months ($\lambda = \frac{1}{3}$).

1. *Probability that the next disruption will occur within 2 months:*

$$P(T \leq 2) = 1 - e^{-\lambda t}$$

$$P(T \leq 2) = 1 - e^{-\frac{1}{3} \cdot 2}$$

$$P(T \leq 2) = 1 - e^{-\frac{2}{3}} \approx 1 - 0.5134 = 0.4866$$

Therefore, the probability that the next disruption will occur within 2 months is approximately 0.4866.

2. *Probability that the next disruption will occur after 5 months:*

$$P(T > 5) = e^{-\lambda t}$$

$$P(T > 5) = e^{-\frac{1}{3} \cdot 5}$$

$$P(T > 5) = e^{-\frac{5}{3}} \approx 0.1889$$

Therefore, the probability that the next disruption will occur after 5 months is approximately 0.1889.

Solution 1.89. *exe93*

To calculate the net present value (NPV) and the break-even point:

1. *Net Present Value (NPV):*

$$NPV = \sum_{t=1}^{10} \frac{400{,}000}{(1 + 0.06)^t} - 2{,}000{,}000$$

Using the annuity formula:

$$NPV = 400{,}000 \left(\frac{1 - (1 + 0.06)^{-10}}{0.06} \right) - 2{,}000{,}000$$

$$NPV = 400{,}000\,(7.3601) - 2{,}000{,}000$$

$$NPV = 2{,}944{,}040 - 2{,}000{,}000 = 944{,}040$$

Therefore, the NPV of the digital transformation over a 10-year period is approximately \$944,040.

2. Break-Even Point: The break-even point occurs when the cumulative savings equal the initial investment:

$$2{,}000{,}000 = \sum_{t=1}^{n} \frac{400{,}000}{(1+0.06)^t}$$

Solving for n using trial and error or a financial calculator gives $n \approx 6.2$ *years.*

Therefore, the break-even point is approximately 6.2 years.

Solution 1.90. *exe94*

To calculate the annual net benefit and the present value of the net benefits:

1. Annual Net Benefit:

$$\text{Annual Net Benefit}$$

$$= (\text{Revenue from recycled materials} + \text{Waste disposal cost reduction}) - \text{Program cost}$$

$$\text{Annual Net Benefit} = (100{,}000 + 150{,}000) - 500{,}000 = -250{,}000$$

Therefore, the annual net benefit is $-\$250{,}000$.

2. Present Value of Net Benefits:

$$PV = \sum_{t=1}^{5} \frac{-250{,}000}{(1+0.05)^t}$$

Using the annuity formula:

$$PV = -250{,}000 \left(\frac{1-(1+0.05)^{-5}}{0.05} \right)$$

$$PV = -250{,}000\,(4.3295) = -1{,}082{,}375$$

Therefore, the present value of the net benefits over a 5-year period is approximately $-\$1{,}082{,}375$.

Solution 1.91. *exe95*

To calculate the net present value (NPV) and the payback period:

1. Net Present Value (NPV):

$$NPV = \sum_{t=1}^{7} \frac{300{,}000 - 200{,}000}{(1 + 0.07)^t} - 1{,}500{,}000$$

$$NPV = \sum_{t=1}^{7} \frac{100{,}000}{(1 + 0.07)^t} - 1{,}500{,}000$$

Using the annuity formula:

$$NPV = 100{,}000 \left(\frac{1 - (1 + 0.07)^{-7}}{0.07} \right) - 1{,}500{,}000$$

$$NPV = 100{,}000\,(5.3893) - 1{,}500{,}000$$

$$NPV = 538{,}930 - 1{,}500{,}000 = -961{,}070$$

Therefore, the NPV of switching to sustainable sourcing over a 7-year period is approximately −\$961,070.

2. Payback Period: The payback period occurs when the cumulative net benefits equal the initial investment:

$$1{,}500{,}000 = \sum_{t=1}^{n} \frac{100{,}000}{(1 + 0.07)^t}$$

Solving for n *using trial and error or a financial calculator gives* $n \approx 15.2$ *years.*

Therefore, the payback period is approximately 15.2 years.

Solution 1.92. *exe96*

To calculate the total economic benefit and the net present value (NPV):

1. Total Economic Benefit:

$$Total\ Carbon\ Benefit = 2{,}000\ metric\ tons/year \times 10 \times 60 = 1{,}200{,}000$$

Annual total economic benefit:

$$Annual\ Total\ Benefit = 250{,}000 + \frac{1{,}200{,}000}{10} = 370{,}000$$

2. Net Present Value (NPV):

$$NPV = \sum_{t=1}^{10} \frac{370{,}000}{(1 + 0.04)^t} - 3{,}000{,}000$$

Using the annuity formula:

$$NPV = 370{,}000\left(\frac{1-(1+0.04)^{-10}}{0.04}\right) - 3{,}000{,}000$$

$$NPV = 370{,}000\,(8.1109) - 3{,}000{,}000$$

$$NPV = 3{,}001{,}033 - 3{,}000{,}000 = 1{,}033$$

Therefore, the NPV of the investment over a 10-year period is approximately $1,033.

Solution 1.93. *exe97*

To calculate the net present value (NPV) and the break-even point:

1. Net Present Value (NPV):

$$NPV = \sum_{t=1}^{8} \frac{150{,}000 + 100{,}000}{(1+0.05)^t} - 1{,}200{,}000$$

$$NPV = \sum_{t=1}^{8} \frac{250{,}000}{(1+0.05)^t} - 1{,}200{,}000$$

Using the annuity formula:

$$NPV = 250{,}000\left(\frac{1-(1+0.05)^{-8}}{0.05}\right) - 1{,}200{,}000$$

$$NPV = 250{,}000\,(6.4632) - 1{,}200{,}000$$

$$NPV = 1{,}615{,}800 - 1{,}200{,}000 = 415{,}800$$

Therefore, the NPV of the traceability system over an 8-year period is approximately $415,800.

2. Break-Even Point: The break-even point occurs when the cumulative net benefits equal the initial investment:

$$1{,}200{,}000 = \sum_{t=1}^{n} \frac{250{,}000}{(1+0.05)^t}$$

Solving for n using trial and error or a financial calculator gives $n \approx 5.3$ years.

Therefore, the break-even point is approximately 5.3 years.

Solution 1.94. *exe98*

1. ***Utilitarianism:***
   - *Overstating inventory levels may lead to temporary financial gain and increased stock prices, benefiting shareholders. However, it may harm the company's long-term reputation, leading to potential legal issues and loss of trust from stakeholders.*
   - *Reporting accurate levels promotes transparency and trust, ensuring long-term sustainability and overall happiness for all stakeholders.*
2. ***Deontology:***
   - *Overstating inventory levels is dishonest and violates moral duties of honesty and integrity.*
   - *Reporting accurate levels adheres to moral rules and duties, maintaining ethical standards regardless of the immediate consequences.*

Therefore, from both utilitarian and deontological perspectives, the company should report accurate inventory levels.

Solution 1.95. *exe99*

1. ***Impact on Society and Environment:***
   - *Implementing eco-friendly inventory management practices, such as reducing waste and optimizing resource use.*
   - *Supporting community initiatives and ensuring fair labor practices in the supply chain.*
2. ***Integration into Business Strategies:***
   - *Incorporating CSR goals into the company's strategic planning and decision-making processes.*
   - *Regularly evaluating and reporting on the social and environmental impacts of inventory management practices to stakeholders.*

Aligning with CSR principles, the company can enhance its reputation and contribute positively to society and the environment.

Solution 1.96. *exe100*

1. ***Identifying Stakeholders:***
    - *Key stakeholders may include employees, customers, suppliers, shareholders, and the community.*
    - *Employees seek job security and fair working conditions; customers desire product availability and quality; suppliers want reliable partnerships; shareholders expect financial returns; the community values environmental responsibility.*
2. ***Implementing Inventory Practices:***
    - *Adopting just-in-time inventory to meet customer demand while minimizing excess stock.*
    - *Ensuring ethical sourcing and fair labor practices in the supply chain.*
    - *Engaging with stakeholders to understand and address their concerns, promoting transparency and trust.*

Balancing stakeholder interests, the company can foster sustainable and ethical inventory management.

Solution 1.97. *exe101*

1. ***Triple Bottom Line (TBL):***
    - *Environmental impact: Reduce carbon footprint through efficient inventory practices and sustainable sourcing.*
    - *Social impact: Ensure fair labor practices and contribute to community well-being.*
    - *Economic impact: Optimize inventory levels to reduce costs and increase profitability.*
2. ***UN Sustainable Development Goals (SDGs):***
    - *Aligning inventory practices with SDGs, such as responsible consumption and production (SDG 12) and decent work and economic growth (SDG 8).*
    - *Implementing measures to monitor and report progress towards achieving these goals.*

Applying TBL and SDG frameworks, the company can enhance the sustainability of its inventory management.

Solution 1.98. *exe102*

1. ***Ethical Trading Initiative (ETI):***
   - *Following ETI guidelines to ensure fair wages, safe working conditions, and ethical sourcing practices.*
   - *Conducting regular audits and assessments to ensure compliance with ethical standards.*
2. ***Global Reporting Initiative (GRI):***
   - *Adopting GRI standards to enhance transparency and accountability in supply chain practices.*
   - *Reporting on sustainability performance, including social, environmental, and economic impacts.*

Integrating ETI and GRI standards, the company can promote ethical and sustainable supply chain management.

Solution 1.99. *exe103*

1. ***Modeling the System:***
   - *Arrival Process: Orders arrive according to a Poisson process with a rate $\lambda = 5$ orders per hour.*
   - *Service Process: Each order takes an exponentially distributed time with mean $\mu = 10$ minutes (or 0.1667 hours) to be fulfilled.*
2. ***Performance Metrics:***
   - *Average number of orders waiting: Use the formula for the average number in the system L in an M/M/1 queue:*

$$L = \frac{\lambda}{\mu - \lambda} = \frac{5}{6 - 5} = 5 \text{ orders}$$

   - *Average waiting time: Use Little's Law $W = \frac{L}{\lambda}$:*

$$W = \frac{5}{5} = 1 \text{ hour}$$

3. ***Impact of Doubling Arrival Rate:***
   - *New arrival rate $\lambda' = 10$ orders per hour.*
   - *Recalculate performance metrics:*

$$L' = \frac{\lambda'}{\mu - \lambda'} = \frac{10}{6 - 10} \implies \text{system becomes unstable}$$

- *The system cannot handle the doubled arrival rate, leading to an infinite queue length and waiting time.*

Solution 1.100. *exe104*

1. ***Modeling the Demand:***
   - *Daily demand follows a normal distribution* $N(100, 15^2)$.
   - *Total demand over lead time L is normally distributed with mean* $\mu_L = 7 \times 100 = 700$ *and standard deviation* $\sigma_L = \sqrt{7} \times 15 \approx 39.69$.
2. ***Monte Carlo Simulation:***
   - *Generate 10,000 random samples from* $N(700, 39.69^2)$ *to represent total demand over the lead time.*
   - *Count the number of samples where the demand exceeds the safety stock of 200 units.*
   - *Estimate the probability of stockout:*

$$\hat{P}(stockout) = \frac{\textit{Number of stockouts}}{10{,}000}$$

3. ***Results:***
   - *Suppose 300 samples out of 10,000 exceed the safety stock.*
   - *Estimated probability of stockout:*

$$\hat{P}(stockout) = \frac{300}{10{,}000} = 0.03 \textit{ or } 3\%$$

Solution 1.101. *exe105*

1. ***Utilitarianism:***
   - *Choose Supplier B because it results in the greatest good for the greatest number by promoting ethical labor practices and benefiting society as a whole.*
2. ***Deontology:***
   - *Choose Supplier B because it adheres to ethical labor standards, which aligns with the duty to treat workers fairly and respect their rights.*

3. ***Virtue Ethics:***

   - *Choose Supplier B because it reflects the virtues of integrity, fairness, and responsibility, which the company should embody in its business practices.*

Solution 1.102. *exe106*

1. ***CSR-based Inventory Management Strategy:***

   - *Social Impact: Implement fair trade policies and ensure suppliers follow ethical labor practices.*
   - *Environmental Impact: Reduce carbon footprint by optimizing logistics and sourcing sustainable materials.*

2. ***Key Performance Indicators (KPIs):***

   - *Percentage of suppliers adhering to ethical labor standards.*
   - *Reduction in carbon emissions from logistics operations.*
   - *Percentage of inventory sourced from sustainable materials.*

Solution 1.103. *exe107*

1. ***Modeling the Stockout Probability:***

   - *Let X be the number of stockouts in 6 months, where $X \sim$ Binomial $(n = 6, p = 0.10)$.*
   - *Calculate $P(X \geq 1)$ using the complement rule:*

   $$P(X \geq 1) = 1 - P(X = 0)$$

   - *Calculate $P(X = 0)$:*

   $$P(X = 0) = (1 - 0.10)^6 = 0.531441$$

   - *Therefore,*

   $$P(X \geq 1) = 1 - 0.531441 = 0.468559$$

Solution 1.104. *exe108*

1. ***Modeling the Demand:***

   - *Demand $D \sim N(500, 100^2)$.*

2. ***Monte Carlo Simulation:***

   - *Generate 10,000 random samples from* $N(500, 100^2)$ *for demand.*
   - *Calculate profit for each sample:*

$$\textit{Profit} = (\textit{Selling Price} - \textit{Cost Price}) \times \textit{Demand} = (50-30) \times D$$

   - *Estimate expected profit:*

$$\hat{\mu} = \frac{1}{10{,}000} \sum_{i=1}^{10{,}000} \textit{Profit}_i$$

   - *Suppose the simulation yields an average demand of 505 units.*
   - *Expected profit:*

$$\hat{\mu} = 20 \times 505 = \$10{,}100$$

Solution 1.105. *exe109*

Key Steps:

- ***Design Phase:*** *Identify products suitable for closed-loop processes and design them for easy disassembly and recycling.*
- ***Collection:*** *Establish systems for collecting used products from customers.*
- ***Reverse Logistics:*** *Develop logistics for transporting returned products to remanufacturing facilities.*
- ***Remanufacturing:*** *Process returned products to recover valuable components and materials.*
- ***Integration:*** *Integrate recovered materials into the production of new products.*
- ***Monitoring:*** *Implement systems to track and monitor the performance of the closed-loop supply chain.*

Solution 1.106. *exe110*

1. ***Challenges:***

   - *High initial costs for setting up collection and remanufacturing systems.*
   - *Technological challenges in disassembling and processing used products.*

- *Customer acceptance of remanufactured products.*
- *Compliance with environmental regulations and standards.*

2. ***Benefits:***

- *Reduced environmental impact through recycling and waste reduction.*
- *Cost savings from recovered materials and components.*
- *Enhanced corporate image and customer loyalty.*
- *Compliance with regulatory requirements and potential tax benefits.*

Solution 1.107. *exe111*

1. ***Digital Technologies in Supply Chains:***

- ***IoT:*** *Real-time tracking of inventory, monitoring environmental conditions, and optimizing transportation routes.*
- ***Blockchain:*** *Enhancing transparency and traceability by providing immutable records of product origins and supply chain transactions.*
- ***AI:*** *Improving decision-making through predictive analytics, demand forecasting, and optimizing inventory levels.*

2. ***Examples of Sustainability Enhancements:***

- *IoT sensors monitoring temperature and humidity to reduce spoilage in perishable goods.*
- *Blockchain ensuring ethical sourcing of raw materials, reducing the risk of conflict minerals.*
- *AI algorithms optimizing delivery routes to minimize fuel consumption and carbon emissions.*

Solution 1.108. *exe112*

1. ***Circular Economy Initiatives:***

- ***Product Take-Back Programs:*** *Companies collect used products from customers to recycle or remanufacture components.*
- ***Remanufacturing:*** *Refurbishing and reusing components to extend the life cycle of products.*
- ***Recycling:*** *Converting waste materials into new products to reduce resource extraction and waste.*

2. ***Impact on Waste Reduction and Resource Conservation:***

- *Reducing landfill waste by diverting products and materials to recycling and remanufacturing streams.*
- *Conserving natural resources by reusing materials instead of extracting new raw materials.*
- *Lowering carbon footprint through reduced manufacturing and transportation activities.*

Solution 1.109. *exe113*

1. A deep understanding of inventory management concepts helps in identifying the optimal inventory levels, reducing holding costs, and improving turnover rates.

2. Techniques such as Just-in-Time (JIT) and Economic Order Quantity (EOQ) models can lead to significant cost savings and efficiency improvements.

3. Best practices in demand forecasting can enhance decision-making, ensuring that inventory levels match market demand, thus preventing overstocking or stockouts.

4. Example: Implementing a JIT system reduced excess inventory by 30% and improved production efficiency by 20% in a manufacturing company.

Solution 1.110. *exe114*

1. Key practical skills include demand forecasting, inventory control, and optimization.

2. Skills can be developed through targeted training programs that focus on hands-on exercises, real-world case studies, and simulation tools.

3. Expected outcomes include improved accuracy in demand predictions, efficient inventory management, and reduced operational costs.

4. Example: A training program on ABC analysis helped inventory managers prioritize their efforts on high-value items, leading to a 15% reduction in stockouts.

Solution 1.111. *exe115*

1. Tools and methodologies such as Six Sigma, Lean Inventory, and Value Stream Mapping can be used to analyze and improve inventory processes.

2. These tools help in identifying bottlenecks, reducing waste, and enhancing process efficiency.

3. Example: Implementing Lean Inventory principles led to a 25% reduction in waste and a 20% increase in process efficiency in a retail company.

Solution 1.112. *exe116*

1. Inventory professionals can prepare for changes by staying updated with industry trends, investing in continuous learning, and adopting flexible inventory management systems.

2. Strategies for developing adaptability include cross-training employees, implementing agile methodologies, and leveraging technology for real-time inventory tracking.

3. Example: A company adopted an agile inventory management system that allowed it to quickly adapt to changes in market demand, reducing lead times by 40%.

Solution 1.113. *exe117*

1. Certification and advanced training validate the expertise of inventory professionals and enhance their credibility.

2. Opportunities for certification include APICS Certified in Production and Inventory Management (CPIM) and Certified Supply Chain Professional (CSCP).

3. These certifications contribute to professional growth by providing advanced knowledge, improving job performance, and opening up career advancement opportunities.

4. Example: Earning a CPIM certification led to a promotion and a 15% salary increase for an inventory manager.

Solution 1.114. *exe118*

1. Effective inventory management ensures optimal stock levels, reduces costs, and improves cash flow, contributing to organizational success.

2. It supports business objectives by aligning inventory practices with strategic goals and enhances customer satisfaction by ensuring product availability.

3. Example: A company improved its order fulfillment rate by 25% through effective inventory management, leading to higher customer satisfaction and increased sales.

Solution 1.115. *exe119*

1. Classroom lectures provide structured learning and allow for direct interaction between instructors and students.

2. Real-time interaction and discussion enhance understanding by enabling students to clarify doubts, engage in debates, and receive immediate feedback.

3. Example: A lecture on inventory optimization techniques included interactive discussions, which helped students grasp complex concepts more effectively.

Solution 1.116. *exe120*

1. Textbooks and study materials provide comprehensive coverage of inventory management topics, including detailed explanations, examples, and case studies.

2. These resources complement classroom learning by allowing students to review and reinforce concepts at their own pace.

3. Features that make them effective include structured content, real-world examples, practice exercises, and review questions.

4. Example: A textbook on inventory management provided case studies that helped students apply theoretical concepts to real-world scenarios, enhancing their practical understanding.

Solution 1.117. *exe121*

1. Big data analytics optimizes inventory levels by analyzing large volumes of data from various sources, such as sales transactions, customer feedback, and supply chain operations.

2. Insights gained include identifying demand patterns, seasonal trends, and customer preferences, which help in accurate demand forecasting and inventory planning.

3. Example: A retailer used big data analytics to analyze sales data and customer behavior, resulting in a 20% reduction in excess inventory and a 15% increase in stock turnover rate.

Solution 1.118. *exe122*

1. Predictive analytics leverages historical sales data, market trends, and other relevant factors to forecast future inventory needs.

2. Statistical models identify patterns and correlations, allowing businesses to predict demand accurately and adjust inventory levels accordingly.

3. Example: An e-commerce company used predictive analytics to forecast peak season demand, reducing stockouts by 25% and overstocking by 30%.

Solution 1.119. *exe123*

1. Data visualization tools present complex inventory data in an intuitive and easily understandable format, such as charts, graphs, and dashboards.

2. These tools help stakeholders quickly identify trends, patterns, and anomalies, facilitating informed decision-making.

3. Example: A supply chain manager used a data visualization dashboard to monitor inventory levels in real time, enabling swift adjustments and reducing lead times by 20

Solution 1.120. *exe124*

1. Real-time data integration provides up-to-date insights by consolidating information from various sources, such as point-of-sale systems, inventory databases, and supply chain sensors.

2. This capability enhances responsiveness and agility, allowing businesses to adjust inventory levels promptly based on current demand and supply conditions.

3. Example: A supermarket chain integrated real-time data from sales terminals and inventory sensors, improving stock replenishment accuracy and reducing out-of-stock instances by 18

Solution 1.121. *exe125*

1. Advanced analytics platforms offer powerful tools, including machine learning algorithms, data mining, and optimization techniques, for comprehensive inventory analysis.

2. These capabilities support advanced inventory management strategies, such as dynamic pricing, automated replenishment, and demand forecasting.

3. Example: A manufacturing company implemented an advanced analytics platform, which utilized machine learning to optimize inventory levels, resulting in a 22% reduction in holding costs and a 17% improvement in service levels.

Solution 1.122. *exe126*

1. Analytical skills in inventory management involve analyzing sales data, monitoring stock levels, and forecasting demand.

2. Tasks such as determining reorder points, identifying slow-moving items, and optimizing stock levels benefit from strong analytical abilities.

3. These skills contribute to inventory performance by enabling data-driven decisions that improve stock availability and reduce holding costs.

Solution 1.123. *exe127*

1. Attention to detail is crucial for accurately recording inventory transactions, counting stock, and processing orders.

2. Precision in these activities ensures that inventory records are accurate, reducing discrepancies and the risk of stockouts or overstocking.

3. Accurate inventory tracking leads to better decision-making and efficient inventory management.

Solution 1.124. *exe128*

1. Problem-solving skills are essential for addressing issues such as stock discrepancies, supply chain disruptions, and process inefficiencies.

2. Common challenges include resolving inventory inaccuracies, managing supplier delays, and optimizing storage space.

3. Effective problem-solving involves identifying the root cause of issues, developing solutions, and implementing corrective actions to prevent recurrence.

Solution 1.125. *exe129*

1. Key organizational skills include managing multiple tasks, maintaining systematic records, and efficiently organizing inventory.

2. These skills help ensure that inventory processes are well-structured, making it easier to track stock levels, process orders, and conduct audits.

3. Efficient organization improves overall inventory management by reducing errors and increasing productivity.

Solution 1.126. *exe130*

1. Technical proficiency is important for effectively using inventory management software and ERP systems.

2. Familiarity with these tools enables accurate tracking of inventory, real-time data analysis, and efficient management of inventory processes.

3. Technical skills enhance the ability to monitor stock levels, forecast demand, and generate reports, leading to better inventory control and decision-making.

Glossary

ABC Analysis: A technique for categorizing inventory items based on their importance and value, typically dividing them into three categories: A (most important), B (moderately important), and C (least important).

Advance Shipping Notice (ASN): A document that provides detailed information about a pending delivery, sent in advance of the actual shipment.

Allocation: The process of assigning available inventory to customer orders or production.

Anticipation Inventory: Inventory held in anticipation of future demand or price increases.

Assembly Line: A manufacturing process in which products are assembled in a sequential manner as they move along a conveyor belt or through various workstations.

Automated Storage and Retrieval System (AS/RS): A type of warehouse technology that uses automated systems to store and retrieve goods, improving efficiency and accuracy.

Backordering: The process of allowing customers to place orders for out-of-stock items, with the promise of future delivery once the items become available.

Backflushing: A method of inventory management where the consumption of components and materials is recorded only after the finished product is completed.

Barcoding: A method of encoding data in a visual pattern readable by a machine, commonly used in inventory management to track and manage products.

Batch Picking: A method of order picking in which multiple orders are grouped together and picked in one pass through the warehouse.

Bill of Materials (BOM): A comprehensive list of raw materials, components, and assemblies required to manufacture a product.

Blanket Order: A long-term agreement between a buyer and supplier to provide goods or services at a predetermined price over a specified period.

Buffer Stock: Additional inventory kept to protect against uncertainties in demand or supply.

Bulk Storage: The storage of large quantities of items in a warehouse, typically used for items that are not frequently accessed.

Carrying Cost: The total cost of holding inventory, including storage, insurance, depreciation, and opportunity costs.

Cash-to-Cash Cycle Time: The time it takes for a company to convert its investment in inventory into cash flow from sales.

Centralized Inventory: A system where all inventory is stored in a single location or a few central locations.

Centralized Purchasing: A procurement strategy where all purchasing activities are handled by a single, central department.

Closed-Loop Supply Chain: A supply chain that integrates forward logistics with reverse logistics, focusing on the complete life cycle of a product.

Collaborative Planning, Forecasting, and Replenishment (CPFR): A business practice that combines the intelligence of multiple trading partners in the planning and fulfillment of customer demand.

Consignment Stock: Inventory that is owned by the supplier but held by the buyer, who only pays for the inventory when it is used or sold.

Continuous Improvement: An ongoing effort to improve products, services, or processes by making incremental improvements over time.

Consignment Inventory: Inventory that is owned by the supplier but held by the customer, who only pays for the inventory when it is used or sold.

Continuous Replenishment: A strategy where inventory is continuously monitored and replenished to maintain optimal levels.

Cross-Docking: A logistics practice where products are received at a warehouse and immediately prepared for shipment without being stored.

Cycle Counting: A method of physical inventory auditing where a small subset of inventory is counted on a rotating schedule.

Demand Planning: The process of forecasting future demand for products and services to ensure that they can be delivered to customers efficiently.

Distribution Requirements Planning (DRP): A method for planning inventory needs in a distribution network.

Dead Stock: Inventory items that have not been sold or used for a long period and are unlikely to be sold in the future.

Demand Forecasting: The process of predicting future customer demand based on historical sales data and market analysis.

Distribution Center: A warehouse or storage facility where products are received, processed, and shipped to customers or retail locations.

Drop Shipping: A fulfillment method where a retailer does not keep goods in stock but instead transfers customer orders to a manufacturer or wholesaler who ships the goods directly to the customer.

Economic Order Quantity (EOQ): A mathematical model that determines the optimal order quantity to minimize total inventory costs, including ordering and holding costs.

Economic Production Quantity (EPQ): A model that determines the optimal production quantity to minimize total production and inventory costs.

Enterprise Resource Planning (ERP): A type of software that integrates various business processes, including inventory management, into a unified system.

FIFO (First In, First Out): An inventory management method where the oldest inventory items are used or sold first.

Finished Goods Inventory: The completed products that are ready for sale to customers.

Fixed Order Period: An inventory control system where orders are placed at regular intervals, regardless of the inventory level.

Fixed Order Quantity: An inventory control system where the same quantity is ordered every time the inventory level falls below a certain point.

Flow Rack: A type of shelving that uses gravity to move products from the back to the front, making it easier to access items.

Freight Forwarder: A company that arranges the transportation of goods on behalf of a shipper, including handling customs and other documentation.

Fulfillment: The process of receiving, processing, and delivering orders to customers.

Group Technology: A manufacturing technique that groups similar parts or products together to improve efficiency and reduce production time.

Heijunka: A lean manufacturing technique for leveling production by sequencing orders in a way that minimizes variations in production volume and type.

Hedge Inventory: Inventory held as a buffer against potential price increases or supply shortages.

Inventory Accuracy: The degree to which inventory records match the actual physical inventory.

Inventory Analytics: The use of data analysis and modeling techniques to understand and optimize inventory management practices.

Inventory Control: The processes and systems used to manage and regulate inventory levels.

Inventory Optimization: The process of determining the most efficient inventory levels and costs for different items and scenarios.

Inventory Shrinkage: The loss of inventory due to theft, damage, or administrative errors.

Inventory Turnover: A ratio that measures how many times inventory is sold and replaced over a given period.

JIT (Just-In-Time): An inventory management strategy where materials and products are delivered exactly when needed in the production process, reducing the need for holding large inventories.

Kanban: A visual inventory management system that uses cards or signals to indicate when new inventory should be ordered.

Kitting: The process of grouping individual items together to create a single product or kit.

Kaizen: A Japanese term meaning "continuous improvement," used to describe a philosophy of making small, incremental changes to improve processes and reduce waste

Lead Time: The time it takes for an order to be placed and received.

Lean Manufacturing: A production philosophy that focuses on minimizing waste and maximizing efficiency in the manufacturing process.

LIFO (Last In, First Out): An inventory management method where the most recently received items are used or sold first.

Logistics: The coordination and management of the movement of goods from suppliers to customers.

Lot Size: The quantity of an item ordered for delivery on a specific date or manufactured in a single production run.

Make-to-Order: A production strategy where items are manufactured only after a customer order is received.

Make-to-Stock: A production strategy where items are manufactured in advance and kept in stock until customer demand arises.

Milk Run: A delivery route that allows for multiple pickups and deliveries in a single trip, often used to reduce transportation costs and improve efficiency.

MRP (Material Requirements Planning): A production planning and inventory control system that manages manufacturing processes.

Obsolete Inventory: Inventory that is no longer usable or sellable due to changes in demand or product design.

On-Demand Inventory: Inventory that is ordered and delivered based on current demand, reducing the need for holding large stocks.

Order Fulfillment: The complete process from receiving an order to delivering the product to the customer.

Order Lead Time: The time it takes from placing an order to receiving the goods.

Outsourcing: The practice of contracting out certain business functions or processes to third-party providers.

Overstock: Inventory that exceeds the demand or storage capacity, often leading to higher carrying costs.

Packing Slip: A document that accompanies a shipment and lists the items included in the delivery.

Perpetual Inventory System: An inventory management system that continuously tracks inventory levels in real-time.

Physical Inventory: A process of counting all inventory items to verify accuracy against inventory records.

Pipeline Inventory: Inventory that is in transit between locations, such as from supplier to warehouse or warehouse to customer.

Point of Sale (POS): The location and system where a customer transaction is completed.

Procurement: The process of acquiring goods and services from suppliers.

Production Planning: The process of organizing and scheduling production activities to ensure that products are manufactured efficiently and on time.

Purchase Order (PO): A document issued by a buyer to a seller indicating the products, quantities, and agreed prices for products or services.

QR Code: A type of barcode that can be scanned to provide information about a product or shipment.

Quality Control: The process of ensuring that products meet specified quality standards through inspection and testing.

Raw Materials Inventory: The basic materials that are used to produce goods in the manufacturing process.

Reorder Point: The inventory level at which a new order should be placed to replenish stock before it runs out.

Reverse Logistics: The process of returning goods from the customer to the supplier for refund, repair, or recycling.

Replenishment: The process of restocking inventory to maintain desired levels.

RFID (Radio-Frequency Identification): A technology that uses electromagnetic fields to automatically identify and track tags attached to objects.

Safety Stock: Extra inventory kept on hand to protect against stockouts caused by demand variability or supply chain disruptions.

Safety Stock Calculation: The process of determining the amount of extra inventory needed to protect against variability in demand or supply.

Service Level: A measure of the ability to meet customer demand without stockouts.

SKU (Stock Keeping Unit): A unique identifier for each distinct product and service that can be purchased.

Sourcing: The process of identifying and selecting suppliers to provide goods or services.

Stockout: A situation where there is no inventory available to meet customer demand.

Supply Chain Optimization: The process of improving the efficiency and effectiveness of the supply chain to reduce costs and improve service levels.

Supply Chain Management: The management of the flow of goods and services, including all processes that transform raw materials into final products.

Third-Party Logistics (3PL): The use of external companies to handle logistics functions such as warehousing, transportation, and distribution.

Transportation Management System (TMS): A software application designed to plan, execute, and optimize the movement of goods.

Turnover Rate: The rate at which inventory is sold and replaced over a given period.

Two-Bin System: An inventory control system where inventory is divided into two bins; when the first bin is empty, a replenishment order is placed while the second bin is used.

Outsourced Manufacturing: The practice of hiring external companies to produce goods on behalf of a company.

Value Stream Mapping: A lean management tool used to visualize and analyze the flow of materials and information through a process.

Vendor-Managed Inventory (VMI): A supply chain initiative where the supplier is responsible for managing and replenishing inventory at the customer's location.

Vendor Rating: A method of evaluating and comparing suppliers based on their performance, quality, and reliability.

Warehouse Layout: The physical arrangement of storage areas, workstations, and equipment in a warehouse.

Warehouse Management System (WMS): A software application designed to support and optimize warehouse functionality and distribution center management.

Work-in-Process (WIP) Inventory: Inventory that is in the process of being manufactured but is not yet completed.

Work Order: A document that provides instructions for manufacturing or assembling a product.

Yield Management: A strategy for maximizing revenue by adjusting prices based on demand and inventory levels.

Zero Inventory: An inventory management strategy where the goal is to have as little inventory as possible without affecting the production process or customer satisfaction.

Zoning: The practice of organizing a warehouse by grouping similar items together to improve efficiency in picking and storing inventory.

Zone Picking: A method of order picking where the warehouse is divided into zones, and workers pick items only within their assigned zones.

References

[1] Jörn-Henrik Thun, Martin Drüke, and André Grübner. Empowering kanban through TPS-principles–An empirical analysis of the toyota production system. *International Journal of Production Research*, 48(23):7089–7106, 2010.

[2] GM Anoop and Vivekanand S Muhammed. A brief overview on toyota production system (TPS). *International Journal for Research in Applied Science and Engineering Technology*, 8(5):2505–2509, 2020.

[3] Jean-François Houde, Peter Newberry, and Katja Seim. Nexus tax laws and economies of density in e-commerce: A study of amazon's fulfillment center network. *Econometrica*, 91(1):147–190, 2023.

[4] Christian Mascle and Julien Gosse. Inventory management maximization based on sales forecast: Case study. *Production Planning & Control*, 25(12):1039–1057, 2014.

[5] Agustin Gutierrez, Ashish Kothari, Carolina Mazuera, and Tobias Schoenherr. Taking supplier collaboration to the next level. *McKinsey & Company*, July, 7, 2020.

[6] Manuel D Rossetti and Ashish V Achlerkar. Evaluation of segmentation techniques for inventory management in large scale multi-item inventory systems. *International Journal of Logistics Systems and Management*, 8(4):403–424, 2011.

[7] Braulio Brunaud, José M Laínez-Aguirre, Jose M Pinto, and Ignacio E Grossmann. Inventory policies and safety stock optimization for supply chain planning. *AIChE Journal*, 65(1):99–112, 2019.

[8] Jagdeep Singh and Harwinder Singh. Continuous improvement philosophy–literature review and directions. *Benchmarking: An International Journal*, 22(1):75–119, 2015.

[9] Taner Arsan, Emrah Başkan, Emrah Ar, and Zeki Bozkuş. A software architecture for inventory management system. In *Innovations and Advances in Computer, Information, Systems Sciences, and Engineering*, pages 15–27. Springer, 2013.

[10] Jaroslava Kubáňová, Iveta Kubasáková, Kristián Čulík, and Lukáš Štítik. Implementation of barcode technology to logistics processes of a company. *Sustainability*, 14(2):790, 2022.

[11] Aurélie Edouard, Yves Sallez, Virginie Fortineau, Samir Lamouri, and Alexandre Berger. Automated storage and retrieval systems: An attractive solution for an urban warehouse's sustainable development. *Sustainability*, 14(15):9518, 2022.

[12] Muhammad Usman Ahmed, Asad Shafiq, and Farzad Mahmoodi. The role of supply chain analytics capability and adaptation in unlocking value from supply chain relationships. *Production Planning & Control*, 33(8):774–789, 2022.

[13] Al-Rwaidan, Rafi Mahmoud, Naimah Aldossary, Mohammad Mousa Eldahamsheh, Majed Kamel Ali Al-Azzam, Ali Zakariya Al-Quran, and Sulieman Ibraheem Shelash Al-Hawary. The impact of cloud-based solutions on digital transformation of HR practices. *International Journal of Data & Network Science*, 7(1):83–90, 2023.

[14] Sundeep Mathur. Continuous multi-eechelon inventory optimization. 2020.

[15] Jože M Rožanec, Blaž Kažič, Maja Škrjanc, Blaž Fortuna, and Dunja Mladenić. Automotive oem demand forecasting: A comparative study of forecasting algorithms and strategies. *Applied Sciences*, 11(15):6787, 2021.

[16] Ioana Popescu and Yaozhong Wu. Dynamic pricing strategies with reference effects. *Operations Research*, 55(3):413–429, 2007.

[17] Amandeep Dhaliwal. The rise of automation and robotics in warehouse management. In *Transforming Management Using Artificial Intelligence Techniques*, pages 63–72. CRC Press, 2020.

[18] Bosco, Mary-Clare. "From Yosemite to a Global Market: How Patagonia, Inc. has Created an Environmentally Sustainable and Socially Equitable Model of Supply-Chain Management" 2017. Pomona Senior Theses. 178. https://scholarship.claremont.edu/pomona_theses/178

[19] Poddar, Prisha. "Integrating sustainability into the Textile Supply Chain-comparing the case of H&M and Patagonia." Scholarly, Aug. https://scholarly. co/wpcontent/uploads/2021/08/Research-Paper-Prisha-Poddar. pdf (2021).

[20] Mojtaba M Shourkaei, Kelsey M Taylor, and Bruno Dyck. Examining sustainable supply chain management via a social-symbolic work lens: Lessons from patagonia. *Business Strategy and the Environment*, 33(2):1477–1496, 2024.

[21] Atiq Zaman and Peter Newman. Plastics: Are they part of the zero-waste agenda or the toxic-waste agenda? *Sustainable Earth*, 4:1–16, 2021.

[22] Giulia Baragiola and Martina Mauri. SDGs and the private sector: Unilever and P&G case studies. 2021.

[23] Atiq Zaman. Zero-waste: A new sustainability paradigm for addressing the global waste problem. In *The Vision Zero Handbook: Theory, Technology and Management for a Zero Casualty Policy*, pages 1195–1218. Springer, 2022.

[24] V Kasturi Rangan, Michael W Toffel, Vincent Dessain, and Jerome Lenhardt. Sustainability at IKEA group. 2014.

[25] Ida Ellesson and Viktor Hultin. A strategic framework for improving inventory management decisions at IKEA, 2016. Department of Industrial Management & Logistics, Division of Production Management.

[26] Lilia Yang and Nanxiang Shao. Sustainability strategies in IKEA with the focus on production and suppliers. 2019.

[27] Maria Lourdes Pahamotang Catalan. IKEA's supply chain: Growth on sustainability. 2022.

[28] Katherine Neebe. Sustainability at walmart: Success over the long haul. *Journal of Applied Corporate Finance*, 32(2):64–71, 2020.

[29] Michelle Van Koert, Deborah Sherwood, and Steve Kennedy. Interface: Creating a climate fit for life through carpet tiles. 2019.

[30] Laasch, Oliver. Principles of Management: Practicing Ethics, Responsibility, Sustainability. 2nd ed. SAGE Publications Ltd, 2021.

Index

ABC Analysis, 46
Activation Functions, 155
Adam (Adaptive Moment Estimation), 155
Adapting to Technological Changes in Inventory Management, 396
Advanced Inventory Optimization Models, 103
Advancements in Inventory Optimization Techniques, 373
Advantages of Safety Stock, 62
Advantages of VED Analysis, 58
Analytic Hierarchy Process (AHP) Method, 264
Artificial Intelligence and Machine Learning, 227
Artificial Intelligence and Machine Learning in Inventory Management, 147
Autonomous mobile robots (AMRs), 161

Backpropagation and Gradient Descent, 155
Balancing Between Overstocking and Stockouts, 10
Basic Terminology, 2
Benefits of Inventory Metrics and KPIs, 251
Benefits of Simulation and Gaming Techniques, 342
Blended Learning Techniques, 360
Blockchain technology, 255
Building a Strong Professional Resume and Portfolio, 402
Bullwhip Effect Reduction, 196

Career Development and Professional Growth, 398
Career Pathways and Advancement Opportunities, 392
Causal Models, 192
Certifications and Training Programs, 393
Challenges and Limitations, 348
Challenges and Opportunities in Inventory Training, 365
Challenges in Inventory Reconciliation, 220
Challenges in MEIO Implementation, 94
Challenges of ABC Analysis, 50
Choosing Reliable and Efficient Inventory Metrics, 235
Classification and Regression, 150
Classifying Risks in Inventory Management, 291
Cloud-Based Inventory Systems, 164
Collaborative Planning, Forecasting, and Replenishment (CPFR), 207
Collaborative Robots (Cobots), 163
Common Challenges in Inventory Research, 379
Community Engagement and Impact, 334
Comparative Analysis of Techniques, 209
Comparison of Inventory Control Techniques, 82
Components of a Risk Matrix, 294
Components of MEIO Models, 93

Continuing Education and Professional Development, 400
Continuous Review (Q,R) Model, 29
Continuous Review Systems, 22
Coordination and Optimal Expenditure on Materials, 90
Costs Associated with Inventory Management, 11
Covariance, 176
Creating a Culture of Responsibility, 332
Criteria for Selecting Appropriate Techniques Based on Organizational Needs, 84
Criteria for Selecting Metrics, 235
Criticality Analysis, 295
Cultural and Market Trends Shaping Inventory Careers, 398
Current Trends in Sustainable Supply Chains, 317
Cycle Counting, 131

Days of Inventory on Hand (DOH), 138
Days Sales of Inventory (DSI), 243
Decision Trees, 149
Definition of Inventory Control Techniques, 41
Delphi Method, 208
Demand Forecast Accuracy, 238
Demand Forecasting, 147, 207
Demand Forecasting and Planning, 130
Demand Risk Mitigation Strategies, 288
Demand Risks, 261
Demand Variability, 14
Descriptive Analytics, 135
Deterministic Models, 26
Developing a Risk Management Plan, 298
Developing Collaborative Research Opportunities, 385
Developing Ethical Frameworks, 331
Digital Tools for Ethical Supply Chain Management, 336
Disadvantages of Safety Stock, 63
Disadvantages of VED Analysis, 59
Dynamic Safety Stock Strategies, 109

E-commerce Industry, 117
Economic Order Quantity (EOQ), 3
Economic Order Quantity (EOQ) Model, 69
Economic Production Quantity (EPQ), 28, 70
Educational and Professional Qualifications, 393
Educational and Strategic Dimensions, 346
Elastic Net, 182
Emerging Areas of Focus, 366
Emerging Topics in Sustainable Inventory Management, 318
Emerging Trends and New Opportunities, 403
Emerging Trends in Inventory Management, 146
Enhancements in Inventory Decision-Making, 379
Entry-Level Positions, 390
Ethical Inventory Policies, 333
Ethical Sourcing Practices, 327
Exercises, 38, 96, 119, 145, 172, 216, 231, 257, 305, 324, 340, 355, 370, 388, 406
Expectation, 175
Exponential Smoothing, 177

Fair Treatment in Production, 327
Feedforward Computation, 154
Financial Risk Mitigation Strategies, 290
Financial Risks, 263
Finished Goods, 18
Flexibility and Collaboration Across Multiple Locations, 168
Food and Beverage Industry, 143

Functioning of Inventory Management Systems Within Organizations, 88
Fundamental Concepts and Practices, 25

Global Market Trends and Their Effect on Inventory Careers, 396
Goals of Inventory Optimization, 20
Gradient Descent, 155
Gross Margin Return on Investment (GMROI), 242

Hands-On Workshops and Practicals, 361
Healthcare Industry, 116
Holding Costs, 11

Impact of Artificial Intelligence and Machine Learning, 382
Impact of Inventory Management on Organizational Operations, 34
Impact of Technological Advancements on Inventory Careers, 394
Impact on Customer Satisfaction and Profitability, 21
Impact on Supply Chain Efficiency, 378
Implementing Blockchain for Transparency, 226
Implementing Risk Management Strategies, 297
Implications for Companies, 300
Implications for Practice and Research, 322
Implications for Practitioners and Researchers, 352, 386
Importance of Continuous Research in Inventory Management, 372
Importance of Inventory Auditing in Inventory Management, 218
Importance of Inventory Control from an Organizational Perspective, 86
Importance of Inventory Control Techniques, 41
Importance of Inventory Management, 233
Importance of Inventory Management in Manufacturing Operations, 123
Importance of Inventory Optimization, 20
Importance of Inventory Professionals in Supply Chain Operations, 390
Importance of Inventory Reconciliation, 220
Importance of Maintaining Optimal Inventory Levels, 21
Importance of Managing Inventory Risks, 260
Importance of Social Responsibility in Inventory Management, 326
Importance of Sustainable Inventory Management, 309
Improving Sales and Merchandising Processes, 250
Innovations in Inventory Forecasting and Planning, 374
Integrating Risk Management into Inventory Processes, 298
Integration of Big Data and Analytics, 374
Integration With Modern Supply Chain Technologies, 225
Internet of Things (IoT), 255
Introduction to Inventory Management, 1
Inventory Accounting and Valuation, 125
Inventory Analytics, 25, 135
Inventory Auditing and Control, 217
Inventory Career Opportunities, 389

Inventory Control and Optimization, 19
Inventory Control Systems, 3, 22
Inventory Control Techniques, 40
Inventory Education and Training, 357
Inventory Ethics and Social Responsibility, 325
Inventory Forecasting, 174
Inventory Forecasting Techniques, 206
Inventory Management in the Manufacturing Sector, 122
Inventory Metrics and KPIs, 233
Inventory Models, 26
Inventory Optimization for Seasonal Demand, 107
Inventory Optimization Methods, 23
Inventory Optimization Using Machine Learning Algorithms, 148
Inventory Reconciliation, 219
Inventory Research and Development, 371
Inventory Risk Management, 259
Inventory Simulation and Gaming, 341
Inventory Sustainability, 308
Inventory to Sales Ratio, 237
Inventory Turnover, 64, 240
Inventory Turnover Ratio, 137
Inventory Value (FIFO), 125

Just-In-Time (JIT), 3

Key Benefits of MEIO, 92

Labor Planning, 204
LASSO (Least Absolute Shrinkage and Selection Operator), 180
Lead Time Demand Forecasting, 142
Lead Time Variability, 15
Leveraging IoT and RFID, 225
Linear Regression, 148, 178
Logistics and Operational Efficiency, 199
Loss Function, 154
Lot-for-Lot (L4L), 28

Maintenance, Repair, and Operations Supplies, 18
Manufacturing Industry, 115
MAPE (Mean Absolute Percentage Error), 178
Measuring Training Effectiveness, 361
MEIO Techniques and Approaches, 93
Methodologies for Teaching Inventory Management, 360
Methods for Calculating Safety Stock Levels, 61
Mid-Level Roles, 391
Minimizing Carbon Footprint, 310
Minimizing Costs Associated With Holding and Managing Inventory, 10
Mitigating Inventory Risks, 288
Models of Sustainable Inventory Management, 312
Modern Educational Approaches, 359
Monitoring and Reporting Ethical Compliance, 336
Monitoring and Reviewing Risk Management Activities, 299
Moving Averages, 208
Multi-Echelon Inventory Model, 103
Multi-Echelon Inventory Optimization (MEIO), 92

Navigating the Job Market, 400
Networking and Professional Associations, 399
Neural Networks, 154
Neuron and Activation Function, 154
Newsvendor Model, 30

Objectives of Inventory Management, 9

Online Learning Platforms and Resources, 363
Operational Risk Mitigation Strategies, 289
Operational Risks, 262
Optimization Algorithms, 155
Optimization Strategies, 48
Order Fulfillment Rate, 139
Ordering Costs, 12

Periodic Review (P,S) Model, 29
Periodic Review Systems, 22
Pharmaceutical Industry, 143
Potential Areas for Growth and Specialization, 403
Potential Areas for Research and Development, 351
Potential for Interdisciplinary Research, 385
Practical Applications, 35
Practical Applications of Inventory Auditing, 222
Practical Applications of Inventory Forecasting, 210, 212
Practical Implications of Inventory Education, 364
Practical Tips for Practitioners, 36
Predictive Analytics, 136
Preparing for Interviews and Assessments, 401
Principles of Inventory Management, 3
Prioritizing Risks, 292
Probabilistic Models, 28
Problems and Challenges in Inventory Management, 13
Production Order Quantity (POQ), 27
Promoting Ethical Sourcing and Production, 311

Qualitative Research Approaches, 376
Quality Control and Assurance, 124
Quantitative and Qualitative Assessment Techniques, 277
Quantitative Forecasting Methods, 208
Quantitative Research Methods, 375

Random Forest, 183
Raw Materials, 17
Real-Time Monitoring and Decision-Making, 164
Regional Variations in Career Opportunities, 397
Regression Analysis, 208
ReLU (Rectified Linear Unit), 155
Reorder Point (ROP), 27, 65
Research Methodologies in Inventory Management, 375
Resource Allocation Ethics, 329
Retail Industry, 114
Ridge Regression, 179
Risk Assessment Models, 263
Risk Matrix, 294
Robotics in Warehouse Operations, 160
Role of Artificial Intelligence and Machine Learning, 363
Role of Inventory Auditing in Supply Chain Coordination, 224
Role of Inventory Forecasting in Inventory Management, 189
Role of Inventory Management, 127
Role of Technology in Enhancing Ethical Practices, 335
Root Cause Analysis, 221

Sales and Operations Planning (S&OP), 207
Seasonal Decomposition, 141
Seasonal Forecasting Models, 108
Sell-Through Rate, 245
Senior and Specialized Positions, 391
Sigmoid Function, 155
Significance of Inventory Management in Manufacturing, 126
Simulation and Gaming Techniques, 361

Skills and Qualifications for Inventory Careers, 392
Skills Required for Technology-Driven Roles, 395
Social Equity and Fairness, 330
Socially Responsible Inventory Management Strategies, 332
Soft Skills and Competencies, 394
Softmax Function, 155
solutions, 407, 425, 428, 429, 432, 433, 438, 442, 450, 453, 456, 461, 463, 464
SSE (Sum of Squared Errors), 179
Stochastic Gradient Descent (SGD), 155
Stock Management, 60
Stockout Costs, 13
Stockout Rate, 139
Strategies for Effective Inventory Education, 358
Strategies to Overcome Demand Variability, 14
Strategies to Overcome Lead Time Variability, 15
Strategies to Overcome Other Common Challenges, 16
Structure of a Risk Matrix, 294
Supplier Relationship Management, 197
Supply Risk Mitigation Strategies, 289
Supply Risks, 262
Support Vector Machine (SVM), 186

Tanh Function, 155
Techniques for Forecasting Demand and Lead Times Specific to Manufacturing, 140
Techniques of Inventory Auditing, 219
Technological Advancements and Their Implications, 227
Technological Advancements in Training Tools, 363
Technological Innovations and Sustainability, 319
Technological Innovations for Transparency, 335
Technological Innovations in Inventory Management, 381
Textile Industry, 143
Theoretical Approaches to Inventory Ethics, 337
Theoretical Framework for Sustainable Inventory Control, 320
Time Series Analysis, 140, 190
Tools and Technologies, 36
Traditional Training Methods, 359
Training and Education for Ethical Practices, 331
Transparency in Distribution, 328
Types of Inventories, 17
Types of Inventory Management Games, 345
Types of Simulation Models, 343

Using Metrics to Monitor Performance, 246
Utilization of Simulation and Modeling Tools, 383

Variance, 176
VED Analysis, 56

Warehouse Management, 124
Warehouse Space Utilization, 201
Waste Reduction and Environmental Impact, 329
Waste Reduction Strategies, 310
Work-in-Progress, 17

FSC
www.fsc.org
MIX
Papier | Fördert gute Waldnutzung
FSC® C083411

Kolibri 360 GmbH
Ferdinand-Jühlke-Straße 7
99095 Erfurt, Deutschland
produktsicherheit@kolibri360.de